John Nichol served in the Royal Air Force for fifteen years. On active duty during the first Gulf War, his Tornado bomber was shot down during a mission over Iraq and he was forced to eject. Captured, tortured and held as a prisoner of war, John was paraded on television, provoking worldwide condemnation and leaving one of the enduring images of the conflict.

John is the bestselling co-author of *Tornado Down* and author of many highly acclaimed Second World War epics, as well as the hugely successful *Sunday Times* bestsellers *Spitfire*, *Lancaster* and *Tornado*. He has made a number of TV documentaries with Second World War veterans, written for national newspapers and magazines, and is a widely quoted commentator on military affairs.

Praise for *Eject! Eject!*

'A fascinating history. Given his extraordinary personal experience, skills and knowledge, *Eject! Eject!* is clearly a book that Nichol and Nichol alone was destined to write. Five stars'
Daily Telegraph

'An absolutely brilliant, brilliant read. Fascinating history and enthralling stories. An absolute corker'
Jeremy Vine

'Catapults you into the heart of the most epic experiences of ejection, escape and survival'
Andy McNab

'The story of the ejection seat's birth and development revealed by John Nichol, is one of genius, bravery and baffling eccentricity, containing thrilling, gripping and dramatic stories of survival'
Daily Express

'Some of the stories are incredible, making for an action packed book'
The Armourer

EJECT! EJECT!

*Escaping disaster in the skies
and surviving what comes next*

JOHN NICHOL

**SIMON &
SCHUSTER**

London · New York · Sydney · Toronto · New Delhi

First published in Great Britain by Simon & Schuster UK Ltd, 2023
This edition published in Great Britain by Simon & Schuster UK Ltd, 2024

1 3 5 7 9 10 8 6 4 2

Simon & Schuster UK Ltd
1st Floor
222 Gray's Inn Road
London WC1X 8HB

Simon & Schuster: Celebrating 100 Years of Publishing in 2024

www.simonandschuster.co.uk
www.simonandschuster.com.au
www.simonandschuster.co.in

Simon & Schuster Australia, Sydney
Simon & Schuster India, New Delhi

A CIP catalogue record for this book
is available from the British Library

Paperback ISBN: 978-1-3985-0943-6
eBook ISBN: 978-1-3985-0942-9

Typeset in Sabon by M Rules
Printed and Bound in the UK using 100% Renewable
Electricity at CPI Group (UK) Ltd

For Sophie

CONTENTS

This book is dedicated to all those involved in the design, manufacture and maintenance of the ejection seats which have given so many grateful aviators – and their families – another chance at life.

ACKNOWLEDGEMENTS

Many people offered their valuable time and considerable expertise while I researched and wrote this book and it isn't possible to thank everyone individually, but I am eternally grateful to you all.

My sincere thanks also go to:

Tony Gaunt at the Martin-Baker Aircraft Company for guiding me around their factory and answering endless questions about ejection seats. And Andrew Martin and Sarah Jeffery for their help and advice.

Fellow authors Dave Gledhill, Mike Napier, Sarah Holt, Santiago Rivas, Rowland White and Ian Black for their knowledge of RAF procedures, conflicts and ejection systems.

Stuart Hadaway at the RAF Air Historical Branch for providing countless copies of official inquiries into many of the accidents I cover.

The wonderful team at my fantastic publisher Simon & Schuster for their encouragement, advice and expertise, and to Tim Bouquet for his time and assistance.

I am always grateful to my friend and agent of over thirty years, Mark Lucas, for his writing expertise and guiding hand of wise counsel.

My wonderful wife Suzannah and daughter Sophie who are always there with much-needed love and support.

Finally, I am truly grateful to the countless ejectees and their loved ones who told me their stories of brushes with death, and joy at survival. I was only able to use a fraction of the accounts I heard, so I hope I have done justice to you all.

* * *

I was assisted by many other serving and retired military personnel, researchers, authors and historians who offered invaluable information and contacts. It is impossible to name them all, but the following provided important leads, accounts, pictures or advice:

Lisa Ambrose, Ward Carroll, Ricci Cobelli, Terry Cook, David Eberly, Tim Ellison, Lesley Gent, Peter Griffiths, Ron Haack, Barrie Hay, Iain Hay, James Heeps, Jez Holmes, Dicky James, Steve Jarmain, Col King, Jenny Lancaster, Graham Lancaster, Mark Lewis, Enrique Lippi, Bob Marston, Pete Marston, Simon Mills, David Morton, Matt Nelson, Andrew Nichols, Rick Peacock-Edwards, Craig Penrice, Tony Pollard, Les Rawlins, Nicola Rees, Mike Rondot, Jo Salter, Andy Saunders, Kath Sherit, John Shields, Ed Smith, William Spencer, John Taylor, Jeff Tice, Glen Turner, Jim Wild, Phil Willsher, Dave Unwin, Mark Wakeman, Dean Wood, Geoff Zuber, Owen Zupp.

FOREWORD

Late summer sunshine highlights the narrow lane winding through the Buckinghamshire countryside. A few miles south, giant aircraft roar skywards from London's Heathrow Airport, holidaymakers strapped safely in their seats. At first glance, the village of Higher Denham, with its community field and recently installed play park, presents a picture-postcard image of English country life. But as the lane narrows to a single track, I'm suddenly confronted by sturdy blue security gates guarding a factory entrance.

An electronic sign on the wall proudly declares that, to date, the Martin-Baker Aircraft Company has saved 7,681 lives. Since it was formed in 1934, Martin-Baker has designed and produced over 90,000 ejection seats for ninety-three of the world's air forces. As I wait for my escort in the security office, I scan the thousands of names inscribed on the wall of aviators granted a second chance at life thanks to the company's incredible devices. After becoming ejectee number 6,089 when I was shot down in combat during the Gulf War in 1991, my own name is on that wall, alongside those of many of my friends.

A host of aircrew find themselves making this pilgrimage. My guide on this occasion is the company's Business Development Executive, Tony Gaunt, whose previous job in the RAF gave him responsibility for the testing and maintenance of ejection seats once in service. Needless to say, all aircrew regard Tony and his fellow armourers with huge respect. And gratitude.

* * *

The machine shop, where giant industrial lathes and presses, some dating back to the 1940s, jostle with state-of-the-art metal cutters, is a hive of activity. I watch a porter drag an old-fashioned trolley loaded with briefcase-sized chunks of solid aluminium towards a huge, computer-controlled grinder which stretches across the entire width of the factory floor. Exactly six hours and seventeen minutes later, they will emerge as perfectly sculpted components of an ejection seat, en route to the 'Rumble Room', where the machine marks are polished out by being blasted with what look like tiny volcanic pebbles. Finally, the finished sections are fed through huge containers of chicken feed. After endless trials, the highly absorbent foodstuff was discovered to provide the best drying medium with which to conclude a process requiring substantial quantities of water.

An engineer sips from a steaming mug of coffee in a nearby workshop as she studies a complex diagram. She then secures two bottle top-sized pieces of tubing with a dab of adhesive and measures the resulting device – which will later form a minute section of an ejection initiator unit – with a micro-meter.

In the sewing room, portable fans cool a dozen men and women hand-stitching the dark-green straps which will make up the seat and parachute harness. Their work is equally precise, and their attention to detail astonishing. Later, I watch Taj lay 45 feet of parachute and rigging lines across a giant table in the packing room. Over the next two days, he will fold them into a black metal headbox the size of two 6-pint milk cartons. Taj has done the same job at Martin-Baker for thirty-five years. 'I have packed many thousands of parachutes,' he says proudly. 'I carry out my work as perfectly as I can, every single time. What you aircrew do up there is always on my mind.' In a surreal moment, Tony Gaunt casually mentions that it is entirely possible that Taj could have packed the parachute that carried me from my burning Tornado 'safely' into the Iraqi desert – the first part of my journey into captivity as a prisoner of war. It sparks our reflections on the fact that while most ejections might be considered 'routine' – a life saved from mortal danger in a dying aircraft – there are occasions when the ejectee's safe return to earth marks the start of a new, sometimes perilous mission to survive. It was a theme which many of those I interviewed for this book echoed.

Elsewhere, sleek, black, fully formed metal seats perch on wheeled frames at various workstations. At each one, a computer monitor guides a technician through the precise fitment of a seemingly endless series of electronic components, wires and cables. 'We all know that these seats will be in aircraft in months to come,' the technician says. 'They could then be flying over Northumberland in training, or Syria in combat. That's really important to all of us. I was taught to think about the person who would sit on this seat. Their life is in my hands. It's why I'm passionate about what we do here.'

We pop into the main headquarters to chat to John and James Martin, twin sons of the company's founder, Sir James Martin. Now in their early eighties, they are still at the helm of the family business. The walls of their office are covered with certificates and awards interspersed with signed photographs and letters from grateful flyers. A portrait of a steely-looking James Martin gazes down over the oak desk where he once worked. In pride of place on either side are industry awards signed by the late Queen Elizabeth II, and her father George VI. John rifles through a pile of paperwork to show me a letter they recently received from a 1950s ejectee expressing his ongoing thanks to the company.

Pride and passion are in evidence everywhere and back in Tony's office, surrounded by models of the many types of aircraft the seats are fitted in, he reinforces the message I had received from those I encountered among the company's 1,200 personnel. 'We have one aim here. To produce something that will hopefully never be used as intended. But, if it is needed in years to come, must work exactly as designed to save someone's life.' And it is not just Martin-Baker saving lives across the globe. American, Chinese and Russian companies design and produce their own ejection systems too. Taking into account the size of their respective military forces, though only limited statistics are available for obvious reasons, it might be estimated that Martin-Baker's tally of 7,681 lives saved could be multiplied tenfold or more to include worldwide ejections.[1]

* * *

The man who sits as *number one* in Martin-Baker's 'Ejection Hall of Fame' is former Second World War bomber pilot, Jo Lancaster. I had the privilege of spending time with Jo in his later years, and over ham sandwiches and mugs of tea in his sitting room, we chatted at length about his own experience of what he and his fellow flyers initially described, with the greatest scepticism, as 'a curious damn contraption'. And, on the odd occasion, 'a bloody dangerous invention'.

When Jo Lancaster ejected in 1949 on one of the first seats produced by Martin-Baker, it took around thirty seconds from pulling the handle for him to be safely under a parachute. As an aviator brought up on fully automatic ejection seats, I was staggered by Jo's description of the interminable process he went through to escape. Jettisoning his canopy, wrenching his ejection handle, blasting clear of the aircraft, waiting for the seat to stabilise, then manually unstrapping and pushing it away before finally locating and pulling his parachute-release handle – all while plummeting earthwards. After my Tornado was blown apart by an enemy missile over Iraq in 1991, a mere 2.5 seconds separated my ejection initiation to automatic parachute deployment.

Jo and I both had to make a conscious decision to pull that handle. Incredibly, today's aircrew benefit from a system so automated that in certain conditions it will take the decision out of the pilot's hands and *auto-eject* them if it considers the flyer faces mortal danger and cannot react quickly enough. But, regardless of the technological evolution of our individual seats, Jo and I – and every other member of our exclusive club – have one thing in common: a visceral understanding and appreciation of what it means to be given another chance. And the sons, daughters and grandchildren of ejection survivors I spoke to during my writing made it abundantly clear that there are countless others who owe their lives to that remarkable invention.

This book is not an exhaustive account of every aspect of the design and development of ejection seats; it is an exploration of the extraordinary journeys travelled by ejecting aviators, and their loved ones.

Because, more often than not, reaching for the yellow and black ejection handle is just the beginning of the story . . .

PROLOGUE

'IT WILL NEVER HAPPEN TO ME'

RAF BITTESWELL, LEICESTERSHIRE
MONDAY, 30 MAY 1949

A day on the ground for John Oliver 'Jo' Lancaster always felt laboured and unfulfilling. This May morning, he was brimming with anticipation. 'I was exploring the limits. There were a lot of unknowns but I really enjoyed this aspect of experimental flying. I was the proverbial pig in whatnot.'[1]

Leaving their rented house just outside Coventry, a city still scarred by the Blitz, he kissed his wife Betty and their 2-year-old son Graham goodbye. At nearby Baginton Airfield he boarded his company's Dragon Rapide, Armstrong Whitworth's 1930s vintage six-passenger short-haul plywood biplane, for the flight over to RAF Bitteswell. The grass runways at Baginton – Armstrong Whitworth's design, administration and production base – were not always serviceable in wet weather, so his employer had moved its test-flying to the three all-weather concrete strips at Bitteswell, 12 miles to the east.

After the short hop, the 30-year-old deputy chief test pilot strode across the tarmac, the sun warm on his face and the blustery wind rippling his jet-black hair. He reflected on how radically aircraft design had changed in his lifetime. Captivated by flying machines as a boy, Jo had sketched them and made wooden models powered by rubber bands. As a young man in the Second World War, he had flown fifty-four heavy bomber sorties for the RAF and won the Distinguished Flying Cross. Flying was his life and now he was heading towards aircraft serial number TS363, an AW52 prototype jet, hence the large 'P' emblazoned on its gleaming white fuselage just below the cockpit. It was hoped it might be as mould-breaking as the world's first powered aircraft: the Wright Brothers' legendary Wright Flyer.

Resembling a 90-foot boomerang with upstanding rudder/fins on each wingtip, the AW52 represented aviation's biggest leap yet into the future. Its stubby, acorn-shaped fuselage measured a little over 37 feet and had no tail. Already dubbed the 'Flying Wing', its ancestry could be traced to two of the great engineers of the nineteenth century, William George Armstrong and Joseph Whitworth, but it boasted more than a hint of the science fiction comic creations of its pilot's youth. Incredibly, it looked remarkably like a smaller version of the iconic B-2 'Spirit' Stealth bomber of today.

Jo – call sign Raider Three – had joined the Armstrong Whitworth test programme in January and this would be his third outing in the AW52. His first, on 23 May, had lasted just over half an hour and helped him get a feel for it. His second, two days later, was a forty-five-minute sortie gradually increasing speed in readiness for today's targets. In the era before computer-aided design and the common-place use of wind tunnels, it was down to the revered test pilots to discover first-hand how a new aircraft would actually fly. 'The only way to establish how these new aircraft performed,' he said, 'was to just climb in and take them airborne.' The flight test department had already given him his instructions for the day: a series of handling checks between 270 and 350mph to establish how the aircraft performed. The ultimate target was 500mph, but that would come later.

Of all the aircraft he had flown, and they were many, he believed this one, powered by two Rolls-Royce turbojets, was by far the most beautiful. Armstrong Whitworth, backed enthusiastically by the British government, hoped that if the testing programme was successful the AW52 would be the stepping stone to a commercial airliner twice its size, powered by four or even six jet engines, with

passengers seated within the wing itself.[2] Their advertisement in
Flight magazine, with an artist's impression of the proposed, much
larger passenger version gliding high over a perfect blue ocean on a
clear and starry night, proudly carried the caption: 'The Shape of
Wings to Come'.[3]

Incredibly, the £200,000 machine (equivalent to £7.2 million
seventy-five years later) came without any comprehensive pilot
manuals or instructions. In limited preparation for the test regime,
Jo had conducted training sessions on tailless gliders and been on
the receiving end of verbal briefings, but that was it. 'There wasn't
much to do. I checked the aircraft over, making sure that everything
was in order and that it all looked okay,' he said. 'The jet was not
equipped with any flight-test instrumentation or data recorders.'
Nor did he have any Flight Reference Cards (FRCs) used by most
aircrew today, containing technical information and checklists
which must be completed before, during and after a mission – and
which spell out the drills that have to be followed in emergencies.
As it was just a routine test flight, this wasn't going to be a problem.

* * *

Thanks to the Pathé News films showing in Britain's cinemas,
which enthusiastically reported their exploits to soundtracks of
stirring patriotic music and clipped-vowel voiceovers, a handful
of test pilots had become household names, not unlike the first
generation of astronauts or the Premier League footballers of the
next millennium. They were an elite band, responsible for taking
aircraft of the future to their absolute limits of speed, altitude and
manoeuvrability, to evaluate the performance of man and machine
under extreme stress.[4]

Most of them knew one another. 'I remember in a single year
after the Second World War, twenty-four British test pilots had lost
their lives,' Jo recalled.

The death of one famous pilot, 36-year-old Geoffrey de
Havilland Jr, three years earlier, had cast a particularly long

shadow. De Havilland had put his prototype jet into a high-speed dive from 10,000 feet over the Thames Estuary, east of London. As it approached 5,000 feet at around 650 miles per hour, there was such pressure and vibration on the front edges of its wings caused by onrushing air that the main spar supporting their weight cracked and the aircraft pitched violently downwards towards Egypt Bay at Gravesend, disintegrating as it fell. De Havilland's body was found on the mud flats the next day. He had suffered a broken neck, almost certainly the result of the aircraft having undergone severe and violent shaking prior to breaking up, causing his head to strike the cockpit canopy with lethal force. Although equipped with a parachute, which had not been deployed, like so many of his generation de Havilland had been killed by a lack of a fully developed aircraft escape system.[5]

'A good friend of mine was also killed when the flying boat in which he was practising low-level aerobatics for a Battle of Britain air show came down,' Jo said. 'It crashed into the sea near Felixstowe Harbour and broke up on impact. He died instantly.' But Jo was no stranger to death, having witnessed many friends killed during the war. 'Bomber Command suffered very heavy losses but you didn't really talk to anyone about the dangers, or that we were obviously facing death every time we flew an op. I was always optimistic about my survival. At the back of my mind I simply thought that if the worst happened and I was shot down, I'd bail out, use my parachute and end up as a POW. As long as the parachute worked, of course!'

Three years after the war, Jo had his own close brush with disaster in an experimental flying boat. The bulky, unconventional beast had floats under its wings, which could be lowered when landing on water. This time only one came down. In spite of his repeated efforts to free the other, it remained stuck fast. 'I had considered running the aircraft into nearby salt marshes but instead I attempted a normal water landing.' With only one float, he risked tipping a wing into the water, cartwheeling over, breaking up and sinking. But after undoing his harness in preparation for a rapid escape, Jo managed a textbook landing, cut his engines and, as the aircraft settled with one wing resting on the water, he slid back the canopy and leapt onto the starboard wing to maintain the aircraft's balance until a recovery crew came out to tow it to safety.[6] He didn't give it another thought. His role as a test pilot was to perhaps risk his own life in the interests of the advancement of aviation. Jo was just happy to be part of it.

* * *

The AW52's cocoon-like cockpit was snug for a man of his imposing, square-shouldered build – like being behind the wheel of a Formula 1 racing car – but his pilot's seat was far from inviting. Climbing in, he eyed it warily. Its rudimentary frame, constructed from welded light alloy tubing and sprayed British racing green, was nothing like the conventional apparatus he was used to. There were two sets of chunky, fawn-coloured canvas straps which met at large buckles and its footrests and thigh guards seemed to belong on a white-knuckle fairground ride. The red handle protruding from the rectangular box directly above his head did nothing to inspire confidence, and nor did the telescopic metal tube fixed to the back of the seat in line with his spine. This was the so-called 'ejection gun', inside which were two explosive charges and the reason why pilots already referred to these strange new devices, somewhat dismissively, as 'bang seats'.[7]

Jo was about to strap into one of the world's first, and most rudimentary, ejection seats. And he was not alone in viewing what he described as 'a curious contraption' with a degree of suspicion which bordered, if he and his fellow aviators were prepared to admit it, on something close to fear. 'I knew that a handful of brave chaps had made a couple of successful trial ejections from other aircraft in "cold blood" tests. But we had not heard of a single one deployed in a genuine emergency.[8] We had very little information about them and I was very sceptical about the whole damn thing.'

Despite his misgivings, Jo sat down and then fastened the two sets of straps around his shoulders and thighs, one for his personal parachute and the other to anchor himself in position on the seat itself. In an emergency, the seat – and Jo – could now be blasted out of the jet together. The integral thigh guards should prevent his legs from being ripped apart on exit.

Most pilots of earlier single-seat aircraft had been accustomed to sitting on some style of bucket seat bolted firmly to the floor, harnessed by two shoulder straps and a lap belt. They would either be wearing a parachute on their back, or be attached to one that hung around their backsides to sit on. Standard practice *in extremis* was to roll your aircraft onto its back, release the harness and fall out, or to grab the edge of the cockpit then roll over the side, opening your parachute when you'd fallen a safe distance. Some tried to clamber out and jump clear. But taking pot luck and hoping for the best was no longer an option. With the advent of the jet age, aircraft were getting faster and faster, and bailing out manually could prove fatal. As one of Jo's colleagues had discovered.

* * *

Squadron Leader William Davie, a 25-year-old test pilot, had already bucked the odds in August 1943 in a stricken Gloster E28/29, a testbed for Britain's new jet engines. The controls had jammed and the aircraft plunged into a high-speed, spinning dive. As one account put it: 'The gyrating plane jettisoned him into a 20,000-feet free-fall, stripping him of his boots, helmet and oxygen mask.' The Scotsman only survived in the rarefied air because he was able to suck on his severed oxygen tube and eventually open his parachute. Less than six months later his luck ran out when he was on a high-speed run in a prototype of Gloster's twin-engine fighter, the Meteor, at 20,000 feet over the Royal Aircraft Establishment at Farnborough. His port engine disintegrated, leaving the aircraft hurtling out of control. As he attempted to bail out, his left arm was severed, probably by the cockpit canopy slamming back shut in the wind blast as he opened it. But his second attempt to escape a falling jet was still successful as, against all odds, he managed to exit the cockpit. Only to then be smashed into the tailfin as he tried to leap clear. Severely injured, he was unable to open his parachute. His body plummeted through the roof of a hangar at Farnborough.[9]

This life and death lottery had to stop. Pilot safety could no longer be left to chance. If the AW52 got into trouble, Jo's ejection seat should blast him up and away from the cockpit at around 400mph. He'd then still have to unstrap his harness and push himself out of the metal seat while he was in free-fall. Before finally opening his own parachute once the seat dropped clear. Would it work? The jury was still out as far as Jo and his contemporaries were concerned. 'I thought it was a rather ingenious, if somewhat curious contraption, but that I would never need it. I *knew* what it did, but hadn't had any particularly detailed instructions on its operation; it was just *there*.' For a man who had spent his flying career doing his best to minimise risks, this new invention seemed to promise a lot more jeopardy than reassurance.

* * *

Safely strapped into the cockpit, Jo surveyed the array of dials on display beyond the handles of his spectacle-shaped control column. To his right and left were more gauges, levers and massed ranks of knobs and switches. The Lancaster bombers he had flown during the war were unsophisticated – even archaic – by comparison with this new generation of flying machine. Their seven-man crews had been crammed into the bomber's fuselage with limited chance of escaping when things went wrong. The AW52 had room for just two. The observer's position directly behind him was not equipped with an ejection seat, and today it was empty. Jo was flying solo in every sense.

He checked his fuel gauges. Everything was just as it should be. He glanced up at the fluffy cumulus floating high above the airfield. Partial cloud cover. A perfect day for flying. Some turbulence had been forecast but nothing untoward. He closed the bug-eyed canopy at the prow, pressurising the cockpit, then smiled and waved to the ground crew. They slid the chocks from the undercarriage and pulled them away at a run as his engines growled and began stirring up the air.

The last thing Jo did before taxiing onto the runway was to reach down, remove a small pin from the ejection gun containing the explosive charge, and insert it into a specifically designed slot in the side of the seat. He'd taken the safety catch off the gun, the pin out of the grenade. His ejection seat was now live.

The noise from the prototype's engines was ear-shattering for the watching ground crew, but inside the cockpit the atmosphere was almost serene. At the head of the runway, he reached out with his left hand and palmed his throttle levers forward as far as they would go, released the brakes and raced down the runway. As the few onlookers around the airfield stared at its curious shape, the AW52 eased up, almost floating airborne at 90mph like some giant white bird or an alien craft heading back to the outer reaches of space.

Jo ran through his allocated testing schedule as he gained height. A fellow test pilot had reported the aircraft's tendency to buck when pulling out of a shallow dive at 280mph. 'I don't recall any great consternation within the company about what he had experienced,' he said. 'And, personally, I wasn't too concerned about any dangers. I was rational. I knew the risks of being a test pilot. Like all of us, I thought, *It will never happen to me.*' His stoicism stemmed from having returned home unscathed from his relentless succession of wartime bombing missions. 'It was a bit dicey at times, but we were lucky to survive each op and get safely back on the ground.'

He could see the town of Rugby sprawling alongside the River Avon through gaps in the clouds. 'I spotted an occasional car; life was going on as usual so this was a pleasurable way to spend my time. I liked being on my own, enjoying the freedom of the skies.'

Unlike tried-and-tested production models, the Flying Wing was a collection of new design concepts. Jo began putting it through its paces at 270mph. 'It might have been graceful to look at but it wasn't particularly easy to fly,' he admitted. 'It had a bit of a flaw as back then there were no power flying controls.' So, the combination of the AW52's huge, sweeping wingspan and short stubby fuselage meant that laterally it handled like a Lancaster, but was as responsive fore and aft as a Spitfire – a technological conflict of interests.

Jo was in effect piloting an experiment. For the next twenty minutes he carried out a series of runs, gradually increasing his speed and monitoring the jet's reactions. He then climbed into bright sunshine at 5,000 feet and began a shallow dive to trigger the next phase of the test. Coming back through the clouds at 320mph, heading south-west towards Bristol, the turbulence increased. 'The first indication something was wrong was instantaneous – a sudden bucking fore and aft like a rollercoaster.'

Every few seconds the Wing climbed 8 feet and then slammed back down. The bucking became increasingly frenetic. Jo was being hurled up and down in his seat. 'It was so violent I couldn't see anything on my instruments. There was an astonishing noise of shaking and rattling as the metal structure was stretched. I rapidly became very disorientated.' He needed every shred of experience to drive away the panic. He tried to throttle back. No response. The noise creased his eardrums. Every rivet, seam and weld was beginning to bend. The Flying Wing was out of control and Jo's head was now being battered around the cockpit as he vainly fought with the controls.

At 3,000 feet and dropping fast, the machine was trying to turn itself inside out. Jo's gut told him that it could break up at any second and he would die. Even if it did stay intact, he feared he would be knocked unconscious by the violent shaking and the jet would plough into the ground with the same result.

His thoughts briefly turned to Betty, who would be working through her Monday chores back home, with no idea she was on the verge of widowhood, or that little Graham was about to be robbed of a father. The new 'contraption' he was initially so suspicious of was now his only means of survival. It was Jo's last chance of exiting the plummeting aircraft and seeing his beloved wife and child again. 'I knew I just had to get out.'

He tried to recall the drills he would need. Get rid of the cockpit canopy. The hood jettison toggle was in front of him, to the left of the control column. Then reach above his head for that red ejection handle. He had been told he should grab it with both hands, wrench it firmly outwards then hard down. This physical action would do three things instantaneously: release a protective blind that would stop his face being torn to pieces by the slipstream when he left the aircraft, get his body into a position that would brace his back and prevent serious injury to his spine, and fire the explosive charges in the seat.

Ground rush. The earth was careering up to meet him.

He had seconds to get out.

Still falling.

The ejection seat was ready to go. All Jo Lancaster had to do was pull his red handle.

Pull it straight out and then hard down.

His hand reached upwards.

But the aircraft was jerking him about with such brutality that he was both physically and mentally disorientated. His vision was blurred. The noise was overwhelming. He felt weak and his perception of light was fading.

One of Britain's finest aviators was rocketing towards the edge of oblivion.

PART I

1930–1951

THE DEATH OF A FRIEND, THE BIRTH OF AN IDEA

CHAPTER ONE

A SHAKY DO

RAF ALCONBURY, HUNTINGDON
24 JULY 1941

Jo Lancaster's illustrious career in the air had nearly been over before it began. He had joined Armstrong Whitworth in Coventry before the war as an apprentice engineer, and also flown as a member of the RAF Volunteer Reserve. He was as near to heaven as it was possible for a teenager to be.

Then, disaster.

After performing some unauthorised low-level aerobatics for a watching friend during a training flight in a biplane, he had pushed his luck too far. While then practising a simulated emergency landing, his engine failed and the aircraft crashed through a fence. Although he escaped injury his aircraft was nearly destroyed, with both port wings shorn off.[1] He was discharged from the RAFVR. His treasured pilot's logbook was endorsed, 'suspended from flying training for a breach of flying discipline. Crashed aircraft thereby causing extensive damage to same.'

The Chief Flying Instructor dismissed his student as 'average'.

Jo's logbook stopped at a blank page on which he pasted a newspaper picture of his biplane with its two severed wings beside his dismissal notice.[2] He had completed fifteen hours solo and, aged only nineteen, his world had fallen apart. 'I was devastated. It was

all I wanted to do; an escape from everything I knew.' Then, in a
dramatic turnaround of fate, war was declared in September 1939.
Pilots were in short supply so the RAF forgave his previous misde-
meanours and recalled him for training. And now here he was, at
the age of twenty-two, on his first tour, an officer and a captain on
40 Squadron ready to fly his Vickers Wellington bomber on a mission as daring as it was desperate.

By early 1941, battle cruiser *Gneisenau*, heavy cruiser *Prinz Eugen*
and battle cruiser *Scharnhorst* had sunk twenty-one British mer-
chant ships between them. Alongside the success of the U-boats, it
looked highly likely that the Germans were winning the Battle of
the Atlantic, cutting off Britain from the rest of the world.

   Scharnhorst had slipped out of Brest and headed south to La
Pallice near Bordeaux, but *Gneisenau* and *Prinz Eugen* were in dry
dock for repairs. They were sitting targets. Or were they? Since the
end of March, the RAF's bombing raids on these ships had involved
hundreds of aircraft with limited success and heavy loss of life.[3] The
clock was ticking. On the far side of the Atlantic, 30,000 Canadian
troops were preparing to sail for Europe. The German cruisers had
to be taken out to protect the convoy from deadly attack. Operation
Sunrise, a daylight raid by seventy-nine Wellingtons from several
squadrons was green-lit for 24 July. It was their last chance.

Jo's crew loved the Vickers Wellington. Its metal carcass of diagonal
braces covered in a tight fabric skin had been developed by Barnes
Wallis, who would go on to invent the famous Dambusters' 'bouncing bomb'. Those who flew the iconic long-range, twin-engine
bomber affectionately nicknamed it the 'Wimpy', after the iconic
cartoon character J. Wellington Wimpy from the *Popeye* newspaper
cartoons, who had an insatiable appetite for hamburgers and cigar
butts.[4] 40 Squadron was supplying six aircraft from its base a few
miles from Huntingdon. It was to be Jo's first daylight raid.

   A small crowd had congregated to wave them off at around 11am.
When they joined the main battle group, they would adopt two

'inverted V-shaped' three-aircraft formations, with Jo's Wellington on the left of his V. 'None of us had any experience of formation flying, but as the route was via the Scilly Isles it gave us a bit of time to practise.' Jo was experiencing a number of 'firsts'.

Just two sorties into his thirty-operation tour in 1941 he had been a member of the funeral party for a rear gunner who had been killed on his twenty-seventh op. German night fighters had raked the tail of the Wellington over Hamburg, and he'd had no chance of escaping. The stricken aircraft had limped back to base. 'I saw the turret where he was killed,' Jo remembered, his voice faltering when he described the experience to the author some seventy-five years later. 'It was a hell of a mess . . . it brought home the reality of war. It was a sobering sight. All very sad . . .'

But like many of the 'Bomber Boys', as they would become known, he was driven by an unshakeable optimism. It did not do to dwell too much on lost comrades. 'We were more concerned with having fun, living life to the full, for the moment.' On this, their fifteenth sortie, Jo and his crew were in their usual good spirits.[5] 'The day was hot and cloudless, with thick smoke haze over southern England; a perfect day for flying. As a crew, we never talked about the dangers, or that we were facing death every time we flew an op. And we rarely thought about the reality of trying to escape from a doomed aircraft.'

Bailing out of a lumbering, stricken bomber was far from easy when early aircraft escape techniques were rudimentary at best: strap on a parachute and jump. While fighter pilots were sometimes lucky enough to climb or roll out of their aircraft, bomber crews had to use cramped escape hatches. And only after those not already wearing them had collected their stowed parachutes, spending precious seconds to fasten an array of hooks, fingers fumbling in frostbite temperatures, as their aircraft plunged towards the ground. Over the course of the war, almost half of the 125,000 men who flew in Bomber Command would be killed on operations or during training. Jo and his colleagues knew none of this, and their focus that July morning was not on statistics – it was on the seventy-nine

Wellingtons now in place as the formation flew south-east towards the Brittany coast.

Flying in at 10,000 feet, they found themselves heading towards cloud of a more threatening kind. Flak. 'It was chaotic,' Jo says. The Wellingtons, not the warships, were the sitting ducks as the German ground defences opened up. 'The sky over the target was thick with flak smoke. We could see, hear, feel and smell the continuous anti-aircraft flak. As we got nearer to the target we were being rocked by explosions. I could smell cordite in the aircraft.'

There was a huge crack inside the cockpit, and Jo jumped when a burning shard of metal punched a hole through his windscreen. 'I tried to make myself as small as possible, hanging onto the controls for dear life.' He gritted his teeth, got his head down and pressed on. He watched the lead aircraft in his V formation intently, holding his position and waiting for them to release their payload; the trigger for his own bomb aimer to drop theirs. He glanced down at the enemy ships in the harbour. 'There was no thought for the men on board. We just needed to hit them and put them out of action.'

A cluster of ME109s shot out from the gloom, engines screaming. The RAF formation was on the verge of turning into a melee. 'A stray Wimpy passed right underneath us with an ME109 hard on its tail. As it broke off its attack and went into a vertical climb, our rear gunner plastered him.' A kill. 'We had destroyed one of the enemy before he could kill us. But I was still relieved to see the pilot's parachute blossom open as he escaped from his doomed aircraft. He was a fellow airman and I was just pleased he possessed the means to survive.'

Jo's pulse started to race when he saw two Wellingtons plummet earthwards in flames and crash into the town. 'I was praying to see parachutes. Would they bail out? There was nothing.' They had no chance of escaping. Ten more deaths added to Bomber Command's toll.

Jo dropped his bombs and turned sharply away from the danger zone. Ten of the seventy-nine Wellingtons involved had gone down, fifty men missing. Jo counted himself fortunate. 'Our little trio of bombers came through without casualty although the aircraft were in a bit of a mess.' In contemporary RAF parlance, Operation Sunrise was what was referred to as 'a shaky do'. There were many more and Jo was lucky to survive that first tour. Very lucky indeed. Between that May and October in 1941, his squadron lost twenty-four crews; 120 men.[6] Like Jo, most of them were still in their twenties. 'I lost a lot of good friends,' he reflected later. 'We grew up very quickly.'

And he never forgot the mornings after a night raid, and the empty seats at breakfast. Nor did he forget the sight of aircraft being shot out of the sky, hoping against hope that he would see their crews escaping, parachutes billowing. Usually, he saw none. Escaping a doomed aircraft was a lottery.

RAF WING, AYLESBURY
12 SEPTEMBER 1942

Still trim in his flying suit at the age of fifty-four, Valentine Henry Baker was about to test-fly the MB3, a camouflaged prototype fighter being put through its final checks. He could not have been more different in appearance or temperament to James Martin, the stocky fellow in a three-piece suit and fedora hat standing beside him.

Born in Wales in 1888, 'Bake', as he was known to his many friends, was everything one could wish for in a flying ace. Dashing and stylishly dressed, Captain Baker had won a Military Cross early in the First World War for, 'conspicuous gallantry and devotion to duty, showing the greatest daring and determination'.[7] For good measure, he had also served in the Royal Navy and the British Army, and survived being shot in the neck at Gallipoli. The bullet was still lodged there, too close to his spine to risk removing it. After the war he had set up a famous flying school; aviation was stamped into Baker's DNA.[8]

James Martin had no such lineage. A farmer's son from County

Down, his forebears had tilled the same 30 acres since the early eighteenth century. But he had no interest whatsoever in following in their footsteps. As a teenager he had tinkered with engines to see how they worked and make them perform better. He'd invented a three-wheeled motor car, a fish fryer, tool sharpeners and bicycles with rain-proof hoods. 'I had the notion that I wanted to design things and not be employed by anyone.'[9]

Having left school at fifteen and been turned down for a place on an engineering degree course at Queen's College, Belfast, he had arrived in England in 1919 aged twenty-six, with £10 in his pocket. He had no qualifications, no job, no contacts, and no workshop. Gathering skills as he went along, he began buying army surplus trucks from a depot in Slough, overhauling and modifying their engines, then selling them on. He also experimented with his own designs for trucks and trailers which he went on to patent.

By turns determined, irascible and single-minded, he was virtually a one-man band: inventor, draughtsman, experimental engineer, toolmaker, fitter, assembly man and salesman. Like many visionaries, he was also obsessively restless. He had bigger plans. For years he had been fascinated by aircraft and how they were constructed and powered. In 1928 he loaded up his machine tools and equipment on a horse and cart and moved into a former linoleum factory in Denham, Buckinghamshire, around 17 miles from London. The sign he hung on the front of the building said 'Martin's Aircraft Works'. He had absolutely no experience of designing or building flying machines, but believed this was a mere detail, which could, and would, be overcome. The following year James Martin booked flying lessons with Valentine Baker.[10]

Although so different in character and background, the two men had hit it off instantly. Blessed with 'a breezy and unfailing good humour', Baker was popular and gregarious, while Martin was teetotal, shunned social gatherings and was happiest when working with mechanics on the shop floor. It was no surprise when they went into business together. By 1934 the sign at the Denham factory had been changed to read: 'The Martin-Baker Aircraft Company'.

Martin would design the aircraft, Baker would be co-designer and test pilot.

Martin's reputation was growing, as were his orders and his staff. As fighter pilot casualties rose during the Battle of Britain in 1940, it became obvious that Spitfires needed cockpit hoods that could be easily jettisoned to increase the chances of escape by parachute. James Martin came up with the answer. In what was to be one of the first forays into assisted escape systems, he mounted a small red rubber ball on the arch of the hood, which was attached to cables connected to unlocking pins. When the ball was pulled, the cables whipped out the pins, leaving the slipstream to tear off the canopy. Countless pilots owed their lives to that little red ball. But his great ambition still consumed him: that Martin-Baker would become a leading aircraft manufacturer.

Valentine Baker had already tested their first prototype in 1935 – the MB1, a two-seater, single-engine monoplane. With war looming once more, the Air Ministry had announced it was looking for a new fighter that could reach speeds of 275mph, so Martin designed the MB2. It could achieve speeds of 350mph and was highly manoeuvrable. But the Air Ministry never pursued the design. Although Martin was swiftly adopting what would become a lifelong antipathy towards Whitehall mandarins, he remained undeterred and produced the MB3. This sleek machine boasted a maximum speed of just over 400mph and six 20mm cannons. On 31 August 1942 Baker carried out the first test flight, reporting that the MB3's handling was excellent. Twelve days later, he was due to take the MB3 airborne again.

Heading out to RAF Wing by car, Martin and Baker stopped to take a pee behind a hedge. Baker's brow was uncharacteristically furrowed. He turned to Martin and said, 'I have a feeling, Jimmy, that something is not quite right. And I don't know what it is.'

When they arrived at the airfield they discovered that the new fighter's engine was not running as smoothly as it should. Unable to find any serious problems, Martin and the engineers cleared it to fly. Baker, typically cheerful and forgetting his doubts of earlier in the day, climbed into the cockpit and strapped himself in. His audience watched him increase speed and hurtle down the runway. All was good.

Until, with no warning, the MB3's engine suddenly cut out. There was no room for an emergency stop. Then, the engine roared back into life. Martin breathed a sigh of relief as the MB3's undercarriage left the tarmac and Baker gained height. He was at 50 feet. Then the engine cut out again.

This time it did not restart.

Valentine Baker disappeared over a line of trees and out of sight. Seconds later there was a huge explosion and black smoke belched into the air. Martin and his mechanics raced towards the crash site.

Reports vary as to what might have happened. He'd hit a tree stump and cartwheeled, it was said. Or maybe it was a haystack.

The sight that greeted James Martin would haunt him for the rest of his days. The MB3 had been destroyed, shorn of its wings and rear half. Some 120 gallons of high-octane fuel was blazing like a bush fire. Steel panels and tubing had virtually melted in the heat. Baker was stuck in the cockpit and the stench of burning flesh caught in the back of the horrified witnesses' throats. As the flames subsided, Martin's men rushed to a fence and wrenched out two panels which they used to prise Baker from the cockpit. It was too late. His remains rolled down the stub of a wing and hit the ground. Valentine Baker's body had been reduced to half its size.

Martin flung himself down onto a grass bank. 'My dear Val,' he sobbed. 'My dear Val.'

In those days, test-flight crashes were not uncommon, but nobody present that day had seen anything as gruesome as this. Sick, distressed and in silence they trudged back to the aircraft hangar. Martin paced up and down, racked with grief and disbelief, unable to banish a flurry of horrific images from his mind, or the fact that the aircraft he'd designed and built had condemned his friend to death. Later, Valentine Baker's coffin, draped in a Union Jack, lay in state in the Denham machine shop, where his many admirers would also gather for his funeral. In a commemorative booklet James Martin wrote: 'He was one of the most loveable souls it has ever been my luck to meet.'

Once back in his Denham office, photographs of Baker grinned at Martin from every wall. There were times when he could almost hear the laughter that had characterised the company's meetings during their fourteen-year partnership. But Martin wasn't just haunted; he was also angry. If the Air Ministry had not forced him to use less powerful engines than he'd wished for, or had agreed to his earlier request that the test flight be launched from an airfield surrounded by open countryside rather than by trees, he believed that Valentine Baker could still have been by his side. He vowed that he would never again let the Air Ministry push him around. Or let the lives of pilots continue to hang by such a slender thread.

There had to be some way of improving their odds of escape. What could he do to assist them? Given James Martin's restlessly inventive mind it would be surprising, following Baker's death, if he were not pondering how he could make that happen.[12]

RECHLIN-LÄRZ AIRFIELD, GERMANY
13 JANUARY 1942

Because of the difficulties in receiving international news, and the ongoing war, James Martin could not have known that several other countries had been investigating the possibilities of assisted escape systems for many years. Indeed, the first record of any form of 'ejection system' reportedly came in 1910, just seven years after the Wright brothers had made the first, heavier-than-air, powered flight at Kitty Hawk, North Carolina. Though little is known about the system, it involved the 'detonation of an explosive charge' in the seat.[13] And in 1912, a French inventor had trialled an explosively deployed parachute attached to a pilot. Fired out of the cockpit by a cannon, it then deployed to haul the occupant clear.[14]

And, despite being in the midst of the Second World War, the Germans were also well ahead of the game.

Seven months before Valentine Baker's death, German test pilot Helmut Schenk climbed into the cockpit of his Heinkel He-280 at the Third Reich's principal testing ground 60 miles north-west of Berlin.[15]

At this stage of its development, his experimental jet was incapable of launching on its own. As the tow aircraft hauled Schenk into the air, the runway beneath them glittered with snow and ice. At 7,875 feet it was time to ditch the cable. He gripped the release handle. It was frozen solid. He tugged it harder but could not budge it. He waggled his wings to alert the crew ahead that he had a problem. They misread Schenk's signal and let go of their end of the hawser, leaving him to wrestle the controls of an underpowered machine with a steel snake dangling from its nose.

There was no way he could land safely.

He had just one chance of survival.

The He-280 was the world's first turbojet-powered fighter aircraft. It was also one of the first aircraft to be fitted with a '*Schleudersitzapparat*': a 265-pound, state-of-the-art, compressed-air-powered 'catapult seat device' escape system.[16] Heinkel had been developing the technology since 1939, firing sandbags and dummies up a rig. They had also completed their first airborne test with a human being. Now it was happening for real.

Schenk discarded his cockpit canopy and pulled the seat release lever, firing the compressed-gas charge and hoping the device would propel him to safety. 'I was thrown clear of the aircraft without coming into contact with it,' he said later. 'During the acceleration I did not lose consciousness or notice any disagreeable feeling.' He had been shot out 19 feet above his jet. 'I realised I was revolving considerably and believe I executed a backward somersault, as I recall seeing the aircraft again. I succeeded in jettisoning the catapult seat, which quickly fell away from me. I then pulled my ripcord and the parachute opened perfectly. The opening shock was more violent than that experienced during catapulting.' Schenk landed safely in the record books, the first man in history to use an ejection seat in an emergency.

While this incredible milestone in aviation history was unknown to most, other leading German manufacturers, such as Dornier and Focke-Wulf, were also working on ejection seats powered by either compressed air or explosive cartridges. Decades ahead of their time, German scientists were even working on a system to convert the entire cockpit into an escape module. Swedish aircraft company Saab had also successfully fired an 80-kilo dummy out of an aircraft travelling at 170mph the same month as Schenk's escape and, by the end of 1942, many new German fighters and fighter-bombers were fitted with basic escape systems. Heinkel's compressed-air system was said to have saved the lives of sixty Luftwaffe pilots[17] before they, Dornier and Focke-Wulf switched to explosive seats to eject pilots.

History suggests that James Martin had not been aware of these developments, or of the work of earlier pioneers such as RAF Flying Officer Peter William Archdale Dudgeon of 208 Squadron back in the 1930s.[18] Dudgeon's colleagues often wondered why he was seen so rarely in the bar of the Officers' Mess. His squadron was switching to a new aircraft, the Armstrong Whitworth Atlas biplane, which could achieve over 200mph in a dive, and Dudgeon had realised that escaping pilots and their observer/gunners could be pinned firmly into their seats by the resulting slipstream, making a conventional parachute bail-out near impossible. He set up a drawing board in his bedroom and went to work with draughtsman's instruments and a box of Meccano.

A few weeks later, he emerged with detailed blueprints and a working scale model for a jack-in-the-box contraption. The Dudgeon seat would be mounted on tubes, each of which contained a powerful compressed spring, held in place by a strong but simple catch. Upon its release, the pilot would be thrust up to the lip of the cockpit, from where he could roll away and release his parachute. In great anticipation, Dudgeon submitted his drawings and model to the Air Ministry, who rejected the idea out of hand and left it to gather dust. It seems that they still believed that aircrew, if given the opportunity, would abandon their damaged aircraft at the drop of a hat rather than go to the trouble of attempting to return it to base. A similar argument had once been voiced against the provision of parachutes to enable crews to bail out of doomed aircraft.

* * *

Attitudes in the defence establishment to assisted escape had to change. And they did following the brutal death of Squadron Leader William Davie in January 1944, who, as we saw in the prologue, had been killed while bailing out of a doomed Gloster Meteor. His lifeless body plummeting through the roof of the Royal Aircraft Establishment (RAE) at Farnborough had been the most dramatic and tragic of wake-up calls.[19]

The RAE had immediately set up a committee to study the challenges of assisted escape, as the pilots who would be flying the new Meteor jets were anxious not to suffer the same fate as Davie. Finally spurred into action in April 1944, the Ministry of Aircraft Production started talking about possible methods of emergency escape. Could pilots be propelled upwards, or downwards? Or maybe they could be saved by part of the aircraft cockpit being blasted away. Then, another challenge. Any device or mechanism would have to cope with the problem of leaving an aircraft at high altitude where a pilot might be disabled by freezing temperatures or killed by lack of oxygen. Another possibility was a brake parachute to slow the whole aircraft down, allowing a pilot to bail out conventionally. The RAF's Physiology Department was also trying to answer such questions as how much acceleration a human body could withstand if ejected.

These complex issues, and the vagaries of assisted escape systems, were being pondered by many military agencies around the world. Development not only continued in Germany and Sweden, where it had originated, but also now in America, and almost certainly

Russia, where news of these ground-breaking systems was slowly filtering through. Britain's Martin-Baker company, however, provides a useful example of the worldwide challenge to save aviators' lives.

In the early summer of 1944 Fighter Command despatched a Wing Commander to Denham to seek James Martin's thoughts on the issue. Although Martin was still focused on aircraft design and had just seen the prototype of his latest fighter, the MB5, embark on its first flight trials, his response was immediate: 'We could fire them out.'[20]

In October 1944 the Air Ministry asked James Martin to come up with a design for a prototype escape system.[21] In spite of his previous engineering ingenuity and successes, Martin's relationship with both the RAF and Whitehall was still spiky at best, especially when it came to aircraft design and development, but he leapt at the challenge. Valentine Baker's death must have been a major factor, but so too was the news that the Air Ministry had given another aircraft manufacturer the same task. The Martin-Baker Aircraft Company was in a race. And James Martin hated losing races.

Presumably still unaware of German and Swedish developments in escape systems, Martin had only a blank page to work from. He had already impressed with his enthusiasm, and the quality and ingenuity of his engineering. In his mind, no problem was insurmountable. In double-quick time he came up with a model and specifications for what he called the 'Swinging Arm'. In simple terms, a long 'pole' constructed of sheet metal would be mounted along the spine of the fuselage, fixed on a pivot just forward of the fin. A hook at the cockpit end of the pole was attached to the pilot's parachute harness like a fishing rod. When a powerful spring was released, the arm would shoot upwards, pivoting from the fixed end and plucking the pilot up and out of the cockpit. Then, with the aid of aerodynamic forces like a medieval catapult, hurl him clear of the fin to safety where he could deploy his parachute. Martin took his model to the Ministry of Aircraft Production on 11 October 1944. The reaction was so positive that he was asked to show it immediately to the minister, Sir Stafford Cripps, who urged Martin to begin development without delay. The head of research and development at the Ministry heaped praise on the design – 'it appears simple and most attractive'.

But Martin's rivals had come up with a similar notion. Martin asked for an aircraft onto which he could fit his Swinging Arm for practical trials. He was told that it would be delivered in two days. Two months later it had still not made an appearance. Martin wrote a stiff letter to the powers that be, demanding to know what was going on. This did not go down well, especially since he had recently aggravated the Ministry with late delivery of his new prototype MB5 aircraft. But after much bridling in Whitehall corridors, Martin's fuselage arrived in December 1944. But then the Ministry began to have second thoughts. Would a Swinging Arm get a pilot out quickly enough and high enough to avoid contact with the tailfin? And how would it work if the aircraft went into a spin? Martin was asked to continue his development work at his own expense. The arm might be a good halfway house, the mandarins reasoned, until something more sophisticated came along.

The Swinging Arm never swung. Not for the first time, James Martin was ahead of the game. Ever the maverick, with a particular passion for explosives, he was already taking another route. In December 1944 he produced his first drawing of an ejection seat.[22]

THE MARTIN-BAKER AIRCRAFT COMPANY, DENHAM 24 JANUARY 1945

Martin decided that the most effective way of ejecting aircrew in an emergency would be to have a seat leave the aircraft with an occupant sitting in it. And the most efficient way of making that happen was to attach an explosive charge to the seat to shoot it upwards and out of the aircraft, safely avoiding hitting the tailfin. But how much explosive would be needed and at what angle should the seat be ejected? Those were challenges that a brilliant engineer like Martin could solve. But he was no medic and he had no idea how much explosive force a human body could withstand. There was no point in saving somebody's life only to cripple them in the process.

Help came to Martin in the shape of a complete human spine, which now sat in his office in a tall glass jar. It was a gift from Peggy Loudon, a general surgeon at Elizabeth Garrett Anderson Hospital in south London, with whom he'd been connected by a twist of fate. Loudon was lodging with the mother of James Martin's then secretary. When they later met, she was immediately interested in his plans for an ejection seat.

Loudon had gathered extensive experience of the human body's structure while operating on horrifically injured air-raid casualties during the Luftwaffe's nightly blitzkrieg bombing of London and other major British cities. Martin was delighted to have a surgeon's brain to pick about the effects of severe trauma on the body. Loudon explained the workings of the spine in detail. She showed him X-rays of spinal fractures. She gave him bones and damaged lumbar vertebrae to examine. Martin soaked it up enthusiastically, preserving various body parts in the refrigerator at the Denham works, causing those looking for the milk during tea breaks no little distress. Martin's secretary turned her nose up at the 'ghastly bits of body' lurking there. Loudon invited Martin to sit in on her operations and post-mortems. Every time he studied the spine in his office it was a vivid reminder of both the possibilities and the dangers of his current project. Loudon had helped him understand how the spine worked and how injuries were caused, but she could not tell him definitively the physical impact of blasting somebody out of an aircraft. As far as James Martin was concerned, nobody had ever tried it before. He would have to find out for himself.

In one of the Denham workshops Martin's engineers were putting the finishing touches to the muscular black steel tripod 16 feet high, which he had designed. One of its reclining legs was a ramp fitted with guide rails to the top. The rig looked not unlike a fairground 'test-your-strength' game but instead of a puck at its base to be struck with a sledgehammer, there was a rudimentary seat like those used by fighter pilots of the day. Fixed to the back of the seat was a DIY explosion kit Martin had constructed using two telescopic metal tubes. Martin referred to this crucial piece of kit as the 'ejection gun'. When the 'gun' was fired by a gunpowder cartridge, it would shoot the seat up the guide rails.

On 20 January 1945 it was time to try it out. Martin loaded up the seat with 200lb of sandbags, equivalent to just over 14 stone. Standing back, he pulled on a length of cable attached to the firing mechanism. His engineers stood a little further back, apprehensive to say the least. There was a substantial explosion. Martin watched as his 'ejection seat' shot rapidly up the rig's guide rails. It was the first small step in what would be its remarkable journey. But he still had no idea how a living spine would be affected. He called for a human volunteer, preferably one weighing around 14 stone.

Enter Bernard Ignatius Lynch.

The burly, husky-voiced southern Irishman, known always as 'Benny', was an engineering fitter in Martin-Baker's experimental aircraft department. He had previously played a similar role in the Irish Free State Air Corps. Some described him as a likeable rogue. In days of rationing he could lay his hands on anything anybody wanted, by fair means or foul, and was prone to getting himself into trouble with the police. Even his luxuriant handlebar moustache couldn't conceal his ready grin. Like all his fellow employees, he was devoted to James Martin for whom he had worked for nearly a decade. But nothing had prepared him for what was about to happen.

Martin was never one to hang around, so just four days after the sandbag test, Benny Lynch had cast aside his work overalls for this very special event, electing to appear in one of his best pinstripe suits, polished black lace-up boots, and a freshly laundered white shirt and tie. His moustache, teased and sculpted, was even more striking than usual.

With a ripple of applause from his assembled colleagues, who were relieved that they were not about to take this particular ride, and bullish encouragement and good wishes from James Martin,

Lynch hauled himself into the seat and adjusted the shoulder harness that criss-crossed his chest. He also tightened a safety strap across his forehead and rested his arms across his raised knees. He folded one hand over the other and closed his eyes. He looked as though he was ready for a meditation class, not acting as a ballistic guinea pig about to risk limb and possibly life.

If there was a countdown, Lynch did not hear it over the sudden and very loud bang. Triggered again by a sharp tug on the cable attached to the firing mechanism, the ejection gun had fired. He felt a heavy jolt in the small of his back as he accelerated up the ramp. The onlookers gazed at him with a mixture of relief, admiration and disbelief. Lynch's ascent did not last long. Martin had set the charge so that he came to rest only 4 feet 8 inches from the ground. Lynch, who had experienced 3½g (three and a half times the force of gravity) on the way up, slid slowly back down to earth to a chorus of cheers. He unstrapped, stood up, adjusted his jacket, and informed Martin that he had 'suffered no discomfort' on his trip. Over the next few days Lynch did two more tests, with Martin progressively increasing the power of the cartridge in the ejection gun. After each one Martin debriefed Lynch on how he felt physically. On his fourth ride Lynch reached a giddy 9 feet 11 inches. This time he did not come down smiling. The explosive kick had been unexpectedly violent. Usually cool and always modest, Benny admitted that he felt like he had been kicked by a horse. He limped off to the pub to recuperate.

Martin went to work on solving the problem with intensive sandbag and Benny Lynch tests and further adjustments to the seat, including modifying the footrests to help the ejectee achieve the correct posture. His ultimate aim was to have a powerful enough explosion to fire a seat – and a pilot – at least 20 feet into the air, safely away from the aircraft's tailfin, without endangering life. In the following weeks news of Martin's pioneering if unbelievable invention spread fast. Many were eager to see the rig and test it out, and none more so than Charles Andrews from *Aeroplane* magazine. He arrived hotfoot at Denham to write an article, and took ride number fourteen. Propelled to a height of 10 feet, he complained of severe back pain. When Martin called the next day to find out if the journalist was okay, he was told: 'He's in hospital.'

'What? Why?'

'He's broken his back.'

Although Andrews had reached a peak force of only 4g, it had happened so quickly that his vertebrae had been crushed. Martin could not understand how his system had produced such devastating results. He was sufficiently alarmed to seek Peggy Loudon's surgical advice. Back in his office he devoted yet more study to the spine in the glass cabinet. How could he square the circle of an explosive charge that would successfully fire a pilot out of a cockpit but without so much force that his life would be put at risk? Back then Martin had no reliable information on how much g-force a pilot could withstand during ejection without causing serious physical injury, especially to his back.[23]

G is the measure of acceleration due to gravity and the effect it has on the body. As humans we are all normally subject to a force of 1g. Rapid increases in velocity or sudden changes of direction are measured in further units of gravity. Speeding around a tight bend in a car might result in a 2g force, a rollercoaster might produce perhaps 3–4g. Most humans could safely withstand around 7–9g, but only for a matter of seconds. At 9g the body feels nine times heavier than usual, so a 12 stone (76.2 kilo) pilot would feel 108 stone (914 kilos). Blood would rush to the feet and the heart would not be able to pump hard enough to bring this heavier blood back to the brain. So vision would narrow to a tunnel and then go black. If the acceleration did not decrease, a pilot could pass out, lose control of the aircraft and be killed. A very rapid onset of high g-force could also crush internal organs and bones.[24]

None of this would have been of any consolation to Charles Andrews, but his injuries did help increase Martin's understanding of the effects of g-force on the spine. He fitted a high-speed camera to the rig, an indicator to the seat and painted the guide rails in black and white stripes so he could measure acceleration by the inch. By studying the high-speed film of many tests and comparing them to the X-rays of Charles Andrews' injuries, Martin was to make a major discovery. It was not the amount of g that was significant, but the speed at which it was imposed.

To reduce this rate of the sudden onset of g, he devised a two-cartridge ejection gun. When fired, the first stage would raise the seat smoothly. The second cartridge, activated by flame from the first, boosted the seat further, building up the speed to 60 feet per second, the maximum needed for safe ejection.

Martin had also placed the firing handle above the occupant's head. When yanked forwards and downwards it pulled the connecting cable attached to the 'sear' – the wedge of metal that held the firing bolt safely back – from the ejection gun. When released, the bolt came down like the hammer on a shotgun, to hit the detonator firing the first cartridge housed in the ejection gun. Importantly, pulling the handle out and down not only brought out a screen like a roller-blind to protect the ejectee's face, the motion also ensured that the pilot assumed the correct posture in the seat.

Modifications made, tests continued with Lynch as principal rig-jockey. The original 16-footer had been replaced by a monster 65 feet high, described by one observer as a 'terrifying tramway on which the test men could experience the full thrust of an authentic ejection'.[25] Lynch was shot up the rail to a height of over 26 feet without mishap. By March 1945 Martin's ejection seat was ready for flight tests.

With his rivals still in the race, Martin carried out a series of successful airborne tests ejecting life-size dummies out of aircraft at varying speeds. But, as with the rig, the ejection seat could only really be assessed in flight with a human occupant. While others suddenly remembered they had something safer to do, the redoubtable Benny Lynch once again volunteered. He headed to RAF Ringway near Manchester for training with the taskmasters of the Parachute Regiment.

In just four months, Martin had achieved a great deal with his new technology and silenced many doubters. In April 1945 the Ministry of Aircraft Production issued Martin-Baker with a £10,000 contract to produce two high-speed pilot ejection units. They would provide a Gloster Meteor jet fighter for his use. There were 180 human and dummy rig tests between April and June and Martin's Denham neighbours were well used to the daily claps of thunder and the sound of steel in motion. Notable RAF figures from Fighter Command and the Institute of Aviation Medicine came to Denham to ride up the rig, as did Royal Navy officers and members of the Ministry of Aircraft Production. Heads nodded in approval, and some relief, when they returned to terra firma in one piece.

On 24 April 1945 Martin had also completed the first ground test ejection of a dummy from a static aircraft for the benefit of the RAF's Physiology Department, which had queried some of the rig data supplied by Martin on injury risks. It was satisfied that ejection at a maximum 9–10g would not cause injury to aircrew. Martin, meanwhile, was working on a new, less aggressive and more effective explosive for his cartridges. But his main focus was still on the flight tests which continued through that year.

* * *

Escape system developments were also continuing apace in America. Since the fall of Nazi Germany in May 1945, the US Army Air Force Intelligence Service had seized a huge amount of German military data and equipment in Operation LUSTY, which its scientists and engineers were analysing eagerly. This included a wide variety of Luftwaffe aircraft and their largely unknown ejection seat technology. The discovery would give a boost to the American entry into the escape-systems race. Russian aircraft manufacturers were also picking up where the Germans had left off, and Saab, too, was still very much in the game.[26] A game James Martin had become even more determined to win.

Here was a man driven by his compassion for aviators, for whom he had enormous respect, and a keen commercial instinct. The

RAF needed him to come out on top. The war in Europe was over, but it was still losing horrifying numbers of aircraft. In 1945, 592 had gone down in accidents, resulting in more than 600 deaths. In 1946 it was 1,014, with nearly 700 fatalities.[27] Other air forces were suffering similar losses.

The ejection seat could not come soon enough.

But the seat was only half the equation. Once it – and the pilot – were safely out of the aircraft, the seat needed to be properly stabilised so that the pilot could release himself, fall free and then deploy his personal parachute and descend to safety. It was time for Martin to invent and fit another 'gun' to the seat. Once the seat was just clear of the aircraft, the so-called 'drogue gun' would then automatically fire out a metal weight known as the 'bullet'. A line was attached from the top of the bullet to a 'drogue parachute' contained in the seat itself. When the gun fired, the bullet pulled out the drogue from its canister and into the slipstream where it instantly deployed on long lines, avoiding becoming entangled with the seat. Once inflated, the drogue would stabilise the seat and prevent it, and the pilot, tumbling.

After trial-and-error experiments, as some in the air establishment saw them, using different sizes of drogue with canopies and cables that had twisted, tangled with the seat or simply snapped free because of the forces they had to endure, Martin had settled on a drogue parachute 2 feet in diameter, with twelve nylon lines and a 7½-inch vent hole to reduce the severity of deployment. It was successfully tested at 350mph.

After two years of extensive research and tests, with countless steps both forwards and back, now came the ultimate challenge. Could James Martin's seat actually save the life of a human being? Operational pilots were not going to put their trust in dummies. Dummies could not be debriefed. Dummies could not tell you how it felt to be shot out of an aircraft in flight, or confirm that a body could survive the stresses that Martin's experiments indicated were possible.

Luckily, Benny Lynch had returned from his parachute training. He even looked a little leaner. Martin was pleased to see him. With Benny back, it was nearly time to reach for the sky.

CHAPTER TWO

A Curious Contraption

Martin-Baker Airfield, Chalgrove, Oxfordshire 24 July 1946

Benny Lynch had swapped his pinstripe suit for mechanic's overalls once more, but this time they were topped off with a Biggles-style leather flying helmet. He acknowledged the small group of observers who had assembled in high expectation that summer's evening. They came from the RAF, the Air Ministry and the Royal Aircraft Establishment. Representatives were also present from the US Army Air Force and the US Navy, which had already placed an order with Martin-Baker for a 100-foot test rig together with a complete ejection seat.[1]

Lynch was beaming. Whether out of bravery or bravado, he was not letting on. Rumour within Martin-Baker had it that Martin would give Lynch a tot of whisky before a test. And possibly afterwards too.[2] James Martin did not appear to share his go-to man's confidence. His brow was furrowed and he was short on conversation. 'I hate the live tests far worse than the fellows who actually make them,' he later said. 'And I hated the very thought of the first one worst of all.' He had accepted sole responsibility if anything went wrong, and still had visions of the day that Valentine Baker didn't make it home.

Meteor EE416 boasted two ejection seats, both untested. After

finishing a final cigarette, Lynch put on his parachute and climbed
into the aircraft where he lowered himself into the seat which
had been installed in the modified ammunition bay behind the
pilot.[3] He pulled the seat harness over his shoulders and strapped
himself in. His pilot for the test, Captain Jack Scott, strapped
himself in too.

The clock ticked towards 9pm. Take-off had been scheduled for a
window during which the meteorologists forecast the winds would
be lightest. With a final wave, they taxied out and were airborne.
Another aircraft, fitted with a camera, followed, recording every
moment of the test. Watching the two climb high into the sky, Martin
knew this summer's evening was the point of no return. Scott levelled
off at 8,000 feet, the starting point for his run over the airfield. Lynch's
mouth was suddenly parchment dry. 'I never felt so lonely in my whole
damned life.'[4] But he had any fear fully under control, unlike some
on the ground who would later confess to having screwed their eyes
shut. As one observer put it: 'The test man is a technician fitted for
his responsibilities by attitude, aptitude and experience. Benny Lynch
was, every nerve and muscle of him, a test man.'

Jack Scott checked his instruments but he only had thoughts
for the courageous colleague sitting behind him, separated by an
armour-plated partition. He increased his speed steadily to 320mph.
In Lynch's compartment a row of coloured lights began to wink
at him in sequence. There was no intercom communication in a
Meteor. Each light signified a different stage in the process. The
red bulb glowed. Five seconds to ejection. The green flashed on.
Lynch pushed a button underneath it. Now Scott knew Lynch was
ready to go.

At 9.15, Bernard Ignatius Lynch reached for the handle above his
head. He pulled it sharply out and then down towards his chest.
The guard was over his face. Lynch was firmly anchored, his back
straight and strong. Almost instantaneously there was a flash of
flame and a puff of smoke as the two cartridges fired perfectly in
sequence. The seat raced up its runners at 60 feet a second and shot
Lynch into the unknown.

'The punch was powerful but not painful,' Lynch recalled. 'The slipstream threw me over on my back.' The air blast was pushing him at 300mph in one direction while the explosive thrust of his seat was forcing him upwards. 'It was pretty confusing. On top of that, the noise in the first few moments was deafening.' Once Lynch had risen 24 feet, the drogue gun fired, blasting the stabilising parachute out from the top of his seat.

So far, so good. The Meteor was gone. All was quiet. Benny Lynch was safely in his seat at 7,000 feet. But his job was only half completed. The worst was still ahead. He now had to manually get rid of the seat so he could then deploy his personal parachute. If he was still attached to the seat when he landed, he could be killed. The seat harness straps were attached to him over the straps for his own parachute. If he confused the two release buckles and jettisoned his parachute by mistake, he was in deep trouble. He had to be careful. He felt for the seat harness release and pulled. He felt it loosen. He kicked his body forward.

Lynch fell out of the seat and into space.

Those on the ground training their binoculars upwards could see nothing. It was purely something to do as the seconds ticked by; Lynch was too far away to be seen. Benny Lynch, now in free-fall, began counting, as the Parachute Regiment instructors had drummed into him.

One second, two seconds, three ...

He must wait eight seconds before reaching for the ripcord on his personal parachute. He needed plenty of space between himself and the seat now floating down on its own recovery parachute, fitted for the test, or he risked becoming entangled as they dropped away.[5]

Slowly.

Don't rush.

Six seconds, seven, eight.

Lynch pulled his ripcord. He was wrenched sharply upwards as his parachute canopy billowed out. It was a beautiful sight. The whole sequence had taken nearly thirty seconds from pulling the ejection handle until his own parachute deployed.[6]

There was a slight breeze. He looked around, taking in the neat patchwork of rural England beneath him. And then the airfield came into view.

Now James Martin saw him. He could unclench his fists. One man on one parachute descending serenely ushered in the era of the Martin-Baker Aircraft Company's ejection seat. Benny Lynch made a textbook landing exactly as he had been taught. He got to his feet, hauled in his canopy and removed his Biggles helmet. He had cemented his place in British aviation history. The icing on the cake was a public house within easy walking distance, where he rewarded himself with a welcome-home pint.

As Scott climbed down from the Meteor onto the runway, there were handshakes all round. James Martin and the RAF brass congratulated each other. Safe ejection from an aircraft was now a reality.

* * *

That same July, 1946, a Gothenburg newspaper recorded a moment of similar excitement. It splashed the headline: 'Charge Under Seat Delivered Pilot from Certain Death'. Following a mid-air collision, a Swedish fighter pilot owed his life to a Saab ejection seat that had gone into production in 1943.[7]

The first live test ejection in America took place in August 1946 over Wright-Patterson Air Force Base in Ohio, when Sergeant Lawrence Lambert used an American seat to punch out of a P-61 Black Widow night fighter. It was a fitting location for making history. Wright Field was named after Wilbur and Orville Wright, who had built and tested two of their Flyer aircraft there in 1904 and 1905.

The year before Lambert's ejection, a recovered German seat had been shipped to Wright Field for study courtesy of Operation LUSTY. Because of the shortages of steel the Germans experienced in the latter stages of the war, the seat was actually constructed of laminated wood. Even so, the captured technology was to prove invaluable in America's early development of ejection seats. Lambert's seat was also explosively operated, as one engineer involved in the trials explained: 'It thrust the seat/man combination up the rail mounted on the rear of the cockpit at a high initial velocity and hurled it from the aircraft.' Like the Martin-Baker seat, Larry Lambert had to physically free himself before manually deploying his own parachute.[8]

It was also in August 1946 that James Martin arrived at the US Navy Yard in Philadelphia. Already eyeing up his potential export market, he was there to supervise the use of his recently installed 100-foot test rig. After successful dummy and human runs, his hosts asked for guarantees of the Martin-Baker seat's performance. 'I won't give you any,' he told them robustly. 'If you don't like it, you can throw it in the Delaware River.'[9] The US Navy duly modified a Douglas A-26 Invader, a twin-engine light bomber, to install a Martin-Baker seat in the rear gunner's position in which a willing volunteer successfully ejected at 5,000 feet and a speed of 250mph on 1 November 1946. The live test was staged to convince doubting US Navy airmen that it was reliable and safe to use.[10]

Like their RAF counterparts, many American pilots still had strong misgivings about trusting their fate to a metal seat packed with explosives.

After Benny Lynch's first live test, Martin had further refined the initial version of the seat, reducing its weight and improving its performance at different speeds and altitudes and in variable weather conditions over the course of thirty-one dummy ejections. After those many seat modifications and redesigns, Benny Lynch completed the second live airborne ejection on 11 August 1947. Having left the Meteor at 6,000 feet and 200mph, he successfully disengaged himself from the seat and was under his fully developed personal parachute just eight seconds after the ejection.[11] While it was still a prototype, he'd put his trust in what would become the Martin-Baker Mk1 – the company's first production seat.

Martin and the Air Ministry were, as usual, moving at different speeds. Now fifty-four, Valentine Baker's age when he died, the inventor was even more restless and keen to forge ahead. The Air Ministry was circumspect and cautious. It had discovered on a visit to Denham that Martin was determined to now design a fully automatic ejection seat, where the whole process, from pulling the handle to the pilot's personal parachute deploying, was completely automated. While not opposed to such a system, the mandarins insisted that he work in stages. They said that any work on a fully automatic seat should be put on hold until a manual seat, in which an ejectee had to open their own parachute, had been put into full production. But Martin was adamant an automatic seat was the way ahead. There was more friction. An angry exchange of letters.[12]

CHALGROVE AIRFIELD
14 AUGUST 1947

James Martin was buoyant. He had received a letter from the Air Ministry telling him that, 'if some outstanding points could be settled satisfactorily, the decision will be in favour of your firm'. The Martin-Baker Mk1 seat, close to completion after the many dummy tests and Lynch's second airborne outing, would be fitted as standard to all new military jet aircraft. But before formal confirmation, Martin was required to carry out a series of human airborne test ejections to prove its success rate to the sceptics. Given the number of tests required, it was obvious that Benny Lynch could not do them all. Martin needed back-up. Peter Page, another mechanic who had joined the company after serving in the Parachute Regiment, was keen to bring his wartime skills to the programme. The day after Lynch's second test, Page had also ejected from a Meteor at 6,000 feet and 200mph.

Two days later Martin decided that he needed to up the ante. On Page's second test, he was to eject at 400mph, the fastest ever attempted. But on this occasion, the guinea pig would be the last to find out.

Page shot out of the aircraft a fraction of a second after he fired the seat.[13] The ejection itself was perfect. But as he rose clear of the jet, the 400mph slipstream hit him as hard as if he had driven into a brick wall. Then, the jolt of the seat's stabilising parachute opening nearly jerked him out of his harness. Page's feet were torn from the seat's footrests. His body slewed sideways. His legs were now dangling out of the seat. He was flailing around, 6,000 feet above the airfield. He still had to locate and unbuckle the seat harness and fall free, ready to deploy his own parachute. His hands clawed to find the seat's release catch. But at some point in the chaos, he also pulled his own ripcord and the bundle of rigging lines and folds of nylon canopy escaped from his parachute pack and wrapped themselves around both himself and the seat.

Page now faced a terrifying dilemma. Even if he managed to release himself from the seat and it floated away, there was a real danger that his own, already released parachute, would not open properly when he dropped clear and he would plummet into the ground. Or he could stay buckled into the seat and take his chances. There was no choice really. He stayed in the seat. As a professional parachutist he would have stayed focused as he fell, trying not to think of what damage he would sustain when the combined weight of him and the seat hit the ground. The seat's own parachute (luckily, as it was a test device, the seat itself had yet another parachute fitted to aid recovery) had slowed his descent somewhat, but he was still hurtling back to earth at an alarming speed. Closing in on the airfield he was desperately trying to shift back into a proper sitting position.

James Martin was horrified. Onlookers hoped for a miracle. Page smacked onto the concrete runway with a sickening thump. When they reached him, he was in a terrible mess. He looked to be paralysed and was finding it almost impossible to breathe. It did not take the hospital doctors long to tell Martin that his man had broken his neck. He was lucky to be alive but, even so, from his hospital bed Page said he was 'satisfied' that he 'had done his stuff by paving the way' in what was a British success story. He spent six months in hospital wearing a special body cast to repair the spinal damage.

Martin was distressed. He needed time to decide if it was safe to continue the programme. He made another simple modification to the seat: a safety strap over the ripcord housing which would prevent another premature activation of an ejectee's own parachute.

On 29 August, fifteen days after Page's accident, the Chalgrove Meteor was back in the air at 12,000 feet and now 420mph. This time, Benny Lynch landed perfectly, except for slight bruising to his legs, a detail that Martin did not share with the Air Ministry. More importantly, he had finally proven the smoothness of the ejection gun that fired the seat, the effective protection afforded by the face screen, the efficiency of the stabilising drogue, and the ability of an ejectee to release themselves from the seat and deploy their personal parachute.

* * *

Martin-Baker was then asked to put the ejection seat through its paces in front of the public at a Battle of Britain air display at RAF Benson in Oxfordshire on 20 September. Martin agreed as long as the test was done at low speed. His dynamic duo took to the skies once more. Scott gave the signal to leave the aircraft. Lynch pulled down the face screen. And nothing happened.

The ejection gun had failed to fire.

Scott had no choice but to return the Meteor to the airfield, both men terrified that the malfunctioning seat might explode at any moment. The closer they got to the ground, the greater the danger for Lynch. An ejection at low altitude would kill him. Even if he managed to separate himself from the seat and pull his ripcord, his parachute would not have time to open properly.

Scott approached the runway gingerly, hoping to achieve the smoothest landing of his life. A rapid inspection of the seat revealed the problem. A modified sear on the firing bolt had been fitted to the ejection gun at the request of the Air Ministry but much against Martin's wishes. It had failed to properly release the bolt and fire the first charge, something for which he would later chastise the desk jockeys of Whitehall. Martin's tried-and-tested sear was quickly reinstalled in its place and the problem was fixed. Scott, meanwhile, was only too aware that several thousand spectators were feeling short-changed and he asked Lynch if he fancied going up again. For once Benny's courage and good humour deserted him. He would – understandably – sit this one out.

In a truly astonishing turn of events, 'Mr Keyes', a spectator at the air show, suddenly rushed out of the crowd, volunteering enthusiastically to take Lynch's place.[14] Three weeks earlier Keyes had met James Martin and told him how much he would like to 'have a go' on an ejection seat. Martin had turned him down flat. He was not prepared to take that risk with an amateur. Especially not one who, it transpired, had extremely poor eyesight. But Martin was not at the RAF Benson display and, bafflingly, the company's on-site engineer decided that allowing an untrained civilian to try out the ejection seat would be a good idea. He gave Keyes the nod.

Scott briefed the excited Keyes on the ejection procedure. Keyes told his pilot that he wanted to delay pulling his ripcord. Scott ruled that out. He must stick to his instructions, or that was the end of it. Keyes promised he would follow the protocol. Scott was not convinced that he had been listening, so took him higher than he would have done with Lynch, to allow his new passenger and ejection-seat-tester more time to get rid of the seat and begin his parachute descent. At 4,500 feet Scott gave Keyes the green light to eject. Initially, everything went as it should have done. Mr Keyes, until a few moments before a mere civilian air-show spectator, pulled the ejection handle and shot out of the aircraft, then jettisoned the seat. He was having the time of his life. Jack Scott looked down, expecting to see a parachute. He was horrified to see Keyes plummeting earthwards, with no hint of a billowing canopy. Surely it couldn't have malfunctioned?

As Keyes had ejected, his goggles had blown off. He could barely see beyond his own nose and only pulled his ripcord when the green blur of the airfield swam into view. Too late. His canopy opened just as he bulleted into the ground in front of the horrified crowd. Keyes lay unconscious in hospital for three days. Though comprehensively battered and bruised, amazingly he had not broken any bones. When Keyes woke up, James Martin was probably one of the most relieved men in Britain.

The Air Ministry was not amused. Martin was told that their Meteor was not to be used for unauthorised joyrides. 'The incidents at Benson were deplorable,' they said, 'particularly as they occurred in the presence of a large crowd and are likely to bring ejection equipment into disrepute.'[15]

Back on the ground, testing on the rigs at Denham continued as Martin sought to perfect his ejection seat. Former Second World War bomber pilot Jo Lancaster was one of the many visitors to try it out there in 1947. Following his first tour in 1941, Jo had converted to Lancaster bombers for his second. Now retired from the RAF, he was working as a civilian test pilot for Saunders-Roe, which made flying boats on the Isle of Wight. 'I went up to Martin-Baker with my boss,' he remembered. 'It was the first time I had ever seen the ejection seat test rig. It all looked most concerning.'[16]

Jo's boss went first.

'He pulled the face screen handle and this damn contraption just shot up the tripod. It looked completely mad, like some out-of-control fairground ride. I thought, *Bloody hell, I want to turn around and go home*.' But there was no way out. Feeling rather apprehensive, Jo strapped himself in. 'I pulled the handle and there was a huge bang. The next thing I knew I was 20 feet in the air and looking down at my feet. It was a harsh ride. My head was thrown forwards and afterwards my backside was quite sore from the upwards blast.'

Still in pain, Jo later went to see his GP and explained what had happened. His doctor looked at him blankly. 'He had obviously never heard of an ejection seat and couldn't understand what I was talking about. I think he thought I was a bit mad.' The pain did not go away for several months but Jo managed to ignore it. There was one thing he could not ignore. 'My initial suspicions about this curious new device had not been calmed.' Like many pilots who had minimal information about the fledgling ejection seat and its airborne tests, he came to a conclusion: 'It was a bloody dangerous invention!'

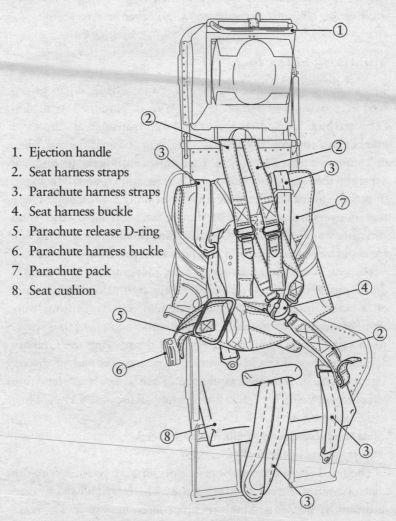

1. Ejection handle
2. Seat harness straps
3. Parachute harness straps
4. Seat harness buckle
5. Parachute release D-ring
6. Parachute harness buckle
7. Parachute pack
8. Seat cushion

PRE-MK1 EJECTION SEAT

On 2 January 1948 the Air Ministry cancelled its contracts with competing engineering firms; Martin had won his race. Dangerous or not, the Martin-Baker Mk1 ejection seat could go into production.

For Jo Lancaster there was a consolation. 'I simply couldn't imagine any circumstances where I would ever need to use it.'

MONDAY, 30 MAY 1949

Jo was flying his boomerang-shaped AW52 'Flying Wing' proto-type jet, partway through his test flight from Bitteswell Airfield in Leicestershire. He had climbed into bright sunshine at 5,000 feet and began a shallow dive to begin the next phase of the test when everything began to go wrong. Every few seconds the jet climbed 8 feet and then slammed back down, hurling him around the cockpit. 'It was so violent I couldn't see anything on my instruments. There was an overwhelming noise of shaking and rattling as the metal structure was stretched.' The aircraft was threatening to break up. He could no longer control it and he was running out of time.

He was nearly down to 3,000 feet, plunging towards certain death. But he was sitting on a piece of apparatus that he and many pilots still viewed with deep suspicion, if not fear, dismissively terming the recently installed ejection seats simply as 'bang seats'. Jo thought briefly of his wife Betty and their young son Graham at home a few miles to his west near Coventry. He had no choice. The 'bang seat' was his last chance of escaping the plummeting aircraft and seeing his beloved wife and child again. 'I knew I just had to get out.'

He tried to recall the drills he would need.

First, get rid of the cockpit canopy.

The hood jettison toggle was in front of him, to the left of the control column. He reached forward with his left hand and yanked it. Instantly, the Flying Wing's cockpit canopy flew away. The noise of the jet and its slipstream was now deafening. Simultaneously he flexed his arms, reached up over his head and felt for the ejection handle. Grabbing it firmly with both hands, he pulled it down in front of his face with all his strength.

A fraction of a second pause then an enormous explosion as the first cartridge in the ejection gun fired, almost instantly igniting the second. The seat was on the move and travelling fast up its rails. It was like being in a weird dream and Jo's thoughts were scrambling – 'it was all totally unbelievable'.

Before ejecting, Jo should have put his feet on the rests at the bottom of the seat. He was lucky he didn't. 'If I had done that, the spectacle-shaped control stick in front of me would have been in the way and seriously damaged my knees as I rose up.' There was some violent tumbling then another jerk. The drogue parachute had fired out to stabilise the seat as it shot out of the aircraft.

Then nothing. The Flying Wing was nowhere to be seen. It was as though it had been airbrushed from the sky.

Jo Lancaster had ejected successfully at 3,000 feet.

In the silence, time appeared to be standing still, while everything was also happening very quickly. Quicker than Jo could comprehend. The next thing he remembered 'was sitting strapped in my seat, leaning forward at about 45 degrees and looking down at the ground. The green fields were approaching me rather quickly.'

His senses kicked back in. He was wearing two harnesses, one for his parachute and another attaching him to the seat. He had to get rid of the seat fast. He felt gingerly for its release buckle across his sternum. 'I knew I had to be careful not to release my parachute harness which sat underneath it.' Now, sure he had got the right one, he pressed and twisted. He wriggled the harness free of his shoulders and, still tilted forwards, literally fell out of the seat. He was alone and in free-fall; the ground approaching even faster. He reached for his parachute's ripcord and pulled it hard. He felt a massive jerk under each armpit and was wrenched upwards. His parachute had fully inflated. For the first time, he felt safe.

From pulling the ejection seat handle in his cockpit to being under his own parachute had taken around thirty seconds.[17]

In the countryside below, life was carrying on regardless of the

unfolding aerial drama, though a local baker loading his van at
the windmill in the market town of Southam did look up when he
heard the then unusual sound of a jet engine. He was astounded to
see what looked like a giant wing losing height then crashing just
a few miles away.[18] Jo's AW52 had carried on without him, skim-
ming across the treetops at 400mph before crashing, ripping out its
engines on impact and breaking into pieces.

Jo was now floating serenely towards the ground. But where was
the metal seat? Still falling and, if it hit him, he would be in serious
trouble. He craned his neck to locate it. Out of nowhere it shot past
him and disappeared.[19] Too close for comfort. With the aircraft
gone, and the ejection seat accounted for, Jo looked for somewhere
safe to land. The breeze had stiffened and he realised it was sweeping him straight towards the Grand Union Canal a few hundred
feet below. He tried to remember his wartime parachute drills. He
tugged on one line after another, trying to change direction, but
only succeeded in swinging alarmingly backwards and forwards.

He was still heading for the water now 100 feet below.

He had no option but to take whatever fate decided to throw at
him. The ground raced up to meet him. He went through a hedge
like a pendulum and crumpled down hard, landing shoulder-first a
few feet from the water. He had come down behind the Cuttle Inn
at Long Itchington in Warwickshire, just 19 miles from Bitteswell
airfield where he had started this eventful day.

For a while he lay there stunned, trying to make sense of the
drama that had engulfed him for the last few minutes. It was all a
blur. He staggered to his feet and freed himself from his harness.
He was gasping for air. The wind had been knocked out of his lungs
and he was sure he had broken his shoulder. He had to get help. He
heard a voice calling to him. A nearby farmer ran up and helped him
gather up his parachute. 'He took me to his farmhouse 100 yards
away where his wife produced a cup of tea.'

Apart from the relief of being alive, one thought continued to
reverberate through Jo's mind. What if somebody had been sitting
behind him in the observer's seat of the AW52 as had been the case

on previous test flights? Only the pilot had an ejection seat. 'If there had been somebody sitting there that day, with no means of ejecting, I could not have left him to die. I would have tried to stay with the aircraft and we would have gone down together. Thank God I didn't have to make that decision.' Jo would not be the last aviator to face the horrifying prospect of ejecting and leaving a friend to die in a stricken aircraft.

Using the farmer's phone, he called his base at Bitteswell. One of his fellow test pilots answered. 'I simply said, "I've ejected."' The tumultuous events of the day, and his increasing pain, robbed Jo of any further words. Armstrong Whitworth sent a truck to collect him, and his ejection seat which was found intact about half a mile from where he landed.

Back at the factory he saw the company doctor and was whisked to hospital for X-rays. Later he rang his wife Betty to tell her what had happened, and to reassure her that he was safe. 'She didn't seem very impressed that I'd been in such danger.' The X-rays revealed that, as well as losing a large chip of bone from his right shoulder, Jo had fractured his second and third lumbar vertebrae. 'They also said it was the second time it had happened – it turned out that I'd actually broken my back on that test-rig ejection in January 1947.'

Jo Lancaster had become the first British aviator to eject successfully from an aircraft in an emergency. Relief and jubilation at the Martin-Baker Aircraft Company were palpable, coursing through the Denham factory from James Martin's office to the engineers and mechanics on the shop floor. History had been made. The technology, although still basic, worked in the real world. The national press had a field day, vividly describing the loss of a 'top secret' Flying Wing aircraft and celebrating Jo Lancaster's dramatic escape thanks to Martin's invention.

Martin was not one to rest on such a notable achievement. He knew his greatest challenge remained: to get pilots out of a doomed aircraft even quicker in a fully automated seat, where the ejection seat itself completed all the actions from firing to deploying the pilot's parachute.

Bones mended and bruises healed, Jo Lancaster was cleared to return to test-flying two months later. His medical report noted he was fit 'but not to be exposed to hazards of Martin-Baker ejector seat'. Which was rather like telling a sailor they could go to sea, but not in a vessel fitted with lifeboats. Jo ignored the advice.

Sometime after his great escape, a small, custom-made wooden box arrived in the post at Armstrong Whitworth. It was stamped: 'Danger – High Explosive!' Jo smiled when he opened it. 'Inside, there was a congratulatory letter from James Martin himself, and a gold Rolex wristwatch very nicely engraved with my name and the date of my ejection: 30/5/1949.'

Jo Lancaster, now officially Martin-Baker ejectee *number one*, had no idea that eventually many thousands of other aviators would, just like him, owe their lives to an invention that he and many other pilots initially dismissed as an impractical and down-right dangerous 'contraption'.

A contraption that was about to be tested to the limit in the post-Second World War era of regional conflicts.

CHAPTER THREE

ALL OR NOTHING

SAN PABLO BAY, SAN FRANCISCO
31 MAY 1949

Although America was still playing catch-up in the post-war development of ejection systems, trials had been ongoing since the initial inspection of captured German seat technology. But US pilots were still just as sceptical as their British counterparts about a device many reckoned that using would amount to 'sheer suicide'. Some official sources maintained that pilots could not bail out at more than 250mph. 'Human flesh could not withstand the force of air at such speed ... Volunteers making wind tunnel tests at 250 miles an hour had nearly parted with considerable portions of their skin.'[1]

But decorated Second World War bomber pilot Captain Vincent Mazza was out to prove his fellow airmen wrong. He had been involved in the development and testing of the systems for some time and was convinced that ejection seats were the future. With a fellow guinea pig – who had once been a steeplejack – he volunteered to conduct a roadshow, demonstrating the new technology at US airbases up and down the country, using specially modified Lockheed P-80 jet fighters.

The day after Jo Lancaster's British-history-making emergency ejection, they took off with camera-armed escorts from Hamilton Air Force Base near San Francisco, and headed for San Pablo Bay, a tidal estuary which would be their drop zone. Mazza was using a seat that he and colleagues had helped develop from the wooden German models. 'The pilot ejection seat, the newest and most dramatic of the Air Force's safety devices, is their baby,' said one commentator.

From 10,000 feet, they were about to risk their lives to prove that their bizarre device actually worked. Mazza was first to pull the ejection handle as his speed tipped 435mph. 'I was amazed at how easy it was,' he said after touching down gently in the water and being fished out by a boat. 'First I was sitting in the plane, and then I was out in the air. There was no feeling of being rocketed out. It was so smooth.' In front of the stunned onlookers, his fellow aircrew who could not believe he had survived this recklessness, Captain Mazza simply dried himself off, took to the skies once more and ejected again, this time at 555mph. 'There wasn't much difference,' he said. As he parachuted down to the bay he waved to 'the boys circled around', to let them know he was okay. He was now 'certain a man can bail out safely at much higher speeds'.

So remarkable was Mazza's ejection that he was immediately booked to appear on American national radio to discuss his experiences. With commendable understatement, Mazza told his astonished audience: 'It really is the easiest way to get out of an aeroplane.' Attitudes to ejection systems were slowly changing and Mazza would later go on to tell his fellow aviators, 'let us hope that the need to eject never happens to you; but if it does, add a little insurance for the wife and/or girlfriend and GET OUT before lack of altitude defeats the best [ejection seat] engineering efforts'.[2]

'Eject in time' would become the mantra all aviators would hold dear in the coming years.

* * *

Despite Mazza's success, no American pilot had been forced to eject in an emergency like Jo Lancaster until Lieutenant Robert E. Farley took off from March Air Force Base in California on

29 August 1949 for an instrument proficiency flight in an F-86 Sabre.[3] Over the desert, he suddenly experienced problems with his controls and the Sabre began a slow, descending turn to the right. With both hands on the stick, Farley used all his strength to try to keep it level. The jet then started to spiral downwards. He had to take a hand off the stick to adjust his controls. 'As soon as I took my left hand off, the aircraft snapped over onto its back and started a vertical dive.'

The moonscape beneath him was 4,500 feet above sea level and strewn with boulders. The peaks of the surrounding hills were level with his wingtips and, hurtling earthwards, he had just one chance to save his life. 'I fired the canopy, placed my feet in the seat stirrups, and ejected myself from the diving plane about a thousand feet from the ground. I immediately released the seat and simultaneously pulled the ripcord.' When his parachute opened 600 feet above the ground, Farley lost his shoes, dog tags, watch and, in his own words, 'everything except my flying suit'. And while the seat may have separated, it was now tangled in his parachute lines and resting on his head. Landing awkwardly, it slammed down on top of him.

Later, in his hospital bed, with injured feet, lacerations to his left ear and cheek and a gash on his face, which required forty sutures, Farley was asked what he thought of the ejection seat. 'I'm sold on it, brother,' he said. 'Without it, I would have been a dead duck. I would not be here talking to you. It would have been impossible for me to have gotten out. Pilots should make sure they know how to use it. Having it and not knowing is the same as bailing out without your parachute.'

Like Jo Lancaster, Robert Farley had made history and suffered some hefty cuts, breaks and bruises in the process. Their survival deserved celebration, but the technology still had far to go. Pilots were still dying.

FARNBOROUGH AIRFIELD, HAMPSHIRE
3 APRIL 1951

When it came to Battle of Britain fighter aces, few were more fearless than 92 Squadron's Trevor 'Wimpy' Wade. Like the famous Wellington bomber, he was nicknamed after J. Wellington Wimpy from the *Popeye* cartoon series. Perhaps because of his shock of black hair, open, oval face and ready smile, he was never far from the centre of a party with his pretty wife, Josephine. But he found himself mixing it with the Luftwaffe within weeks of their wedding in 1940. Wade destroyed some nineteen German aircraft, including three on a single day. Such was his relish for a dogfight, he survived being shot down an astonishing five times.

On one occasion he made a forced landing on Lewes racecourse after being hit by a German bomber. He was trapped upside down in the cockpit when his aircraft overturned and thanked his lucky stars that it did not catch fire. He was even more blessed when he crawled out from under the wreckage of another Spitfire, seconds before it exploded. After dicing with death so nonchalantly in combat, Wimpy Wade was all too aware of the need for a timely escape from stricken aircraft. After the war, Squadron Leader Wade became chief test pilot at Hawker Aircraft and was now participating in the development of the P1081, an experimental turbojet capable of breaking the sound barrier. Hailed by its makers as 'the shape of things to come', Wimpy had taken the swept-wing fighter on its first test flight in June 1950, and then to the Brussels Air Show four days later, where it aroused huge interest.[4]

But something in him had changed. The fearlessness had gone. He had recently visited his close friend and former RAF fighter ace Tony 'Bolshie' Bartley in Hollywood. 'One evening, Wimpy confessed to me that he had lost his nerve test-flying,' Bartley recalled. 'He was one of the most skilful pilots I knew and had flown with. He was also a very nice fellow and I was fond of him. I told him, "For God's sake, quit while you're ahead. It could happen to any of us."'[5]

Back in the cockpit of the P1081 on the runway at Farnborough,

Wimpy had clearly failed to take Bolshie's advice. He was strapped into a manual ejection seat produced by one of James Martin's engineering competitors. Although Martin-Baker seats had been ordered for all RAF jets, some earlier seats were still in use in civilian development aircraft. That very day the Ministry of Supply had inspected the jet and issued a certificate of airworthiness. Waiting to take off, engine running, Wade had completed his checks and was also satisfied everything was in order. He was his normal cheerful self as he chatted with a Hawker engineer with whom he had just taken lunch.

The engineer climbed down and Wimpy was soon airborne, heading south. Nearing the East Sussex coast, he cranked up for a high-speed run. As he hit 757mph over the village of Ringmer 3 miles east of Lewes, an electrician and a slaughterhouse manager were going about their daily business far below when they heard an ear-splitting roar. Rushing out and looking to the sky, they saw a dark object falling from a plummeting aircraft.

A catastrophic failure of some kind had prompted Wimpy to release his cockpit canopy.[6] His helmet and goggles were stripped from his head by the airflow. He lost his oxygen mask as he fought to recover from the kind of near-vertical dive that would have made ejection perilous. At 2,500 feet he had managed to stabilise the P1081 sufficiently to pull his roller-blind ejection handle down hard over his face. The seat fired successfully, but, for some reason, he failed to complete the process to release himself. Imprisoned in its grasp, he was now somersaulting at high speed, end over end, at more than 120 revolutions per minute. His parachute was never deployed.

A police constable found his shattered body, still in the ejection seat, embedded in a coppice amid a trail of broken branches about half a mile from the wrecked jet. Both the seat and parachute harnesses were still firmly fastened. Had he blacked out and not been able to release them, or was there a mechanical fault? Either way, he hadn't stood a chance. Aged just thirty-one, Squadron Leader Wimpy Wade's luck had finally run out.

At the inquest, the Ministry of Civil Aviation investigations officer announced that the ejection seat had fired normally. Close

examination of the wreckage failed to show what had happened to
the jet that caused Wade to leave it. There were rumours of an explosion. But no evidence. The coroner suggested that he might have been
knocked out as he ejected, or rendered unconscious by the rapidly
rotating seat, and was then unable to free himself and open his parachute. He had suffered multiple injuries, including a deep contusion to
his scalp. What had made Wade attempt to eject remained a mystery.
But could some sort of automatic ejection seat have saved his life?

In an era where flying, especially military flying, tested both man
and machine to the limits, aircraft accidents continued to blight aviation. In 1950, the British military *alone* had lost 380 aircraft, resulting
in 238 fatalities. In 1951, that number would increase to 490 aircraft
lost and 280 deaths.[7] The rapid development of escape systems was
desperately needed to reduce these horrific figures as ejections – both
successful and unsuccessful – continued. In July 1951 an RAF airman
escaped following a collision between two Meteors at 30,000 feet.
Like Wimpy Wade, the first pilot was unable to locate his seat harness
release to extricate himself from the seat. Luckily, unlike Wade, he
somehow managed to deploy his own parachute and made a successful, if heavy, landing, still shackled to the metal seat. The other pilot
had also ejected when his damaged aircraft flipped upside down as he
was attempting to land. Too close to the ground for his parachute to
open fully, he was killed.[8] In those early days it seemed a case of 'all
or nothing' on the ejection front. And the seats had never been tested
in aerial combat, where they were about to be most urgently required.

KIMPO AIR BASE, SOUTH KOREA
29 AUGUST 1951

For 77 Squadron of the Royal Australian Air Force (RAAF), the
Korean War was a hazardous assignment from day one. Its pilots
had just left the age of the propeller and were still coming to terms
with their new twin-engine British Gloster Meteor fighter jets when
hostilities broke out between American-backed South Korea and
the Soviet- and China-sponsored Democratic People's Republic in

the North. Tensions between the two states and their conflicting political ideologies had been bubbling closer and closer to the surface since the end of the Second World War, when the country had been divided along the 38th parallel. All attempts at reunification had failed and, on 25 June 1950, 90,000 Communist troops had crossed the border without warning in an attempt to take over the entire peninsula.

That morning, the phone in the Sergeants' Mess at 77 Squadron's base at Iwakuni in Japan had rung with the news: an operations officer relaying the message that their Squadron might be called upon to 'help do something about it'. Two days later, the United Nations Security Council declared the North Korean action a breach of the peace agreement and called on member nations for assistance. The Australian government instructed 77 Squadron to shift to Korea and police airspace under the command of the US 5th Air Force. It was the first Commonwealth unit to join the conflict, flying under the UN flag.[9]

Warrant Officer Ronald Guthrie was one of the young Australian pilots now flying out of the Kimpo Air Base, still a muddy construction site of spartan tents and makeshift huts, a few miles south of Seoul. Eager for action, Ron had joined the RAAF two years shy of his eighteenth birthday. 'I loved flying from the word go,' Ron said. 'I wanted to get as many aircraft and flying hours as I could to my credit. That was the thing for a young fella.'[10]

Fourteen combat missions in, he was comfortable with his jet. But, like many of his contemporaries, he wasn't so sure about his Martin-Baker Mk1 ejection seat. It had been completely redesigned from the prototype tested by Benny Lynch back in 1947, and was now fitted as standard in all new British and Commonwealth service jets. Like most pilots, Ron was suspicious of sitting on something designed to explode. He had been unnerved a few months earlier as he watched a friend flying a Meteor during a test flight. 'I was standing in the flight line when I heard a thump. Looking up, I saw a parachute blossoming and the aircraft it had parted from circling aimlessly.' Even though he hadn't pulled the firing handle, his friend's ejection seat had mysteriously fired without warning, blasting him unceremoniously out of a perfectly serviceable aircraft. A lug securing the firing mechanism had failed.

'The Meteor crashed into a nearby hill and the pilot landed nearby,' Ron says. 'We sat with trepidation for some time afterwards. The loss of a Meteor had been startling and had given my friend a fright and an early, and unexpected, experience of ejection.' Ron and his fellow pilots were far from happy. 'We simply didn't have enough information to truly trust these strange new devices,' he says. 'I didn't remember ever hearing of anyone using one in anger.'

Near the end of August 1951, Ron was out on the runway, about to start his fifteenth Korean sortie. 'It was just another mission,' he remembers. He was in one of eight RAAF Meteors flying top cover for US B-29 bombers en route to another strategic target on the far side of the 38th parallel, the 'Iron Curtain' dividing the two Koreas. Ron's formation's role was to protect them from attack by high-flying Russian MiG-15 fighters.

It was an unforgiving landscape for a pilot, and Ron and his colleagues were forced to adapt quickly to stark, barren hills that towered over paddy fields in the valleys. Taking off and landing at the airbase in a deep, rugged bowl was particularly hazardous in the dense summer fog and unrelenting rainfall.

Kimpo was only 25 miles from the front line, and 77 Squadron's eight top-cover jets quickly took to the skies. They climbed to 39,000 feet before forming two flights of four jets, well-spaced in battle formation, test-firing their guns en route. Ron brought up the rear in 'Dog Flight'. 'Because of the height of the back of the ejection seat you had a blind spot between four and eight o'clock, so everybody relied on others to look after their backs,' he says. 'The MiGs liked to come in at six o'clock.' As 'the Squadron's senior jet man', with aerobatic experience and 100 hours under his belt, Ron was tail-end Charlie, call sign Dog Four. Nobody was watching his back.

Eyes peeled for enemy raiders, he watched another eight Meteors flying lower-level protection on the fleet of US bombers 5,000 feet beneath him. They were all heading for an infamous segment of Korean airspace known as 'MiG Alley'.

The MiG-15 was a specialised high-altitude fighter with one powerful jet engine, swept-back wings, a high, horizontal tailplane and a pair of 20mm guns. Its 37mm cannon was slow-firing but deadly. Ron was very fond of his twin-engine, straight-wing Meteor, which boasted four 20mm cannons. But he was equally frank about its shortcomings. 'It was well-armed, but too slow. At 20,000 feet, it was more nimble than the MiG and could hold its own, tight turning and diving. At 40,000, it handled like a bag of manure. I likened us in Korea to the Australians who fought at Gallipoli. We were ill-equipped to do the job that needed to be done.' More worryingly, most of the MiG crews were not North Korean. They were highly experienced, battle-hardened Russian fighter pilots; formidable foes.

On the morning of 29 August, high above the Yalu River, which runs along the border between North Korea and China, Senior Lieutenant L. K. Shchukin of the Soviet Air Force was piloting one of twenty-four MiGs primed to intercept any attacks on the dam and hydroelectric power stations located there. He spotted eight Meteors sweeping the river in figures of eight in advance of the bombers. 'It was the first time we had encountered these aircraft,' he recalled. 'Unsure of what they were, we dived down and our squadron split so we could attack them from different angles.'[11]

MiGs flew in formations of six, known as 'trains', each one led by a Russian ace the Americans codenamed 'Casey Jones' after the American folk hero and rail engineer known for his speed, who sacrificed his life to save those in danger from a train collision.[12] An American early warning controller near the mouth of the river came on the radio. 'Heads up! Four trains airborne. Casey Jones at the throttle.'

Ron says, 'I knew we were in for one hell of a time . . .'

Squadron Leader Dick Wilson, leading the first flight of four

Meteors, spotted a train of MiGs sweeping down fast from 40,000 feet. Keeping them in sight, he diverted his formation to meet them. Two more MiGs appeared a few thousand feet below them, scenting bombers. Wilson and his number two dived down to attack them, but as they levelled out at 600mph, his fellow pilot's aircraft suddenly flicked into a spin. Plummeting downwards, he managed to recover, but he'd left Wilson exposed. Shchukin pounced out of the sun to begin his own attack. Two other Meteors swooped to protect Wilson, but the Russian's 37mm shells punched a 3-foot hole in his port wing and punctured a fuel tank. Wilson limped home to base as Shchukin disappeared back into high cloud, having just won his fourth kill of the war.

The dogfight was well and truly on. Opposing jets mobbed each other in a rotating maelstrom of fire. Bringing up the rear of the second Meteor flight, Ron Guthrie suddenly had trouble keeping up. He was losing power in the turns. He watched what looked like 'glowing ping-pong balls' stream under his left wing. They were actually white-hot tracer rounds, 'doing around 700mph to my 640'. MiGs were sliding into the battle in an avalanche, belching 37mm and 23mm shells across the sky with deadly accuracy.

Senior Lieutenant Nikolai Babonin had Ron in his sights. 'We closed to within 900 feet and opened fire,' the Russian former test pilot recalled. His wingman, Aleksei Svintitsky, was behind him, just off to one side, 'so no one caught us off guard'. He saw Guthrie's jet was in trouble. 'Fragments of metal and parts went flying,' he said.[13]

Ron threw his jet into a hard left-hand turn and punched his mic button. 'Break to port,' he yelled to the other Meteors. Too late. 'Two shells had hit me behind the armour plating at the back of my ejection seat and blew out both my radios. I was talking to myself.' Two more MiGs shot past Ron's nose. 'Instinctively, I turned back sharply to the right, hoping to get one of them in my gunsights.' He saw a red star on a silver fuselage and the pilot's head in the cockpit. Although damaged, Ron went after him. 'My finger curled around the trigger of my four cannons. The guns rattled. I was excited as pieces flew off his aircraft. It rolled upside down and dived out of sight.'

'The target continues to fly,' cried Svintitsky, astounded both by Ron's bravery and by his aerobatic skills. Why hadn't he ejected? He was not only still flying, but fighting for all he was worth. Not to be outdone and eager for a kill, Babonin raced to within 160 feet of the rear of Ron's jet. Any closer and they would have collided. 'He fired and fired,' Svintitsky said. And then he fired some more. The heavy cannon cranked out a relentless stream of 37mm rounds. The impact was catastrophic. 'I felt as though a big load of bricks had fallen onto the rear end of my aircraft,' Ron says. The Meteor started to shake convulsively. Babonin's shells had destroyed its tail. The Russian was now so close he got caught in the Meteor engine's jetwash. It flipped his MiG over and sent it into a dive. Babonin managed to right it and headed for home, completely out of ammunition. Ron was his kill.

Ron was in such a state of shock it took him a moment to realise his guns were still firing. He released the trigger and read the speed off his vibrating instrument panel. Nudging 640mph. His clock told him it was six minutes past ten. His speed in the dive was heading inexorably for the sound barrier. 'The aircraft began to shudder and continued to roll.' He urgently grasped the canopy jettison handle on the right-hand side of his cockpit. 'It came away clean.' A gigantic swirl of freezing air flooded in. He grabbed the face-blind firing handle on his Martin-Baker ejection seat with both hands, waited for the jet to finish a roll, and then pulled it hard down over his face. Nothing happened.

To the accompaniment of engine scream and air blast, alarm and panic invaded Ron's head.

Desperately trying to stay calm, he waited for another rotation to end and pulled the handle a second time. Surely the firing mechanism wasn't going to let him down again. There was no explosion. The ejection seat remained firmly fixed in the diving jet.

Though now frantic, Ron suddenly realised why. 'Being

left-handed, I had my pistol holster under my right armpit and my elbow was hitting it when I had pulled the handle. A Red Cross pack on my left side had also prevented me from pulling it down properly.' He spread his arms and elbows wider and tugged it a third time in a desperate effort to escape. His altimeter needles were unwinding below 39,000 feet. There was an almighty and startling bang. 'She worked just like advertised.' An immense upward thrust shot him out of the aircraft.

As he collided with freezing air, Ron blacked out.

* * *

Ron, now Martin-Baker's first combat ejectee, had ejected at 38,000 feet, over 7 miles above the earth. 'As I left the cockpit, my chamois gloves had been ripped off by the 600mph slipstream, along with the knee pockets of my flying suit, which contained spare socks and extra ammunition. Suddenly, I couldn't breathe.' He thought he had also lost his goggles and oxygen mask. His throat was constricting. As his awareness returned, he felt the goggles around his neck. The mask had slipped to his chest. Both of Second World War vintage, they had repositioned themselves when he had pulled the face-blind handle. His ice-cold fingers fumbled both back into place. He sucked the oxygen supply from the emergency bottle in his ejection seat deep into his lungs.

After the cacophony and frenzy of the dogfight, he was still in the seat, stabilised by the drogue parachute, heading downwards in silence, high above the clouds. A pilot flying an armchair instead of a jet. He had the strange sensation that he was stationary, detached from everything. 'It was like I had slipped into another time zone. No more engines, no more gunfire.'

He had a decision to make. The air temperature at that altitude was around -50°C. He was wearing a thin 1942 summer-issue flying suit and a flimsy oilcloth helmet. If he got rid of the seat and opened his parachute now, there was a real danger of frostbite. On the other hand, altitude might give him the opportunity to steer his canopy out over the North Korean coastline, where he stood a chance of being picked up from his life raft by a rescue aircraft. In for a penny, in for a pound. He unfastened his harness, kicked the seat away then pulled his parachute ripcord. The canopy opened at 35,000 feet.

Still 6.6 miles high, he could see both of Korea's coastlines. The panorama of the peninsula was spread out below him, a patchwork of grey, brown and vivid green. Determined to evade capture, Ron tugged at his parachute lines to steer himself towards the sea. Almost instantly, he heard the sound of something flapping above his head. He looked up. His canopy had lost its umbrella shape and was dangerously close to collapsing. 'Not being a parachutist, it frightened the life out of me.' He stopped pulling and prayed. A few moments later it billowed full again.

He accepted bitterly that he was not going to make it to the ocean. A westerly wind was now blowing him inland, his life raft dangling uselessly from his harness. Now it really sank in. He was going to end up in enemy territory.

Down to 15,000 feet. Mountains, rivers, townships, roads and railways were coming into sharper focus. The countryside looked more hostile with every passing minute as he continued his long descent. What would be waiting for him down there? Would he be shot on sight? Tortured? From his intelligence briefings, he knew that the Geneva Convention and the humane treatment of prisoners meant nothing to the North Koreans. His thoughts turned to home. How would his poor mother take the news of his ejection and capture? 'Since her divorce she did not even have the support of a husband. And the death of my only sister, in 1945, would now come back to haunt her.'

The earth was rushing up to meet him. 'In the last 2,000 feet I heard strange sounds, like *fittt-fittt*.' He was being shot at from the ground. He looked up to discover his parachute was now full of holes. He could almost feel the bullets zip through it. 'This was what we had been warned about in our briefings. In alarm, I attempted evasive action, by pulling on the cords to swing myself from side to side.' Instead of moving like a pendulum to avoid the rounds teeming upwards, 'I spilt the chute again and had to stop pulling. It was a frightening experience as I was so near to the ground. Just in time my parachute re-erected.' The last few hundred feet passed rapidly. He was heading towards a paddy field in a valley, 25 miles from the coast. It was twenty-eight minutes since he had bailed out. 'It had been a long, cold ride,' he remembers.

Beneath him two women were stooping over their work. Ron landed bolt upright between them in the paddy, his feet planting into spongy, waterlogged earth. 'They leapt into the air in fright and ran. When they saw I was no threat, they came back as I was unstrapping. The smiling girls took one of my hands. One said, "*Russki da?*"'

'*Russki da!*' Ron replied, thinking quickly. If they assumed he was Russian, perhaps he might have a chance to escape? Satisfied, they began to lead him away. Had his luck turned?

'Then bullets really started flying and the girls took off like rabbits.'

Ron crawled behind a paddy wall for cover and drew his 0.38 Smith & Wesson revolver. Thanks to losing his spare ammunition during the ejection, he only had the six rounds in the chamber to defend himself. 'I saw three army patrols approaching my position, two from the top of the valley and one from the bottom. The nearest soldier began shooting from the hip as he ran.' A clatter of submachine-gun bullets hit the ground perilously close to where Ron was crouching. He raised his revolver, lining up the Korean soldier in the sight and fired two rounds back. His attacker fell forward, bleeding heavily and dropping to his knees.

There were now troops swarming everywhere. Minutes before, the paddy field had been dotted with farmers tending their rice. They had now vanished. Before Ron could fire again, a teenage soldier leapt over the mud wall, wrenched the Smith & Wesson from his grasp and ran around firing it in celebration. A large group of North Korean fighters crowded excitedly around their novel catch, completely ignoring the comrade Ron had shot, who was now bleeding to death, semi-submerged, in the mud. They forced Ron to his knees and bound his wrists with wire. His captors stripped him of his flying suit, life jacket and his RAAF wristwatch. They also robbed him of his signet ring, a treasured gift from his mother.

Martin-Baker ejectee number six was the first pilot to use their seat in combat. Many, many more would follow. He had made the highest ejection on record. But none of this offered Ron Guthrie any consolation.

'I rose to my feet – surrounded.'

It would be several months before anybody knew what had happened to Ron Guthrie. His war in the air was over, but his post-ejection journey was only just beginning.

PART II

1952–1963

THE RACE TO SAVE LIVES

CHAPTER FOUR

SIMPLE SOLUTIONS

For Gabe 'Jock' Bryce, chief test pilot at Vickers-Armstrongs, the brief was simple: flog the aircraft to its limits, and beyond. As one of his fellow pilots said, 'If I can't break the airplane, then neither can the customers.' Like the first British ejectee Jo Lancaster, Jock had served in the RAF, and had a distinguished wartime record.[1] At Hurn Airport in Bournemouth in early 1952, he was now at the controls of the prototype of a new generation of post-war heavy bombers. The Valiant first flew in 1951 and was superior in performance and strike power to almost every other military aircraft in the world. It would later be joined by the Vulcan and the Victor to create the UK's nuclear weapon-armed 'V-force'. Jock was ready to take aircraft WB210 on what he described as 'a rather unusual test flight, to measure and estimate the noise levels likely to be experienced in a four-jet aircraft with the two inboard engines stopped'.

Once inside the fuselage, Jock climbed the short ladder to the cockpit and squeezed through the narrow gap between the two ejection seats. Co-pilot Squadron Leader Brian Foster, RAF Bomber Command's man on the project, followed. 'Behind and below us was the crew compartment, which stretched the width of the fuselage,' Jock says. Facing the rear of the aircraft and a bank of instruments were two navigators and the air electronics officer. Sitting in a row, they were not supplied with ejection seats. Nobody had yet come up with a rear-facing seat for bomber crews. If anything went wrong they would have to bail out the old-fashioned way, through the small oval door on the port side.

After taking off on a fine winter afternoon, Jock headed towards Wisley in Surrey. All was going well. 'We were turning over the river near Kingston upon Thames with the two inboard engines stopped as we had been instructed.'

Jock went to restart them. The starboard engine would not relight.

He was not concerned. 'The Rolls-Royce engines, like the aeroplane, were in the proving stage.'

He tried a second time. 'There was no reaction.'

Jock was not one to panic.

He tried the starboard engine again. 'On the third attempt it lit up all right. I turned away from the built-up area around Kingston and headed south-west for the coast.'

It was time to check the bomb doors. A few weeks earlier during a previous test flight, he had opened them for the first time. One had broken off, plummeting into the New Forest below. Over Portsmouth he was opening and closing them, when, 'I noticed that some of the instruments on the starboard side were not registering.' There were duplicates in the rear-crew compartment. He called down to the flight engineer.

'How are you doing on the starboard engine?'

'That's funny,' the crewman replied. 'We've lost a lot of the instruments on that side.'

The engines were still running and there was no indication anything was wrong. Even so, Jock was not going to take risks and terminated the test flight. 'I headed west at once, towards Hurn, only about three minutes away.'

Crossing Southampton Water, 'there was a dull, muffled explosion, which I heard clearly in spite of my helmet'. Jock couldn't imagine what it was. Still in control, he sat at 7,000 feet, trying to fathom what was wrong. High above the airfield, he reckoned the problem was electrical. It was time to land. 'I moved the joystick from port to starboard. There was no response. I suddenly realised I had no lateral control. It was a terrifying sensation.'

What Jock and his crew did not know was that during the engine shutdown test sequence, unexpended fuel had been collecting around the jet pipes. When the engine had finally been successfully restarted, the pooled fuel ignited and a ferocious fire began to spread through the interior wing structure. The Valiant had been spilling a trail of burning fuel and molten wreckage in its wake; the heat was starting to warp and twist the starboard wing and main spar.

'The aircraft began to force its way into a turn to the left. I could not hold it. I knew in my heart of hearts that I didn't have sufficient control to get down.' The Valiant was swinging violently from port to starboard. 'It was rapidly running away from me. I called the crew. "We're in big trouble. I can't fly the aeroplane."'

Co-pilot Brian Foster looked back over his right shoulder towards the starboard wing. He could see what Jock could not. 'Gee, it's on fire!' he gasped. 'It's coming out of the leading edge. Smoke and flame.' If the outer wing was now ablaze, then the interior, where the engines were and the fire started, would be an inferno.

For the first time in his life, Jock Bryce was going to have to issue an order he dreaded. 'Abandon the aircraft,' he shouted over the intercom. 'Bail out! Bail Out!'

He and Brian Foster blew the hood off their compartment. Foster looked back to make sure that the rear crew had clambered out. Jock knew that if he and Brian didn't evacuate soon, it would be too late to use their ejection seats. He punched his co-pilot on the arm. 'Come on! Get out.'

Brian Foster reached up for his yellow and black ejection handle and dragged it downwards as hard as he could. 'The next moment, I smelt a cordite charge, which was Brian's seat ejecting as he pulled the blind down in front of his face. Almost simultaneously, just as he ascended, there was an explosion behind us as the starboard wing blew off.' Three thousand gallons of fuel had gone up. The bomber was rocked over onto its side. Jock reached for his own handle but before firing his seat he paused momentarily, trying to remember the drills. 'I was using one of the early seats so had to do the sequence manually.' Wasting precious fractions of a second, he rehearsed the drills in his head.

Pull the handle.

Exit the aircraft.

Unstrap and then jettison the seat.

Open the parachute.

'It occurred to me that I would only have one chance to get all this right.'

If Jock didn't sort himself out, he would soon be plunging to his death, strapped into a blazing, hundred-foot coffin. He braced himself, reached above his head and pulled the ejection face-blind handle downwards.

'I must have blacked out, because the next I knew I was in a brilliant blue sky, with England laid out neatly in brown and green below me. Then it was the ground that was above me and the sky below, the whole panorama interchanging as I somersaulted through the sky.'

Tumbling down, he was unable to complete his manual drills to free himself from the seat. It was the stuff of nightmares. 'Again and again, the horizon was rising up at me as I toppled over and over.' Jagged chunks of Valiant were cartwheeling dangerously close, coming at him from every direction.

Finally, the seat began to stabilise. He now had to get rid of it.

But could he remember the process?

'Two sets of straps encased my body.' One for his seat, one for his parachute. One white, the other blue. But which was which? His mind went blank. Get it wrong and he would jettison his parachute and plummet to his death. He traced each buckle back to its source with his fingertips.

More vital seconds wasted.

He thought he had the answer. Seat buckles blue. Parachute buckles white.

Still he hesitated, before finally releasing the blue. He had expected the straps to spring open and the seat to fly away instantaneously. It didn't.

'Nothing happened. Was there some other fastening?' He couldn't find one. He would have to somehow throw the buckle off and climb out of the seat. 'I had nothing to push against, no floor to press my feet on and lift myself out. I kept trying to push the seat away.' Jock was increasingly desperate. 'The ground seemed frighteningly close.' He gave up trying to squirm out and, close to panic, simply pulled his own parachute's ripcord.

His gamble paid off. With just seconds to spare the parachute deployed and he was dragged clear of the seat which finally fell away. It was almost too late; a copse of trees was now filling his vision. He crashed through the branches which flipped him over, then cannoned into a field where he lay spreadeagled on his back, gasping for breath. He was terrified he had broken his spine. He lay quietly, not wanting to face reality.

Jock was suddenly aware of another person. He blinked his eyes open. 'She looked about 30 feet high.' A woman in a kitchen apron was peering down at him intently. 'She asked the question that all English women ask when a parachutist falls into their back garden: "Would you like a cup of tea?"'

The three rear crewmen had all landed relatively safely after bailing out of the side door. But there was a final, grim discovery.

A local doctor had been standing in his garden when he saw the flaming Valiant pass overhead and then explode. He watched as it crashed, then jumped into his car, racing towards the billowing tower of smoke. When he neared the wreckage, he discovered Brian Foster lying in a lane underneath a hedge. The co-pilot was clearly dead. 'He had been bleeding profusely from the mouth and nose,' the medic reported. There was severe facial bruising, blood was coming from his ears and there were 'multiple serious injuries to the lower part of the body'.

When Foster had ejected, the downward force of the exploding wing countered the upward thrust of the ejection gun in his seat.

With an acceleration of 60 feet per second, it was simply not powerful enough to fire his seat clear of the Valiant's large tail and he was cannoned into its towering structure. The cause of death was eventually deemed to be a fracture of the base of the skull caused by impacting the tailplane. It was unlikely that Foster knew anything about it.[2]

This was a problem that James Martin had already foreseen the previous year as the development of these massive new bombers continued. He had produced a modified Mk1 seat powered by an 80 feet/sec gun. It comprised three tubes telescoped into each other, consisting of one primary and then a further *four* secondary charges arranged in two pairs. When the primary cartridge was fired, the inner tubes began to rise, uncovering first one, and then the other set of secondary cartridges, which in turn were ignited by the flame from the cartridge just fired. Airborne tests had produced trajectories of around 90 feet. Enough height to clear a V-bomber's tail. He had delivered two of these modified seats to Vickers. But for some – unrecorded – reason, the company had not fitted them to the jet Jock Bryce was testing.[3]

Jock was horrified to hear his co-pilot had been killed. In deep shock, he and his surviving crew went to see the smouldering wreckage the following day. 'In a cavity in the ground, about 4 feet down, and three times the width of the flight deck was the spot where I had been sitting.' It chilled him to the bone. 'It was easy to imagine myself still in that cockpit and I felt slightly sick. I discovered I was not quite such a tough Glaswegian as I thought.'

Jock Bryce was the eleventh person whose life had been saved by a Martin-Baker seat. If Brian Foster had been sitting on the more powerful seat, that number would have been his.[4]

* * *

Bryce's and Foster's individual ejections – one successful, one fatal – demonstrated the increasing challenges faced by ejection seat manufacturers. Firstly, the ejectee needed to be fired safely out of an aircraft, *then* a more efficient, automatic system to rapidly release them from the falling seat was needed. Until now, James Martin had focused on the seat as an engineering challenge. It was only after the escapes of the likes of Ron Guthrie – propelled into further danger after his life was saved, unsure what terrors his future held – and now Jock Bryce, that he began to realise the true importance of what he had invented. Controlled testing was valuable, but learning about ejection from those who had done it for real was priceless. He invited Jock Bryce to his Denham factory to talk through his experience in painstaking detail. Both he and Jock knew that Foster's 'death by misadventure' could and should have been avoided. He later claimed that there had initially been 'no interest' from the authorities in the improved version of his early ejection seat.

The other three crewmen owed their survival to good old-fashioned luck. They had got out of the escape hatch just in time to open their parachutes, in the same way as many of their forebears had done during the war. But all ejection seat manufacturers, James Martin included, were already looking into multi-crew ejection systems for the coming generation of multi-crew bombers. Martin believed that every man should have an equal chance of escape. The Air Ministry didn't seem to share his conviction. He was informed that because the design of the V-force bombers was now finalised, it was not possible to fit ejection seats for rear crew. Martin was incredulous. He said the 'decision was criminal'. And he wasn't going to accept it without a fight.[5]

Over in America, where the attitude seemed very different, his counterparts were working not only on the 'standard' upward ejection seats, but also on developing downward-firing seats for the bomber crew sitting in the lower section of an aircraft. If they worked, they would offer a clear path for ejection, avoiding the fate suffered by the likes of Brian Foster.[6]

North Korea

After being shot down and captured, Ron Guthrie had been stripped to his underwear, bound with a length of telephone cable and paraded through a series of baying, hostile crowds for interrogation. His successful ejection was the beginning of a long road. He had no idea of his destination, but in his gut he knew it would be perilous. 'I had injured my back and legs during the bailout,' he said. 'Shock was taking effect as my miserable imprisonment began. I was convinced there would be no decent treatment. My overriding thought – *Will I be shot?*'[7]

At her home in Kingsgrove, New South Wales, his mother pored, line by line, over the letter she received from Australia's Minister for Air.

'It is with regret that we have learned that your son, Warrant Officer Ronald David Guthrie, has been reported missing in Korea,' it read. 'You may be assured that every possible effort is being made to locate him. I express the sentiments of the whole nation, when I say that we all join with you in the hope that he is safe. All members of the Government express sympathy in your present anxiety, and hope that some good news of him will soon be received.'

In North Korea, where his captors were accusing him of being an 'imperialist war-mongering American', there really was no 'good news' about her son's predicament. He found himself being interviewed by two foreigners, who were 'helping the Korean people in their just struggle'. They were Russian pilot Nikolai Babonin, who had shot him down, and his wingman, Aleksei Svintitsky. They were scornful of their former foe. His Meteor was inferior to their MiGs. His flying suit was second-rate.

They laid a pistol on the table. He stared at it in fear. It was his own, the one he had used to shoot one of his captors. It was still caked in paddy field mud. 'They told me that it was in a disgraceful condition. In their air force it would lead to severe disciplinary action.'

Their tone then softened, and they relived their dogfight from every angle. They respected Ron as a fellow pilot and slipped him a food parcel. And a surprise. 'They told me that a MiG had been downed by my guns.' The news boosted Ron's morale. Though not for long.

He was moved on, again past more crowds of jeering peasants, lunging and grabbing at him, through a succession of rat-infested, lice-ridden cells, sharing the tiny spaces with thieves, murderers and deserters. The guards were brutal, dealing out vicious beatings at the slightest provocation. Speaking was forbidden. 'One man caught talking had his arm broken by a guard,' Ron recalled. 'For lesser infringements the guilty party was forced to stand with his arms fully extended above his head for hours until he collapsed with exhaustion. Everyone was controlled by terror and the fearful possibility of similar persecution.' Even the slightest movement was punished. 'The guards crept silently along the passageways. Any prisoner who scratched, or dozed and slumped sideways was beaten. There was endless scope for reflection; the tedious passage of time was a torture in itself.'

He was hustled onto the back of a military truck, which jerked into motion. By now he had no idea where he was. 'Two guards manhandled me roughly. They forced my hands behind my neck and secured my wrists with wire. They passed more wires around my waist and down to my ankles and then yanked my legs back as far up as they would go.'

Trussed up like a hog and unable to move, Ron was dumped on a city-centre pavement for more urban humiliation. 'It was not easy to take; passing civilians, especially children, poked, kicked and spat on me.' Others beat him with sticks. 'In a fit of utter despair and desperation, I cursed them all.' After an hour, the torture was ended. Next stop, Ron was thrown into a deep hole in the ground and the ladder was removed, his lodgings for the night.

In the morning he was brought to the surface and hauled in front of a regally attired North Korean Air Force chief. He demanded to know information about Ron's squadron and base. 'My refusal to answer was greeted immediately by a frenzied attack from one of the other officers.' Wielding a sturdy truncheon, the interrogator repeatedly battered Ron's head, upper body and arms. The assault was relentless. Numb with pain, limbs a vivid red and blue, 'I was hustled outside, stumbling between two soldiers who propelled me up a narrow path to a barren hillside.'

They handed him a spade.

'A guard ordered me, "Dig hole. Dig grave!"'

'I was shocked, knowing what was going to happen. I refused to co-operate, throwing down the spade.' The guards attacked Ron again, beating him viciously until he collapsed on the ground.

The violent interrogator looked into his eyes. 'You think again! You co-operate. Or die now!'

Boscombe Down Airfield, Wiltshire
29 August 1952

The first anniversary of Ron Guthrie's ejection was also the day that civilian test pilot Alexander 'Ben' Gunn strapped himself into the cockpit of a small, matt-black delta-wing 'P120' research jet at Boscombe Down Airfield. At around 5,000 feet he ramped up the speed to 517mph, the fastest the experimental aircraft had ever reached. 'Instantly there was a high-pitched buzz,' he recalled, 'followed by the loudest bang I have ever heard.'

He became disorientated as the jet went through a series of rapid rolls. He was out of control, diving towards the ground. At 3,000 feet, he hit severe turbulence. The jet turned into a rollercoaster. He needed to get out.

By the time Ben released his cockpit canopy and pulled the ejection handle on his face blind, the jet was upside down. He was blasted out downwards, spiralling and twisting wildly. The drogue parachute was unable to stabilise him.

There was worse to come.

His legs and arms were flailing like a rag doll. He couldn't reach the seat buckle. He had only seconds to manually release himself, push free and open his parachute.

Hurtling downwards, he somehow found a seat strap on his right thigh and tracked it across his body until he reached the buckle and released it. He was feet from the ground. As the seat fell away, his parachute yanked him upwards. Not far enough. He plummeted through the upper boughs of a tree, somersaulting and crashing onto the hard ground.

With a cracked wrist and ankle, Ben hobbled to a nearby cottage. It belonged to a retired test pilot who had worked for the same company. He didn't need his host to tell him that, as Martin-Baker ejectee number twenty-four, he was incredibly lucky to be alive. By the time help arrived, they had consumed a bottle of brandy and the world looked a little better. As he had done with Jock Bryce, James Martin invited Ben to the Martin-Baker factory to describe his near-death ejection. He was even more eager to discuss the requirements of an automatic ejection seat for these ever-more complex flying scenarios.[8]

* * *

It was becoming increasingly apparent to all ejection manufacturers that the rapid development of jet-age aircraft, flying faster, higher, or sometimes at very low levels, meant the old system of 'manual ejection' was no longer viable. Their ejection systems needed to keep pace with the changing demands of modern aviation. A more efficient, fully automated system was needed. Each manufacturer was looking into this problem, and James Martin's solution is representative of the process. In 1952 he introduced the first major evolution of his ejection seat.

His first seat had saved a number of lives, but he also knew that 30 per cent of ejections still proved to be fatal. The Mk1 seat – like all early devices – was a basic piece of kit. Although a remarkable invention, it did nothing more than get the occupant away from his aircraft. He was then left to his own devices to ditch the seat and pull his parachute ripcord. Some escapees hesitated too long, fumbled or were disorientated during the blast of their ejection into the high-speed airflow. If an ejection happened at a thousand feet a pilot would have mere seconds to release himself from the seat and deploy his parachute, or he would hit the ground and be killed.

Martin redesigned the Mk1 to make it fully automatic. It was a ground-breaking step. On that first seat the small drogue parachute was attached to the top by a rigid fastening; its sole purpose was to stabilise the seat and allow the pilot to manually release himself. On this new Mk2 seat, the drogue's role was changed. It firstly deployed as before, to stabilise the seat, but then it needed to release *itself* from the seat, transferring its pull to the occupant's *own* parachute, dragging it out into the airflow in order to deploy. So the seat's fixed drogue connection was now replaced by a 'scissor shackle' which could be mechanically opened, to release the drogue parachute.

When the pilot pulled his face-blind handle and the seat rose, a static line attached to the cockpit floor activated a new device – a clockwork time-release unit bolted to the right side of the seat. The drogue initially deployed to stabilise the seat as normal, but, at the same time, the new release unit was ticking down five seconds, after which its gears and wheels released a spring-loaded plunger in the shackle mechanism, automatically opening the new fastening like a pair of scissors and releasing the drogue from the seat.

At the *same* time, yet another new mechanical device fixed to the seat pulled a cable, rotating the pilot's seat buckle and unlocking his harness. As the drogue rose away from the seat, its upward momentum pulled another connector, withdrawing the pilot's personal parachute, deploying it and pulling him free from the seat.

Yet another new part of the process was needed to stop the automatic sequence kicking in at high altitude, where lack of oxygen and freezing temperatures could be fatal. Martin fitted a further new invention. A pressure-sensitive mechanical device – the barostatic unit – determined what height the ejection seat was at, delaying the automatic separation operation until the seat and its occupant had fallen, still stabilised by the drogue, through the cold, rarefied air to below 10,000 feet where they could breathe unaided. An emergency oxygen bottle had also been fitted to the seat and was turned on automatically during ejection.

It was an astonishing development in seat technology, allowing a pilot to pull his ejection handle and, even if knocked unconscious, be floating down to safety around six or seven seconds later. Jo Lancaster's landmark thirty-second escape, just three years earlier, suddenly seemed like ancient history.

Martin's reliable seat-tester, the redoubtable Irishman Benny Lynch, completed two successful live tests of the new automatic seat. 'It is possible to be ejected as low as 450 feet,' Martin said, 'and have 200 feet to spare when the main chute is fully developed, and at speeds of up to 600mph. A very great improvement on the manually operated seat.' Even better, all manually operated seats in service aircraft could easily and cheaply be converted to fully automatic.

Although seat mechanisms were being improved to reduce the time from ejection to parachute deployment, a pilot still had to jettison the cockpit canopy manually before pulling the face blind. Valuable seconds were being wasted and if the canopy did not leave the aircraft cleanly the seat could collide with it. There had also been cases where pilots had forgotten to get rid of the canopy before ejecting and were injured. And, to compound all these problems, those ejecting at high speeds could be buffeted by the powerful slipstreams while still in the cockpit after getting rid of the canopy, which made locating and gripping the face-blind handle a real challenge.

Martin believed it would be possible to link the whole system together, connecting a canopy release device into the automatic ejection sequence. When the pilot pulled the face-blind handle it would trigger a cartridge on the canopy to force it cleanly away. There would be a one-second delay for this to happen before the seat fired on its time-release mechanism. He went to work.[9]

NORTH KOREA

Ron Guthrie's post-ejection ordeal was showing no signs of easing. After weeks of torture and brutality he had managed to escape from an interrogation centre staffed by a bunch of sadistic fanatics, along with his friend Jack, a fellow POW. With the help of sympathetic North Korean villagers, they made it 35 miles to the coast and commandeered a small fishing boat. But in the hostile landscape, there was little chance of success and they had eventually been recaptured. Ron knew he was now in serious trouble. They were taken to an infamous torture camp where the floors were stained with blood.

'As we stepped inside, I was blinded by an electric light then clubbed across my left cheek.' He was knocked unconscious, crumpling to the floor as the blows continued to rain down. 'When I woke up, two Korean interrogators were kicking Jack as he lay on the floor. One was wielding an iron bar, I guessed that was what had knocked me out. The looks of pleasure and excitement on their faces indicated they were enjoying the exercise. They turned on me and the merciless bashing continued. I attempted to curl up in a ball but was now receiving a bout of savage kicks. I passed out again.'

Ron came to, squirming in pain, to find the guards laughing. Another frenzied assault followed until he passed out for a third time.

Broken in body and spirit, their wounds left untreated and infected, Ron and Jack were eventually transported to another camp near Pyongyang and dumped in a wooden hut. 'We were horrified by what lay before us,' he says. 'There were about forty prisoners who hardly looked up as we entered. They were in a worse condition than us. Frail and sick, a mass of injuries and bruises. One unfortunate fellow lying on a bunk seemed to be wearing a cream jumper. I was horrified to discover it was in fact khaki, but totally covered in lice.' Sadly, during the night, the unfortunate man died from the maltreatment, dysentery and lice infestation.

Weeks turned into months and the brutality continued through the freezing winter of 1951. With thirty other prisoners Ron survived a ten-day march in temperatures of -30°C to a Chinese-run prison camp near the border. It was home to 350 UN POWs. Amid the pain and degradation of their continuing suffering, Ron had no idea that peace talks were grinding on through 1952, and in the monsoon humidity of summer 1953, without warning, the twelve hours a day he had to sit cross-legged, listening to political lectures, stopped. Then he realised that he could no longer hear combat aircraft overhead. Only the bullfrogs croaking in the paddy fields beyond the barbed wire.

Rumours spread through the camp. Electrifying news. The Korean War was over. A truce had been signed on the morning of 27 July 1953. That evening, leaflets fluttered down from an American B-26 bomber. Ron rushed to grab one, and there it was in black and white. The peace was real. POWs were going to be exchanged.

Eventually, on 2 September 1953 Ron was handed over to the Americans. 'We pulled up in sight of the helicopter pad,' he says. 'As I placed my foot unsteadily on the steps I was approached by two of the biggest American military policemen I have ever seen. These kind men took me under each arm as my legs buckled. Later, they carried me down the steps and through a beautiful archway bearing the words, "WELCOME GATE TO FREEDOM". Tears rolled down my cheeks.'

On 12 September, Ron and a small contingent of 77 Squadron prisoners boarded a flight to Australia. Two years after ejecting, and after two years of fear and brutality, he was finally on his way home. 'As I came down the aircraft steps at Sydney Airport, I caught sight of my mother, my father and a number of relatives and friends lined up by the barrier, all waving. It was a truly emotional moment. A staff car took us back to my mother's place. It was certainly a stylish arrival home.'

Ron had been one of the very first airmen to use the ejection seat in combat and it is estimated that there were around 2,000 ejections over the course of the Korean War; a huge proving ground for this fledgling lifesaving invention. Later investigation highlighted many of the problems that were already known and being solved, such as difficulty in manually releasing from the seat, ejecting too low for the parachute to open, and severe tumbling in the slipstream.[10]

Ron was just grateful that a device he had initially been very suspicious of had actually worked as advertised. Speaking to the author many years later in 2021, he reflected on his life, saying that during his time as a prisoner of war, he had often considered the irony of the seat saving his life, but then catapulting him into those years of unimaginable suffering. He would not be the last military flyer to reflect on that conundrum.

Ron died in October 2021 aged ninety-six, just a few months after talking about his experiences with the author. His obituary said:

> Ronald David Guthrie represented that spirit of the joy of life. His had been in peril, but throughout his, at times brutal, captivity in North Korea, he seems never to have given up hope. Vale Ron Guthrie; a Good and True Australian.[11]

EGLIN AIR FORCE BASE, FLORIDA
7 DECEMBER 1953

In the UK, there was still no enthusiasm in government circles for designing rear-crew ejection seats. However, over in the US, attitudes were very different and the world's first test of a downward ejecting seat for the latest multi-crew bombers beckoned.[12]

The long-range, high-altitude Boeing B-47 Stratojet was a fearsome machine. A masterpiece of revolutionary aerodynamics, America's first swept-wing, multi-engine bomber had initially flown as a prototype in the late 1940s. Weighing 51 tons, its silver fuselage was 107 feet long. The slender wings spanned 116 feet, with three jet engines suspended from each. Now the mainstay of the US Air Force, and progenitor of all subsequent large American military jets, it had a crew of three. Pilot and co-pilot were equipped with standard ejection seats, but, in an emergency, the navigator had to climb through an escape hatch in the belly of the aircraft and bail out conventionally. The two pilots would fire their own seats once he was clear.

Until now. The giant jet was the obvious choice for the first test-firing of a downward ejection seat, which replaced the standard non-ejecting seat of the navigator, hunched down in the nose, with the pilot and co-pilot sitting in tandem in a large bubble canopy above him. USAF chiefs had approved the development of a down-ward seat; they were not sold on the increasingly powerful guns pioneered by Martin and others to shoot aircrew clear of tailfins, fearing the effects they would have on the human body. They also believed that the downward trajectory from multi-crew bombers would minimise the risk of the seats colliding on ejection.

After months of testing on dummies, they needed to see how it would work with a human.

USAF Colonel Arthur 'Chic' Henderson was the guinea pig. One of a small group of men across the world prepared to risk everything for an infant but fast-developing technology, Chic was a Master Parachutist with a record of 127 jumps. Even so, he was still nervous. Through the window of the briefing shack, he could see the B-47 glinting in 80 degrees of Florida sunshine. 'It was poised like some violent beast ready to strike,' he recalled. 'The usual squeamish feel-ing was with me and my mind was racing. What will the first human test be like? Is there something we could have slipped up on?'[13]

A voice snapped him out of his reverie. 'Let's get ready, Colonel.'

Henderson slid down into the seat and strapped in, clipping together his shoulder harness, his lap strap and his crotch straps, which fastened around the tops of his thighs. He was wearing a bright red flying suit so the cameras of the accompanying air-craft could track him as the seat, and he, fell. After he'd stared down through the open hatch for what seemed like an age, it was slammed shut and the B-47 gracefully left the runway, heading for Choctawhatchee Bay on the Florida Panhandle. He kept his eyes on the altimeter and speed indicator as the Stratojet climbed to the test altitude of 10,000 feet at 230mph. With thirty seconds to go, Chic pushed his head back firmly against the rest and grasped the D-ring between his thighs with both hands.

'Come what may, I was ready to go.'

'We are on the final run,' the co-pilot told him.

'Five, four, three, two, one. FIRE!'

He pulled the ejection handle. The hatch beneath his feet blew clear and, as the seat shot downwards on its guide rails, a mechanical

timer was activated. After Chic left the belly of the aircraft, a cartridge was triggered, which would automatically release his harness.
Two seconds later it fired. 'I left the seat by the combined forces
of gravity and drag. The greater the speed, the more forceful the
separation.' An automatic parachute-release timer was also armed.

Seven seconds after ejection, his parachute opened. 'A glorious
and welcome swirl of white. I found myself swaying in the breeze
and sighed with relief on a gentle descent.' Splashing down into the
sea six minutes later, Chic unsnapped his oxygen mask and inflated
his underarm life floats. 'I felt swell,' he remembered.

Chic Henderson had made history; he was the first man in the
world to complete a downward ejection.

'Swish is the easiest way of describing it,' he said later. 'There
was no jolt. The instant you fire the seat, you're gone. A kaleido-
scopic sense of colour unfolds before your eyes. There is no
blackout, merely a few moments of confusion, free-falling through
emptiness.'[14]

Seven ejections were made in that first series of tests, firing the
occupant out at ever increasing speeds, up to nearly 500mph. Four
of Chic's colleagues were successful; the others suffered broken
limbs caused by flailing in the faster airflow after leaving the jet.
This issue of 'flailing limbs' was becoming a major headache for
seat manufacturers and various methods were being tried to minimise the problem. On Chic's B-47 seat, modifications had been
made to harness straps and leg-guards to minimise flailing and in
1954 he led off the second round of tests. This time his ejection
wasn't quite so smooth and he described the moment he left the
B-47 at high speed as being like 'a Saint Bernard dog shaking a
rabbit in its mouth'.

Initial teething problems sorted, the downward seat was
approved and entered service. Chic Henderson, the veteran skydiver and ejection seat pioneer – who would go on to be known as
'the jumping grandfather' – recalled, 'we had been inspired by the
certain knowledge that reliable automatic ejection equipment is an
absolute *must* for jet aircraft'.

* * *

James Martin was pleased with his new, automatic seats, but was more than sceptical about the American concept of downward ejecting, describing the process as 'utterly ludicrous'.[15] But he too was learning that during ejection at higher speeds, the windblast lifted feet off the rests and legs over the existing thigh guards, risking severe injury to knees and hips.

His solution was simple. Two reinforced nylon cords were attached to the cockpit floor. When strapping in, the pilot, now wearing a set of 'garters' clipped around his legs above the calf, routed the cords via the front of the seat, then through small metal D-rings in the garters before finally attaching them, via clips, back onto the seat. During flight the cords were slack to allow for leg movement, but in an ejection, as one end was fixed to the cockpit floor, they would simply tighten as the seat rose up and, utilising a 'non-return' fixing at the seat connection, they would hold the legs against the front of the seat. The floor fixings were engineered to simply shear off under the force of the rising ejection seat when the cords reached their maximum length, leaving the legs held firmly in place. The automatic harness unlocking system was now redesigned to also release the leg restraints.[16]

With ejection speeds increasing rapidly, seats were now also fitted with a mechanical sensor, measuring g-force, which delayed the pilot's parachute opening until his post-ejection speed through the air reduced. If deployed too soon, the g-force experienced at high speeds could cause the parachute canopy to burst. All these new systems needed constant testing and the answers would not come from the workshop or testbed data. Martin needed to know what ejecting at high speed and high altitude felt like, and how his new automatic seat would operate there. Who better to tell him than the man who had made such a major contribution to the testing and development of his very first seats in the late 1940s? He called yet again for his reliable friend, Benny Lynch.

Lynch took off in a Meteor from Chalgrove Airfield in Oxfordshire on 17 March 1954, strapped into the open cockpit behind the pilot. This would be his seventeenth ejection.

Back in 1947 he had blasted out of the same aircraft on his prototype Mk1 and, having disengaged himself from the seat, was under a fully developed parachute some thirty seconds after initiating ejection. Now this latest seat he was attached to would do it all for him, sensing the ejection forces he was experiencing, then height, determining when to release both itself and his personal parachute.

The pilot took the Meteor to 30,000 feet. A green light flashed on Benny's instrument panel telling him it was time to go. He pulled the face blind and rocketed out of the jet in a perfect exit. On the inside of his wrist, he was wearing an altimeter and a stopwatch. He kept an eye on both as he plunged earthwards, still strapped in the seat but stabilised by its small drogue parachute . . .

25,000 feet.

20,000.

15,000. The seat would soon release him and his parachute would deploy. Benny braced for the jolt.

10,000 feet. Now?

8,000. Still nothing.

At 6,000 feet he knew he was in real trouble.

Automatic separation should have happened by now. Something had obviously failed and he was still strapped in the seat.

The cause of the failure was simple: when Lynch ejected he had lost his grip on the face-blind handle which began blowing around violently in the airflow, hitting a key component of the clockwork release mechanism. A small metal rod in the system had come out as designed, arming it ready to start the countdown to release the harness and parachute at the correct altitude, but now it was knocked back into the 'pre-armed' position so the mechanism couldn't begin its process. The seat was never going to release Benny.

The rapidly approaching Oxfordshire fields filled Benny's view. Ignorant of the cause of his predicament, and with just seconds to spare, the seasoned veteran kept his cool and operated the emergency lever to manually separate from the seat. He then pulled the reserve parachute on his chest.

But his woes were not yet over.

As he floated down, he spotted another problem looming fast. Unless he did something, he was going to crash into a wire fence. He desperately pulled at his parachute cords to try to steer clear. The fence was getting closer and, as he hit the ground, he was side-slipping hard to avoid it. A sudden sharp pain stabbed his lower leg. He had severely fractured his ankle. Even though it was Benny's first and only ejection injury, it was the end of his days of being shot out of aircraft. Where would Martin ever find another guinea pig as brave and dependable, or as relentlessly willing and cheerful?

Lynch limped out of the ejection story for some well-earned R&R in his local pub, with a British Empire Medal in recognition of his service.

James Martin's solution to the problem was, yet again, relatively simple. He applied a small tweak to fix the tiny rod, re-engineering it so that it needed to be rotated before it could ever be pressed back into the 'safe' position. A remarkably easy solution to a potentially deadly problem. And a problem which would never have been discovered without human testing.[17]

* * *

Other ejection firsts were equally unplanned.

US test pilot George Smith had been enjoying his weekend off work but he stopped by his office at Los Angeles International Airport on Saturday, 26 February 1955 to drop in a report. He was returning to his car when the company's flight despatcher called him back. A brand-new F-100 Super Sabre was sitting on the flight line and needed to be put through its paces before being turned over to the Air Force. Did George want the job?[18]

An aviator to his bones, George was more than happy to take up the challenge. His task was to chase two other supersonic Sabres that were taxiing out to the runway. He pulled on a lightweight flight suit, parachute and life vest over his street clothes, and rushed to his aircraft. Taking off, he climbed rapidly to 35,000 feet over the Pacific Ocean.

So far, so good.

Suddenly, he felt the flight control stick 'go sticky'. Then, without warning, his hydraulic system failed. The Sabre pitched into a dive. The 15-stone pilot pulled back the stick with all his strength but the dive steepened to near-vertical. The aircraft's speed increased rapidly, passing Mach 1, more than 760mph. He realised that this was going to be a short ride. The accompanying F-100 pilots were yelling into his headset. 'Bail out, George! Bail out!'

He had plunged down 30,000 feet and was now doing 785mph. He jettisoned his cockpit canopy. The roar of the airstream was unlike anything he had ever heard. He pulled his seat ejection triggers. Fired out of his cockpit to be hit by the near 800mph airflow, he was subjected to an estimated 40g deceleration, all while wearing his weekend casuals. He was immediately knocked unconscious.

His limp body was battered by the ferocious blast, so powerful it ripped off his helmet, oxygen mask, footwear and wristwatch. Blood was forced into his head, which became grotesquely swollen and made his facial features unrecognisable. His nose was torn away from his upper lip and his lungs were inflated like balloons. His whole body was being horribly battered and bruised as he flailed end-over-end in the seat.

The still-unconscious George parted company from the seat automatically as programmed, followed by the deployment of his parachute, but the pulverising wind ripped out a third of the panels as though they were made of tissue paper. The other Sabre pilots saw their unconscious colleague dropping like a rock. The crippled jet crashed into the sea close enough to a fishing boat to almost sink it. The crew thought they were being bombed. Then they saw George hit the water. Luckily, the skipper was a former navy rescue expert. They cut him out of his harness and pulled him aboard. He had blood flowing down his legs, dripping from his feet.

Hovering close to death, George was rushed to hospital where he was in a coma for six days. When he woke up, he was blind in both eyes. The force of the ejection had caused them to haemorrhage. Every joint in his arms and legs had been dislocated, his liver and kidneys were damaged, and his gall bladder and seventeen feet of intestines had to be removed. After four operations and seven months in hospital, he eventually recovered from one of the world's first supersonic emergency ejections.[19] Not only did George return to good health and regain his sight, a hypnotist helped him redis-cover his confidence and, incredibly, he returned to test-flying a year later.

Seat manufacturers would study his ejection in great detail, recre-ating the circumstances during extensive testing, further advancing survival equipment and seat design. For those complex, life-defining calculations, they used mechanical adding machines and slide rules – pocket calculators still being twenty years in the future.

George was just thirty-one when he was given a second chance at life thanks to his automatic ejection seat. He lived another thirty-nine years, later saying, 'I can remember that flight very well up to the point I ejected. After that, my memory is a blank. I guess I'm pretty lucky to be here.'

As jets got faster, ejection seats were being tested at ever-higher speeds and altitudes to keep up with the astonishing rate of devel-opment of aircraft. And those who might have to use them were increasingly aware of their significance. Ejection seats were no longer feared. The days when they were regarded with suspicion, as 'curious contraptions', were long gone and aircrew expected them to be fitted as standard to every jet they flew.

But, as ever in military aviation, expectations were not always matched by reality, and, for those on the front line, the results could be disastrous.

CHAPTER FIVE

A SECOND CHANCE AT LIFE

Reporters, assorted MPs and high-ranking dignitaries peered through the driving rain for the first sight of the RAF's brilliant white aircraft of the future. Nicknamed the 'Tin Triangle', the delta-wing Avro Vulcan contained enough sheet metal to cover one and a half football pitches and was constructed from over 100,000 different components, including more than a dozen miles of electric cable and 430,000 nuts, bolts and rivets. Its four hungry jet engines, fed by fourteen fuel tanks, produced as much power as eighteen railway engines. Another heavy bomber destined for the growing V-force, she could deploy a single atomic weapon more powerful than all the bombs dropped on Germany in the Second World War.

The spectators waiting at Heathrow Airport in west London glanced at their watches. It was shortly before 10am. Vulcan XA897, the first of the new breed to be delivered to the RAF, was returning from a highly successful showcase tour to Australia and New Zealand.

Strapped into their ejection seats, barely a hip-width apart on a surprisingly cramped flight deck, Squadron Leader Donald 'Podge' Howard and his co-pilot, Air Marshal Sir Harry 'Broady' Broadhurst, Commander-in-Chief, Bomber Command, were at the controls. Two navigators, the air electronics officer and a civilian adviser from Avro occupied high-backed, rear-facing seats in an even more claustrophobic compartment behind and several feet beneath them. Natural light trickled onto their banks of instrument panels through two small, circular windows. There were still no rear-crew ejection seats fitted to this new generation of RAF jets; in an emergency, they had to dive out of an escape hatch and deploy their parachutes manually by tugging on the metal handle.[1]

At 10.04, the Vulcan was at 1,500 feet and 5 miles from touch-down. Howard began his descent through the atrocious weather. The welcoming party still couldn't see the giant jet. And Howard still couldn't see the ground. The airfield approach controller used radar to guide him down the glide path towards the landing point. Howard searched for a first glimpse of the runway lights through the three small cockpit windscreen panels, which some pilots likened to staring out of a letterbox. He had agreed that he would descend until his altimeter read 300 feet above the ground. If he still could not see the runway and it was not possible to land, he would simply overshoot and come around for a second attempt. He had ample fuel to divert to a different airfield if he had to, but neither he nor his co-pilot wanted to miss the red-carpet moment. The press reception was designed to bolster the image and standing of the RAF's nuclear bomber force.

As the weather worsened, he reached 1,000 yards from the touchdown point, realised he was 80 feet above his glide path, then overcorrected. Now he was 100 feet too low. A minute later, after a critical misunderstanding about altitude between Howard and the ground controller, the bomber ploughed into a field a quarter of a mile short of the runway perimeter track. Howard shoved his four throttle levers forward hard, winding up the engines to harness the Vulcan's enormous power to get airborne again. It was hopeless. The rough ground had ripped off both main undercarriage units. The huge jet reared up 200 feet and surged forward, now irretrievably out of control.

Howard ordered the crew to abandon the aircraft but, at that height, it was a pretty pointless command. He and Broadhurst reached for the firing handles on their ejection seats and blasted out. They had just enough height for their parachutes to open. Within seconds the nose and starboard wing dropped and the Vulcan barrelled into the ground beside the runway, breaking up, the hot engines and sparks igniting the spewing fuel. It was a tangled inferno. Broadhurst landed on the runway, breaking his leg, and was ferried to an RAF hospital. Howard hit the grass and suffered minor bruising to his head. He was seen leaving the airport in a car with his wife who had watched the horror unfold, a cigarette dangling from his lip.

Howard and Broadhurst had been given a second chance at life. Their four crewmates perished; without the benefit of ejection seats and so close to the ground, they were doomed. They died where they sat, strapped into their conventional seats, as their crippled bomber ended its highly publicised journey in an inferno of flame and molten metal. The RAF Board of Inquiry opened the next day to investigate the accident and reported in mid-December. The Secretary of State for Air told MPs at Westminster that the crash had not been caused by a technical fault; the ground controller and the pilot were equally to blame. 'It was the duty of the pilot to give the order to abandon the aircraft and to eject when they did. Both realised the other occupants had no chance of escape.'[2]

Their deaths had brought the issue forcibly to public attention. Concerns over the lack of rear ejection seats in Britain's V-force bombers had begun years earlier when it was first decided they would not be provided. The *Daily Express* was now asking: 'Why no ejection seats for three of the crew?' It launched a campaign. Most of the giant American bombers had ejection seats for all crew members, why not Vulcans, Valiants and Victors?[3]

V-bomber pilots knew they faced an agonising decision if things went wrong. Eject and abandon their crewmates, or stay and risk adding themselves to the death toll. Those who survived thanks to ejection seats would face the unenviable prospect of returning to face the families and friends of rear crew who had perished. To many, it seemed morally wrong that the Air Ministry had decided that only the front-seaters would have the technology that had already saved many hundreds of lives.[4]

The *Daily Express* continued pressing the case, demanding ejection seats for rear-facing bomber crews. The arguments and accusations would go on for many years, but James Martin needed no encouragement. With zero interest from government or the RAF, he invested his own money in the development of a rear-facing version of the Mk4 seat which he could retrofit to the giant bombers. Determined to show the pen-pushers that ejection seats could easily be provided for rear crew, he was a man on a mission.[5]

* * *

In July 1958 he welcomed a new aspiring ejection seat-tester to Denham. William Tulloch Hutton Hay was a stocky, sociable Aberdonian, known to one and all as 'Doddy'. Relentlessly cheerful, he had a taste for the finer things in life. During the Second World War, the Scotsman had been a bomber tail gunner – a very dangerous role in itself. 'It had not all been honey and roses,' Doddy said. 'Sometimes, cooped up in my little glass gazebo with ammunition boxes chafing cold against my knees and the acrid stink of cordite and glycol in my nostrils, I wondered if my choice had been a wise one. My interest in aircrew survival had begun at first-hand.'

After the war, Doddy became an RAF parachute instructor and represented Great Britain in the 1954 World Skydiving Championships. He had also visited the Martin-Baker factory where he watched a film of his predecessor, seat-testing pioneer Benny Lynch, in action. He'd seen first-hand reports of the earliest emergency ejections. And read moving letters of thanks from the flyers whose lives had been saved by them. Doddy was determined to get involved.

And Martin realised Doddy was just the type of chap he needed. A talented parachutist who could jump out of an aircraft blindfolded, and once did so from a balloon while riding a bicycle. And possessing exactly the type of adventurous free spirit Martin needed to test his rear-facing seats. 'Well, Mr Hay,' he said. 'Since you seem so determined to get yourself an almighty kick in the arse, and since you're so remarkably well qualified to receive one, I reckon the least I can do is to give it you. I promise you it won't be any nice little love-pat.'[6]

RAF FINNINGLEY, YORKSHIRE
1 JULY 1960

When Doddy first set eyes on the Valiant he was overawed by its size. 'It was vast. Stark and silver, cold and operational. I walked around it, scenting, assessing, and stroking it. I gazed up at the towering steeple of the blade-shaped tail unit.'

As part of James Martin's personal quest to provide a viable escape option for non-pilot crewmen, the aircraft had been fitted with a specially adapted Mk4 ejection seat, optimised and mounted for rear-facing ejection. After a series of static ground tests, there had been three dummy ejections, one on the runway during a take-off run, and then two at an altitude of 200 feet. Now it was Doddy's turn to undertake the first human test. He had been weighed, measured and suspended facedown from what looked like a meat hook to establish the exact centre of gravity of his 5-foot-6 frame. He was ready.

A small group of V-force test pilots at Finningley looked on with interest and wished him well. The pilot who would be at the controls leaned towards him. 'I hope to heaven it works. I really care about the outcome.' They all knew just how much was at stake. Sporting the zippered white parachuting overalls he had worn at the World Championships, with the Union Jack emblazoned on the left breast, white sports socks and shiny black boots with heavy cleats to give himself traction when landing on slippery or uneven terrain, Doddy climbed the ladder into the fuselage. Draped around his neck was the lucky silk scarf that he had worn throughout the war.

The pilot and co-pilot were already in position as Doddy crawled back and down towards the new seat, mounted on its slender guide rails. A tiny aperture had been cut in the fuselage, through which he would be blasted. He reckoned it was barely 2 inches wider than his shoulders. If the seat did not rise perfectly, he would impact the inside of the aircraft and undoubtedly be killed. When he'd first been shown the interior, he had been surprised to see a sheet of curved steel between his ejection seat and the rest of the crew space. It was an emergency blast screen, an engineer explained. 'You may be the star of this show, Doddy,' he was told, 'but you've got a supporting cast, and when you leave in a sheet of flame, they've got to have some protection.'

He squeezed himself onto the seat. 'Several pairs of hands at once went to work on the harness, clipping on the leg-restraint lines, pulling each strap and buckle tighter and tighter still until they bit into me, holding me fast, blending me almost into the seat itself.' Doddy gave them the thumbs-up, then lowered his bone-dome over his soft inner helmet. He immediately felt uncomfortable. The seat back-pad wouldn't allow him to comfortably position his head. 'My spine will be bent forward,' he told them. They reassured him he should be okay.

What the hell did 'should' mean?

The ground crew came to say farewell as the final minutes ticked away. 'The safety lock is out now, Doddy,' they warned. 'For God's sake don't pull the handle until you really mean to. You're sitting on her, and she's alive.'

Some 150 miles to their south, at Chalgrove Airfield in Oxfordshire, Martin was hosting top brass from the Ministry of Aviation and Bomber Command, along with Doddy's wife, Jenny, who was less than convinced that her beloved husband should be risking his life like this. But the Valiant had already taken off and was heading south; the drop was due in ten minutes. Flying in at 1,000 feet Doddy felt no tension. Hands steady and cool. Then, just five minutes to firing, he was passed a note. Some of the VIP observers had been delayed. The ejection was put back an hour. Still strapped in his rear-facing, experimental, live ejection seat, Doddy reached inside his suit and pulled out the *Daily Telegraph* sports pages and a barley sugar. The Valiant flew around in circles. Martin's latest guinea pig was dying for a cigarette. He had just completed the crossword when the co-pilot told him, 'We're clear to start the run-in, ten minutes from now.' Doddy calmly focused on his surroundings. In front of him was a row of lights. Suddenly one flashed green. The pilot's signal to prepare to go.

The Valiant was travelling at nearly 300mph.

'Ninety seconds from Chalgrove.'

The green light flashed again and Doddy gave the handle a huge haul downwards. 'The gun roared out beneath my backside. A shaft of sheer agony seared up my back and into my brain, but the seat and I were moving up through the dark fuselage and into daylight like an express train from a tunnel.' As he rolled and tumbled in the slipstream, he saw the huge tail of the Valiant slice through the sky beneath him. Unlike Jock Bryce's co-pilot who had been killed impacting it, the improved ejection gun charges meant he was well clear of danger.

The seat stabilised. Doddy counted the seconds. The seat parted company with him as advertised, and his main parachute canopy unfurled and jolted him upright. His upper back felt like it was on fire; as he suspected, his seating position had not been ideal and he'd clearly been injured. But the adrenaline was coursing. 'I looked down and could see I was directly over the airfield. I'd done it. I was very much alive, the new seat was a success, and so, by God, was I.'

Doddy spreadeagled his arms and legs, a prearranged signal to Jenny watching below that he was okay. He saw James Martin's black Bentley leave the crowd of spectators and follow his drift. He landed perfectly, right in the middle of the airfield, and was running around his parachute to collapse it even before the ground crew reached him. He pulled a small tin from his suit, took out a cigarette and lit up. A jubilant Martin rushed up to congratulate him. 'It was a lovely shot, Mr Hay. Now come and tell me all about it.'

The RAF observers were ecstatic, the Men from the Ministry less so. It had classified the test as secret and barred all press; there was to be no publicity for this development. They did not even stay for the celebration lunch at a local country pub, no doubt aware that they could no longer rule out rear-facing seats on the grounds they were impractical. Martin, who had invested close to a quarter of a million pounds of his own money in developing the rear-facing seat, had proved them wrong and claimed a world first.[7]

Doddy was in his element, glass in hand, joshing with the dignitaries, but he couldn't fool his wife. She knew his grin was glassy. 'You're hurt,' Jenny said. 'Is it bad?'

It was very bad. When one of the RAF officers had slapped him heartily on the back in congratulation, he'd 'damn near fainted'. Jenny could see the pain in her husband's grimace, telling him, 'They are all so wrapped up in what's happened they haven't even counted your legs. I know your face, you're putting on a good show but you're as white as a sheet.' Doddy agreed: 'The pain was setting in hard and horrid. I knew I would have to do something about it, and soon.'

* * *

The following day the *Daily Express* ran a big splash with a long-lens picture of the Valiant taken by a photographer who had been lurking by the Chalgrove perimeter fence. The headline read: 'First Man Out'.

'First Man' Hay was already on his way to Harley Street, to see a back specialist. The medical verdict was not good. The surgeon studied the X-rays and pointed out the series of crush fractures that had crumbled his fifth, sixth and seventh thoracic vertebrae caused by the surging power of the rear-facing ejection seat. The consultant told Doddy that his testing career was over.

He was determined to prove the medics wrong. He was not done with ejection seats. He loved the work, the pioneering spirit and playing a part in developing a technology that saved lives. Over the coming weeks and months, he pushed his body through punishing rehabilitation exercises. There were daily visits to the London clinic 'for proddings and pullings and pushings'. He was sustained, over those exhausting and mind-numbing weeks, by the knowledge that he had proved the worth of a rear-facing seat. 'The job had been done. Very soon, surely, the wheels of authority must start to turn.' Then he picked up a copy of the *Daily Express*. The banner headline rocked him.

BOMBER MEN STAY IN PERIL – BY ORDER

It went on: 'The men in the back of Britain's V-bombers are NOT to be provided with lifesaving ejector seats, although their installation has been proved technically possible.'

The Air Ministry had issued a statement:

In view of the experience and the low accident rate over the five years that the V-bombers have been in RAF service, and of the considerable time, effort, disruption and cost which would be involved in modifying the aircraft, it has been decided that the provision of ejection seats for rear-crew members cannot be justified.

'Cannot be justified'. The three words darkened Doddy's mood. The *Express* called the Air Ministry's decision 'monstrous'. Money had won out over morality. More questions were asked in Parliament. If American bombers could have ejection seats for all crew, why not British?

The Secretary of State for Air was not inclined to answer.[8]

The *Daily Express*, now at the centre of the battle to provide rear-crew ejection seats, ran another headline.

IT'S STILL 'NO' TO THE MEN IN PERIL

'A plea to end the life-and-death dilemma for RAF V-bomber crews was turned down in the Commons yesterday.'

But there was no retreat, the Secretary of State merely emphasising that the large costs involved, plus having V-force jets away from the front line for extended periods to be modified, meant they could not be justified. The *Express* reported that the minister had been advised: 'That the greatest number of accidents in V-bombers took place at high altitude, where all crew had time to escape, but he admitted with a shrug: "At low level this is not so easy."'

All those rear crew affected by this decision would agree with the sentiment that it was 'not so easy' to bail out manually at low level. Though none of them 'shrugged' at the reality of what that would mean.

The arguments over the possibilities of retrofitting ejection seats into the V-force jets would rumble on for many years. It is difficult to accurately assess how many lives were lost due to a lack of them. But there were certainly a number.

In March 1963 a Victor bomber stalled after a night take-off from RAF Wittering in Northamptonshire, the only option was to bail out. The four-man rear crew were unable to open the escape hatch. The pilot could hear them shouting over the intercom that they were trapped. At 2,000 feet he ordered his co-pilot to eject, so at least one person could tell the world what had happened. The pilot kept fighting to save his aircraft until he perished with his crew as it spun into the ground, setting trees ablaze and enveloping a village in a black cloud of smoke from the burning fuel. At the later inquest his surviving co-pilot said, 'I think the captain very courageously stayed at the controls when he could have used his ejection seat, because of the knowledge that the rest of the crew could not get out.'[9]

There were certainly other occasions over the life of the V-bomber force where pilots survived by ejecting and rear crew perished because they couldn't. One Vulcan pilot later reflected on the controversy, and how it affected some of the crews, writing, 'this highly controversial policy was maintained despite a practical scheme to fit ejection seats for the rear crew and despite the tragic cases'. And, commenting on the earlier Heathrow crash where the four rear crew died in the burning wreckage, he demonstrated that many pilots were very aware of what leaving their friends behind meant:

'Those pilots chose to eject. Perhaps as the four rear crew

members on that flight heard the firing mechanism of the pilots' ejection seats, then a sudden and terrible instant [realisation] of doom. Maybe as they sat in their non-ejection seats ready to be thrust helplessly, inevitably towards the abyss, perhaps time twisted to turn seconds into an eternity. And, even with eyes tight shut, colours of red, white, blue and green flashed. Flashed across the screens of those closed eyes. Perhaps their final thoughts were less of mounting panic in the midst of reckless endangerment, and more, one can only hope, of an ultimate, mysterious sense of concord beyond, in the imminence of death.'[10]

It mattered not, costs trumped benefit and rear-crew seats would never be provided for the V-force.

Both James Martin and Doddy Hay were appalled by the ongoing situation, but they eventually had to move on. And, in the meantime, Martin was already working on the next major, ground-breaking development in the ejection seat story.

'Forget about recriminations,' he told Hay. 'I've got something very tasty bubbling up in the pot, something very important, and you're the man I'd like to sample it for me. That's, of course, if you are still willing.'

Doddy shot forward like an eager schoolboy. 'Willing? What's the recipe?'

Martin paused, tapped his pencil on his desk and replied with a single word. 'Rockets.'[11]

* * *

At the start of the 1960s, 377 airmen owed their lives to a Martin-Baker ejection seat. Many more hundreds, possibly thousands, must have been saved around the world on other manufacturers' seats.[12] About 93 per cent of ejections were successful, but that figure shrank to 60 when pilots ejected at low altitude or from aircraft descending fast. In most cases they had hit the ground before their parachutes could fully deploy.

The first ejection 'bang seats' used explosive charges to propel the user up and out of the aircraft. At higher levels, there was plenty of time for the parachute to deploy and inflate. That said, the force of the large explosive charge needed could certainly damage the occupant, as Doddy Hay and others had already discovered. But it was better than dying in a flaming wreck.

Those early 'bang seats' relied on an element of forward motion provided by the aircraft's movement to help stream and deploy the parachute. But at the lower levels some jets were now flying at, perhaps, just a few hundred feet above the ground, and, more importantly, just after take-off and before landing, the ejection system simply could not provide enough height for the main parachute to fully deploy. And, as was becoming increasingly important with advancing aircraft technology, what happened if the aircraft was stationary on the ground when some sort of catastrophic emergency occurred – perhaps an engine fire – when the jet was not moving? Could the seat be developed to cater for those occasions? The next major step change in seat technology beckoned, so-called 'zero-zero' (zero height, zero speed) rocket-assisted ejection seats, where an occupant, still on the ground but fully strapped in, could eject while motionless, gaining enough height for his parachute to safely deploy before a reasonably controlled, albeit very short, descent to terra firma.

The Americans were already experimenting with rocket-powered seats and James Martin was determined not to fall behind in the race. He had been quietly working on rocket-seat design for a number of years. Hundreds of tests had already been conducted using dummies of varying weights on ground rigs, ejecting from slow-moving aircraft, and then on high-speed runs. He used a specially adapted truck mounted with a cockpit rig on which his experimental seat could sit, firstly driving it along the runway as a launch pad, and then simply sitting it stationary in a corner of the airfield. It was a risky period of development, and on an earlier ground test from a fast-moving aircraft with another of his human guinea pigs, Martin had been so concerned about the dangers involved that he had slipped away, fearful of what might unfold. Alone in the corner of a hangar, he had dropped to his knees and prayed.

His prayers, and faith in his technology, were rewarded – his ongoing test ejections proving that his zero-zero rocket seat could save the life of a pilot even if he was stationary on the ground. It had been trial and error until he came up with the technology he was convinced would work. Now, Martin showed Doddy Hay into a room at the Denham works. Always kept under lock and key, it was where Martin stored his new inventions.

He whipped off a dustsheet.

Doddy was bemused. It looked just like any other ejection seat.

'But it isn't,' Martin said. It was yet another modified Mk4. Martin had retained the ejection gun but had added a pair of steel combustion tubes mounted either side of the rear of the seat. These contained the solid rocket motors which fed into an efflux chamber, which then discharged downwards below the seat, providing the required thrust. The gun would initiate the ejection but, as the seat began to rise up the rails and out of the aircraft, a static line fixed to the aircraft floor now pulled another sear from the rocket packs, igniting the propellant, thrusting the seat and occupant hundreds of feet into the air, way higher than the first ejection seats.

'Mr Hay,' Martin said to Doddy. 'We are just about ready to leave the past, skip the present – it's well catered for – and move right into the future. We are going to provide the answer to ejection problems for years and years to come. I'm going to shoot you faster and further than any man has ever been shot before. How do you feel about that?'[13]

'Great. I had no idea it was possible,' Doddy said.

'It isn't yet,' Martin told him. 'But very soon it will be, and you're the man who's going to prove it.'

Doddy asked him which aircraft would be used, and at what altitude.

'No speed, Mr Hay,' Martin replied. 'No altitude. And no aeroplane.'

He didn't look like a man who was joking.[14]

CHALGROVE AIRFIELD
1 APRIL 1961

Doddy Hay tried to banish from his mind that it was April Fools' Day. The new rocket seat was sitting on a static rig in the middle of the airfield. It seemed to be beckoning him on an intriguing but perilous mission.

'You still feel you want to take it on?' James Martin had asked him. 'It's just you and the seat.'

Doddy took a cigarette from his pocket and pondered for a short while. He was keenly aware that just a handful of years earlier, Benny Lynch had opened the era of live airborne seat testing. Now he was standing alongside the next generation of pioneers on the brink of a new epoch, where words like rocket and space had a bold, brave ring.

He told Martin, 'Whatever the barrier, I want to be the first man through it.'

It had been a different story when Doddy watched an early test-firing of a rocket seat from the platform with a dummy. 'With an ear-splitting bang merging into a roar, the gun triggered off the rockets and the seat streaked upwards in a sheet of orange flame,' he recalled. At 150 feet, something went wrong. The parachutes flapped like washing in a high wind. Seat and dummy crashed down, gouging a 'deep and ugly pit on the airfield. I didn't want to watch. I wanted to be sick.'

But it was too late to back out now; Chalgrove was buzzing with excitement. Scientists, technicians and officers from several air forces had come to watch. Journalists and TV crews jostled for the best vantage point. Doddy's long-suffering wife Jenny had gripped his wrist when she heard somebody say, 'Hay must be quite, quite, mad.' Another joined in: 'He is really asking for trouble this time.' Doddy still worried that a rocket ejection might pulp him. Equally, he knew this test could lead to safer seats for pilots of the future. His sense of adventure and challenge was winning out. 'I was committed, and so was the girl beside me. Now, more than ever. There simply had to be faith.' Jenny, as usual, was less than enthusiastic.

As the onlookers were ushered back, he looked at the test seat. 'It was stark, grim and black, lodged securely in the gaunt iron framework of the little launching pad, the whole contraption bearing a faintly disturbing resemblance to an electric chair with its straps and buckles.' He stepped onto the pad. While he was being strapped in for the world's first zero-zero ejection he smiled for the cameras. He looked like a man about to face nothing more hazardous than a haircut.

It was time to go.

Doddy pulled the face blind. The ejection gun fired beneath him and, as the seat rose, the rocket sear was pulled. As they roared into life, he felt their colossal thrust. Earlier seats had shot a pilot up to around 70 feet. Travelling through 100mph, Doddy went beyond this before he'd had time to register it. A 20-foot streak of orange rocket flame blazed from beneath his backside. 'There followed the most extraordinary split second of my life, a fantastic flashing image that will never be repeated. Climbing on the crest of the rockets I was rushing upwards, as though to pierce the sky. The sensation was unbelievably thrilling.' The drogue parachute deployed and at 380 feet he reached the pinnacle of the ride. 'I swung across the sky in a gigantic cartwheel, while the seat fell away.' His parachute inflated fully and he gave his spreadeagle signal to show Jenny he was safe, then floated down to massed applause. Doddy had soared around five times higher than any previous ground ejection.[15] As he said, 'The whole future of air survival had changed.'

Martin, in a three-piece suit and fedora, was soon posing for pictures with his tester, dressed as always in his gleaming white parachute overalls.

Doddy and Jenny returned to their four-star country hotel, relieved there was no further damage to his spine. Things were looking up so he headed straight for his favourite destination, where the barman was already waiting with a bottle of chilled champagne.[16]

LE BOURGET AIRPORT, PARIS
2 JUNE 1961

Doddy Hay became quite the celebrity. The headline writers had enjoyed themselves after his zero-zero triumph: 'Amazing Rocket Bale-Out'; 'Whoosh Goes the Pilot to Safety'.

When he and Jenny went for their normal Sunday lunchtime snifter at their local, they found themselves at the centre of an impromptu party to celebrate his achievement. At his favourite Soho eatery, the waiter knew exactly what his customer had been doing. 'Careful, sir,' he said with a grin, as he pushed Doddy's seat towards the table.

When James Martin announced that he would demonstrate his incredible new rocket seat at the Paris International Air Show, Doddy was determined to be the one on board. This time he was not going to be fired from the ground with crowds close by. The seat and its rig were mounted on a flatbed truck which would be driven into place shortly before the launch. Climbing aboard, Doddy stood by the seat, helmet in hand as the vehicle crawled its way to the designated blast-off point. When it got there, Doddy was horrified. It was less than 30 yards from the main runway. It wasn't safe, he protested. But the authorities would not let him move further back, in case rocket flames incinerated a line of expensive helicopters nearby.

He had just over two minutes to complete the test before a low flypast was conducted by a group of fighters. He needed to be safely back on the ground before they arrived. Sweating and swearing, he buckled into the seat. He had been told that the wind speed was just over 5mph, running parallel with the runway. He would be quite safe.

Doddy pulled the firing handle. Just as in the first test, flames shot out beneath him as the seat's rockets powered him above the airfield. The big crowd drawn from all over the world had never seen anything like it. It looked like some crazy fairground ride. As soon as he reached the crest of the ride Doddy knew he was in serious trouble. The wind was much stronger and blowing him towards the runway.

As a parachutist, he had made plenty of hard landfalls before. He didn't panic. His trusty boots would not let him down. Coming in to land, the wind kept him airborne, speeding him across the runway so the boots' metal cleats touched first, causing a trail of sparks along the concrete. As he eventually crashed down, his limbs were flailing in all directions. The crowds gasped. 'With a horrible, crunching snap, my left leg shot up between my shoulder blades.' His parachute was still inflated and dragged him on. Doddy fought to collapse it as waves of nausea crashed over him. The pain was excruciating. His tibia and fibula had fractured. Bone splinters embedded themselves in his ankle.

In near delirium, Doddy was rushed to a local hospital then to University College Hospital in London forty-eight hours later, where one of Britain's finest orthopaedic surgeons went to work piecing him back together. Unable to walk and confined to bed for weeks, Doddy relieved the tedium between operations by reviewing books for *The Observer*. He went home to the aching boredom of more twice-daily remedial sessions. As ever, his spirit was unbroken and, despite dire warnings to the contrary, he was eventually able to regain some mobility and move around on crutches. After five months, he discarded these in favour of two walking sticks. Yet again, he was defying the odds.

Seven months later Doddy was walking unaided, albeit with a heavy limp, when he and Jenny arrived for a special evening in London's Dorchester Hotel ballroom. The black-tie dinner in January 1962 was hosted by James Martin to celebrate the saving of 500 lives. Some 115 very grateful Martin-Baker ejectees were there, many with their equally grateful wives.

Guests were thrilled to see Doddy in one piece. After the dinner, Martin spoke. 'When you read about someone bailing out, there does not seem much to it, but the end result is rather different. There are now 515 [Martin-Baker] ejectees, and if one saw them all together with their wives and children, it would make a fairly large assembly. It is one part of the ejection seat business that a person like myself does not think about, as we are hard pressed to try and make the seat better and better. All we want to read about is successful escapes, and apart from that, it is just an ordinary job with lots of problems. However, when you see the end product, it makes you think that after all it has been worth it.'[17]

After the speeches, the orchestra swung into the 'Gay Gordons' Scottish dance. Doddy took Jenny's hand. 'Let's show them, shall we?'

'We can't,' she said. 'You'll fall.'

'In front of this crowd?' Doddy replied. 'I wouldn't dare. It's a calculated risk.'

United States Air Force Academy Colorado Springs, 1962

Florida-born Guy D. Gruters was about to witness a different kind of entertainment, and a stark reminder of the calculated risks he would be required to take if he graduated as a military pilot. His father had been a lieutenant colonel in the Signals Corps and spent two and a half years on General Eisenhower's staff in the run-up to D-Day during the Second World War. Service and sacrifice were in Guy's blood. The 20-year-old was halfway through his four-year officer training when he and his fellow cadets were called to the US Air Force Academy lecture theatre. He settled into his seat as the opening titles of a USAF training film started to roll.

'*B-58 Escape Capsule Flight Testing . . .*'

At this early stage of their service, Guy and his friends had certainly heard of ejection seats, but knew virtually nothing about them. And, anyway, this was a very different escape system. The race to preserve the lives of airmen like Guy, by means of increasingly advanced systems, was ongoing, as manufacturers in America, Europe and the Soviet Union continued developing improved versions.

As Guy watched the film, the commentary continued. '*Since the beginning of aviation history at Kittyhawk [with the Wright brothers' first powered flight], preservation of life has been the airman's chief concern. Many methods have been used since the early days of parachutes.*'

Guy was about to learn about the very latest method.

A silver B-58 Hustler, an American four-engine supersonic strategic bomber, streaked across the screen. '*Its speed was Mach 1.3 [around 900mph],*' the voiceover announced. '*Altitude 35,000 feet.*'

An emergency crew escape capsule shot out of the rear cockpit in a ball of flame. It was a stark contrast to a standard ejection seat. The bright red pod, the size of a large wheelie bin, totally enclosed the occupant, protecting them from the potentially lethal high-speed blast. It was just one of many new developments in aircraft escape systems. The tiny capsule soared some 225 feet above the aircraft before its parachute deployed. Waiting onlookers followed its track, shielding their eyes from the desert sun. After a descent lasting seven minutes and forty-nine seconds, the capsule drifted gently down onto the arid Californian desert. As it hit the sand, a posse of technicians, many of them incongruously wearing shirts and ties in the searing heat – more than an echo of Benny Lynch being shot up ejection test rigs in his Sunday best – raced up in estate wagons to check the capsule. And its occupant.

'*All hardware and instrumentation functioned perfectly,*' the commentary continued.

An aero medical officer and several specialists clambered out of a nearby helicopter ready to liberate the 'pilot'. They glanced through the hardened glass panel on the front, wary about raising the capsule's clamshell-style folding door. They were right to take precautions; the occupant of the test capsule was no human tester like Doddy Hay or Benny Lynch – it was a large black bear and, understandably, it was not too happy.

Taking care not to dislodge its medical sensors, four technicians each grasped one limb of the heavily sedated bear and carried it away, upside down, tongue lolling from its mouth. The soundtrack swelled. *'An on-the-spot examination, combined with the usual regime of tests conducted later, showed the bear to be in good shape. The ejection was a complete success, to give great encouragement to all concerned.'*[18]

The young Guy Gruters, who at that stage had not even started his pilot training, was certainly concerned about how any ejection system he might need to use would operate. But as he left the auditorium, discussing the astonishing film with his friends, he was by no means convinced by this particular escape device. 'The bear looked really miserable,' he said later. 'It seemed a really strange thing to be doing. I couldn't imagine what the poor bear must have endured, but I guess they were different times back then.'[19]

Humans had tested the crew escape capsule, but only from a subsonic B-58. The USAF subjected six bears and a chimpanzee to supersonic testing before risking aircrew at those high speeds. One bear with 'a previously undetected brain condition' died. Another American black bear was found to have a laceration of the liver which was attributed partially to an overdose of anaesthesia. The other animals survived their ejection, sustaining a few small fractures and bruising. But they did not live much longer as the boffins ordered most of them to be euthanised so that autopsies could be carried out on the unfortunate creatures, to be absolutely sure they had no hidden injuries. One science writer would later say, 'On one hand, using bears for these tests was an extremely practical solution to the problem. The Air Force was working on a fix for something that had already caused human deaths. On the other hand, it's hard not to cringe when you imagine the terror and confusion these animals experienced. Luckily, the testing programme didn't last very long.'[20]

Tests using unwilling animals may not have lasted long, but the development of totally enclosed crew ejection capsules was now well underway.

Chalgrove Airfield
6 December 1963

Back in the UK, James Martin had his own further tests in mind, involving willing humans. He was consumed by rocket-seat development, giving aircrew the best chance of survival in all circumstances. He had designed a 200-foot-long testing track supported at one end by an 80-foot tower. High-tensile cables were attached to the top of the tower to steady the structure, anchored by concrete blocks in the ground.

After further successful tests in the wake of Doddy Hay's Paris accident, his rocket seat had proved it had zero-zero capability stationary on the ground, and worked perfectly racing down the runway, or flying level at very low altitude. But what if the aircraft was very close to the ground *and* descending nose-down? In this situation, the seat had to be able to counter the downward motion of the aircraft – and hence the downward vector of the seat – generating enough upward force, without damaging the occupant, to fire seat and pilot high enough to allow automatic separation and parachute deployment.

Martin's new rig, the first of its kind in the world, was able to mimic an aircraft in a low-level, nose-down dive.

A small sled, mounted on the track with the seat attached, was winched to the top of the tower and then propelled down the slide at around 100mph, while a dummy was ejected at predetermined heights. The force created by the rocket motors was adjusted to ensure they could overcome the descent motion of the aircraft.

Putting rocket seats into a two-seater aircraft presented a further challenge, particularly if they were side by side, as they would be in many jet trainers. As one pilot ejected, the flames from his seat could incinerate the other. Martin's solution was to delay firing the rocket one-sixteenth of a second, until the first seat was up and back far enough to ensure it wouldn't ignite the cockpit, or the co-pilot. It had worked well in tests. Now he needed a human to give it a go for real.

Step forward yet again, Doddy Hay.

The Scotsman, now fully recovered from the catastrophic injuries he suffered in Paris, drove into Chalgrove on 6 December 1963 to see the stark grey Meteor test-jet waiting for him on the tarmac. Now forty-one, he wondered if he was still up to it. Over the long months of tests – and injuries – doubts and hesitations had begun to creep in. Did he have the same speed of reaction? Was his physical peak behind him? Even so, he felt a crackle of intensity and excitement as he climbed into his famous white overalls and went through his usual pre-flight rituals, right down to the barley sugars and cigarettes.

He climbed up the ladder, into the rear cockpit and strapped into the latest version of the rocket seat. The Meteor shot down the runway, climbing over Oxfordshire to the test altitude of 2,000 feet. As he levelled out heading towards Chalgrove, the first green light flashed in Doddy's cockpit. It meant the run was good. He began to relax, waiting for the five-second signal. It was good to be back in the zone. 'My God, I thought, this business is becoming routine.' The next warning light flashed, and then the 'Go'. Doddy yanked down hard on the blind. He felt the first smooth surge of movement as the gun fired him up and then the time-delayed rockets exploded beneath him. Then he instantly realised something was terribly wrong. 'I was catapulted high and twisting sideways, far out of the cockpit,' he recalled. The increased rocket power had taken a terrible toll. 'I felt, I *knew*, my body was broken and I cared not for what might lie ahead. I only wanted everything to end.' His parachute opened safely but he could not physically look up to see it. His arms and legs were dangling uselessly. He was unable to steer his parachute with his hands. 'I tried, God I tried, but there was left in them neither strength nor control. From my toes to my fingertips my body was screaming in protest.' His back and groin convulsed in pain. He was at the whim of a wind that would slap him down on the airfield. 'Somewhere down on the ground was my wife, going through God knows what, waiting for my all-safe signal.' There would be no spreadeagle this time.

'The earth was coming up at me. I could not lift a finger to ward it off.'

Doddy smashed down helplessly on the airfield. The ground crew ran urgently towards him, collapsed his billowing parachute and hauled him out of his harness. After regaining control of his faculties, a Martin-Baker driver took him home. 'Jesus Christ, Doddy,' he said, 'we've really done it to you this time.' Later, he looked at himself in the mirror. His legs and arms were swollen and mottled purple from the force of the blast.

Initially, he was determined to do more tests. Whatever the cost. But he soon realised what traumas he was putting his devoted wife through, never knowing if he would survive the next test. 'One day, I looked up to find Jenny watching me silently. And weeping.'

Doddy Hay accepted that his testing days were truly over. Bravery had taken him a long way, but no further. His was the last Martin-Baker human test ejection. Fully instrumented dummies would soon come in; those pioneers of early ejection testing were no longer needed to demonstrate what could be possible. The doughty Scotsman with a love of adventure, fine dining and cocktails, not always in that order, was the last of those magnificent men to test their ejection machines.[21]

PART III

1964–1980

VIETNAM'S KILLING FIELDS

CHAPTER SIX

EJECT IN TIME

UNITED STATES AIR FORCE ACADEMY
COLORADO SPRINGS, 1964

The images of red escape capsules containing reluctant bears being blasted out of giant bombers were now far behind Guy Gruters. And even the possibility of a 'conventional' ejection from a fighter at some point in the future was not at the forefront of his mind. Passing out of the US Air Force Academy in Colorado Springs most certainly was, preferably with the best grades possible. A physically imposing man, Guy was on the USAFA judo team and a talented boxer. Raised in New Jersey, he had spent his childhood camping, hunting and becoming an Eagle Scout, the rarely achieved highest rank of the Boy Scouts of America. He devoured history books, especially those with dramatic accounts of aerial combat during the Second World War. He knew his future was in the air, and that it would be in the growing shadow of a very different conflict.[1]

Vietnam had been sucking the United States into an increasingly violent military whirlpool since the mid-1950s. The Korean War was the first time ejection seats were used in combat, but they would come of age in this latest attempt to stem the Communist tide. Guy Gruters had no doubt which side he was on, and where his duty lay. 'I had read many books on Communism and I believed it was a blight on the world and a threat to the United States.' Vietnam was now divided along the 17th parallel, with the North supported and armed by China and the Soviet Union, and the South trained, equipped and bankrolled by the United States.

While keenly aware of events unfolding in South-East Asia, Guy was busy enjoying his time at the academy. 'I made good friends and learned a huge amount about other people, their lives and backgrounds.' One of them was Lance Sijan, a young man from Milwaukee in Wisconsin, in the academic year behind Guy. Both were members of the academy's skydiving club, and Sijan was also 'a great football player and athlete, but quiet and humble. A good man. He left a great impression on me because of his abilities. He was one of those people who stood out from the crowd of ambitious young men; I had no doubt he would go far, both as an Air Force officer and in his personal life. Military service was such that you never really knew when you'd see a friend or colleague again, but I had a feeling that our paths in the Air Force would cross at some point in the future.'

Guy graduated in 1964 and the two friends bade each other goodbye and wished each other good luck as the growing conflict in Vietnam edged ever closer to this generation of young aviators. As his training continued, Guy's other focus was on starting a family with his new wife, Sandy. Sandra Lee Hancock, a bubbly, dark-haired graduate from the University of Florida, was born in Jacksonville in 1941. She first met Guy on a blind date at the end of 1962. 'I had no real experience of the military and no concept of what being married to an Air Force pilot would mean,' she said. 'But I knew I wanted to spend my life with Guy. He was clever, interesting, polite, and he challenged me.'[2]

Their first daughter, Mari, was born in 1964.[3] The following year Guy completed his pilot training, finishing second in a class of thirty-two. Sandy was in hospital having their second daughter, Sheri, when 'a nurse came in, crying, to tell us that one of Guy's colleagues had been killed in a plane crash. I knew that flying could be dangerous but this was a reality check. I began to understand the dangers Guy and his colleagues faced.'

Guy was training on a North American Aviation F-100 Super Sabre, the US Air Force's first supersonic fighter, which had entered service in 1954. Just over 52 feet long and predominantly silver, its swept-back wings and the oval air intake in its nose were its most distinctive features.[4] 'I enjoyed flying it, but it was not the easiest,' Guy remembers. 'When I looked at the statistics, I was shocked to see that we had lost hundreds of aircrew in fatal peacetime accidents. At that point in the development of flying, we had little previous experience of, or training in, the demands of supersonic flight.' Guy was right to be concerned. In the course of its service between 1954 and 1979, an astonishing 889 USAF Super Sabres were destroyed in accidents and, even though they were fitted with ejection seats, 324 pilots still lost their lives.[5]

Guy read all those early accident reports carefully, 'so I knew what I was getting into'. One thing struck him forcefully. 'I felt many of the pilots who died had left it too late and might have survived if they had taken the tough decision and ejected earlier. The ejection seat was still a relatively new concept, but it was saving lives and I just trusted it would work as designed when I needed it. Much like I trusted and relied on all my aircraft equipment, like the engine fire extinguishers or the warning panel. The ejection seat was just *there*. And after reading all those reports of unnecessary deaths, I vowed to keep one thought in mind every time I flew: *Eject in time*.'

* * *

Guy's Super Sabre was fitted with a conventional 'open' ejection seat, unlike the totally enclosed escape capsules he'd seen demonstrated using wild bears on the B-58 bomber film years earlier. However, when the Hustler had first gone into production in 1956, it too had initially been equipped with upward-firing 'normal' ejection seats, one for each of the tandem-seated, three-man crew. But punching out from a supersonic aircraft travelling at more than 1,000mph, fully exposed, could be seriously hazardous to health. The wind-blast at those speeds could easily break and dislocate limbs, tear ligaments, or even kill the seat occupant. The early seats were fitted with side-mounted headrests to protect the ejectee. Straps attached to hands and feet tightened automatically on ejection to minimise arm and leg flailing when being hit by a windspeed equivalent to almost ten times the power of a category-5 hurricane. All this helped, but it simply wasn't enough.

1. Clamshell door folded up during normal flight
2. Seat harness
3. Integral aircraft control column
4. Leg retraction mechanism
5. Ejection handle
6. Three-piece clamshell door closed during ejection sequence
7. Viewing window

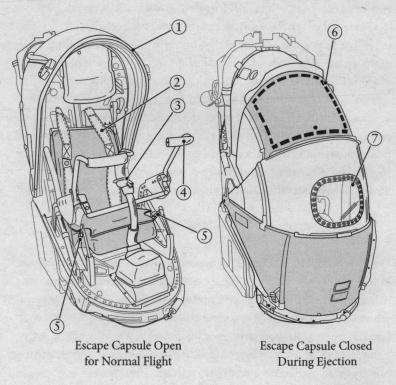

Escape Capsule Open for Normal Flight

Escape Capsule Closed During Ejection

STANLEY AVIATION COMPANY ESCAPE CAPSULE

Something revolutionary was needed and it came from the Stanley Aviation Corporation of Denver, Colorado. They designed a fully enclosed escape capsule, or pod, for each crew member which could be fitted directly into the body of the aircraft. As the corporate marketing video trumpeted, 'it exemplified the concept of *shirt-sleeve flying*'.[6] Apart from a light flying helmet for communications, no extra protection equipment would need to be worn; the occupant would be completely protected within the capsule.

Its clamshell door folded up and back above the pod and remained open during normal flight. Raised up like a visor on a flying helmet, it enabled access to aircraft equipment as normal. The control column was an integral part of the capsule, positioned between the pilot's legs. Internal lap, chest and shoulder straps restrained the crews within the pod as normal.

If they had to bail out, each had two ejection triggers on either side of the seat; pulling either sharply upwards initiated the sequence. The shoulder straps tightened automatically and a bar descended across the lower shins, forcing the legs back against the seat and raising the knees up so the occupant adopted a foetal position, while still sitting upright in the seat. The folding door would snap downwards to slam shut, and the overhead aircraft canopy was blown clear to create an opening. The rockets fired and the crewman, safely wrapped in his capsule, was blasted up, out and away.

A drogue parachute stabilised the craft in the same way as with a standard ejection seat. Then the giant main parachute deployed. The airtight capsule, a complete environmental protection system with a small porthole at the front and its own independent oxygen supply, was pressurised in high-altitude bailouts. Each pod was fitted with shock absorbers to lessen the impact of hitting the ground, and flotation devices were at each corner so it would stay upright after landing on water. Because of its size, much more survival equipment could be stored, including a three-day supply of water and fourteen-day supply of food. A hunting rifle could be carried, presumably to ward off hostile wildlife, like bears. And a change of clothing to aid survival on the ground was added.

Now, thanks to the human testers, but perhaps more importantly to the pioneering gang of bears and one chimpanzee – the unsung heroes of ejection history – escape capsules had been fitted in all B-58s by late 1962, allowing aircrew to eject safely at twice the speed of sound, around 1,300mph, and from as high as 70,000 feet.[7]

* * *

That Stanley escape capsule was the first in a series of enclosed, pod-type crew ejection systems. And they would be soon saving lives thousands of miles away in Vietnam, where the United States was ramping up its military presence.

As Guy Gruters' pilot training continued, the anti-war movement in America was gathering force as the body bags came home, day in, day out. The nation that had initially supported the first major conflict to be televised and covered extensively in the world's media was now divided. But Guy Gruters had no doubts about his duty. 'I certainly understood how terrible war could be for civilians. However, I was prepared to protect my country, and our freedoms, in whatever way was needed. If that meant sacrificing my own life fighting Communists in foreign lands, so be it.'

At the conclusion of his training in the summer of 1966, he was offered the opportunity to complete a doctorate back at the Air Force Academy. He turned down that chance to avoid war. The only place the 24-year-old pilot wanted to be was on the front line, fighting alongside his colleagues. He volunteered for a year-long tour in Vietnam.

Sandy did her best to suppress her fears. By 1967, as Guy prepared to deploy, some half a million American troops were in Vietnam. Just over 15,000 had been killed and more than 109,000 injured.[8] She had to be strong for her two very young daughters. Mari was only two and a half. 'I remember so vividly the day in March when Guy left for war. He would be gone for a year and, in reality, there was little to be said. I told him that the children and I would miss him terribly. I made him promise to write as often as he could and that I would write to him every day. To be honest, I don't think I actually understood what the reality of war would mean. This was the way it was going to be and I had to make the best of it. He had his job to do as a military pilot and I had my job to do as a military wife and mother. So there was no point in worrying about it.'

She still felt a huge emotional wrench as she watched her husband go. 'I really admired his sense of duty. He was so afraid that the war would be over before he got the chance to do his bit. I just presumed that with God's help he would come home safely to me.'

* * *

While aircrews faced daily mortal danger in the skies over Vietnam, projects to save their lives with ever-more advanced escape systems rolled on at home.

Oscar W. Sepp was widely regarded as a genius. He trained as a mechanical engineer and in 1954, concerned that people were losing the old-fashioned screw tops from their toothpaste tubes, he designed and patented the world's first flip-top cap – an invention still very much in evidence today. But he earned his most meaningful accolades for his lifesaving work with the US Air Force, where he designed and modified ejection components and parachutes in several US aircraft. Undoubtedly aware of the B-58's ground-breaking single-person escape capsules, he and his colleagues turned their minds to helping to produce the next iteration of escape systems – the means of ejecting a complete cockpit from a stricken jet.[9]

The F-111 'Aardvark' was one of the USAF's latest combat aircraft, a swing-wing, supersonic, multi-role combat jet. Named after the long-nosed nocturnal mammal, it was equipped with a huge, rocket-powered ejection capsule comprising the entire cockpit in which the pilot and his Weapon Systems Officer (WSO) sat side by side. Weighing in at 1½ tons, it was about the same weight and size as a Ford Focus. Unlike the smaller pod on the B-58, there were no clam-like doors slamming shut in order to protect the crew. And there was no reason to haul back legs to protect them from impacting equipment. There was no hole in the top of the fuselage. The crew sat in their

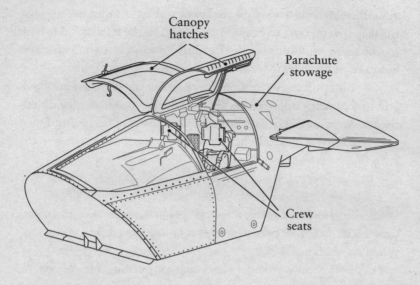

F-111 COCKPIT CAPSULE DETACHED FROM AIRCRAFT

relatively comfortable cockpit in normal flying kit. If they had to eject, their complete cockpit, with them safely strapped inside, would detach from the fuselage and blast them to safety. The engineering needed to successfully deploy such an enormous module from its supersonic parent jet was certainly challenging for Sepp and the inventors.

On 19 October 1967 test pilot David J. Thigpen and engineer Max Gordon lifted off from Carswell Air Force Base at Fort Worth in Texas on a routine proving flight in an F-111. Fifty minutes into the mission, the Aardvark approached 39,000 feet at 1,300mph. Thigpen glanced in alarm at his instruments, which showed he was rapidly losing hydraulic pressure. He put the jet into a gentle turn to return to Carswell. More systems failed and the aircraft began to twitch, then vibrate violently. 'It was pretty obvious we'd have to leave the ship,' Thigpen remembered. Because both crew had ejection triggers, he turned to Gordon as he wrestled with the controls. 'Go ahead, Max! Pull the handle.'

Gordon pulled the cockpit ejection handle at 27,000 feet.

Oscar Sepp's intricate electrical circuit immediately set off a sequence of explosive charges that pulled their harnesses tight and began blasting apart the mechanical fastenings holding the cockpit section to the airframe. The charges also guillotined the cables, lines and tubes connecting the capsule's systems to the Aardvark's hull, slicing the entire cockpit clean away from the jet.

To the crew, the sequence seemed to take for ever.

In fact, it was completed in a third of a second.

The huge rocket motors fired through the outlets at the back of and beneath the module, producing an astonishing 27,000lb of thrust. About the same amount *both* of their aircraft engines produced *together* in normal maximum power.[10] As it rose from the fuselage, sensors detected the speed at which the aircraft was flying and adjusted the amount of thrust needed to blast it clear. For 1960s technology, it was a truly remarkable system.

Thigpen gave the thumbs-up to Gordon who leaned across, grabbed his hand and shook it. 'We felt a slight jerk as the drogue chute anchored in the roof opened.' At 15,000 feet it stabilised and slowed the capsule to 345mph. It was still pointing towards the ground and rotating slowly. As they passed through 14,000 feet the massive recovery parachute, attached by bridle cables to their front and rear, began to play out. In seven seconds it bloomed to its full 70-foot diameter. 'I'll have to admit I felt some relief when the big chute opened, straightening the capsule and slowing the descent.' Thigpen had a talent for understatement.

Meanwhile, their aircraft – minus its cockpit – had crashed and exploded in north Texas. As they floated gently earthwards at 31 feet per second, they opened their side windows to ventilate the cabin, which smelt of gunpowder from a spent detonator. They even took off their helmets. Max Gordon unbuckled his harness to reach for the aircraft's checklist. They shook hands again. It was an incredible sensation to be still sitting *in* their cockpit, but no longer *part* of an aerodynamic jet bomber.

Nearly three years earlier, Thigpen had suffered a catastrophic engine fire shortly after take-off. His life had been saved by ejecting on his 'standard' Martin-Baker seat at 400 feet. Now, this first use of the capsule in an emergency took ejection to a whole new level. 'During the fall, it was almost like still flying the airplane, since we were still in the cockpit.' Cushioning airbags fitted to the underside of the heavy module inflated. Another airman described a capsule landing as 'like jumping into a pillow'. It was also watertight. If they had come down in the sea, self-righting flotation balloons would have bobbed it to the surface. After a gentle landing, they opened their cockpit canopy hatches, climbed out and walked calmly to a farmhouse 500 yards away, where they put a call through to their home base. Not only had they escaped, but so had $400,000 of reusable cockpit equipment.

However, a later F-111 cockpit ejection would demonstrate how easily matters could go very wrong, very quickly, with this particular type of system. During a low-level run on the bombing range at Nellis Air Force Base in Nevada, an Aardvark appeared to develop a will of its own at 5,000 feet. It started to buck and jink violently, as though trying to turn itself inside out. At 1,800 feet, both crew members pulled their ejection handles in unison. Because of the Aardvark's attitude and speed, when the rocket motors fired, the force of the thrust pushed the capsule straight off the aircraft, which was in a nose-down position. Blasting free of the doomed jet, the enclosed crew were now looking straight down at the ground, which was rushing towards them at breakneck speed. The WSO remembers thinking, 'There's no way this is going to stop. I'm a dead man.' He waited for the inevitable.

Suddenly, the giant recovery parachute deployed, snapping them upright and slowing their fall just 30 feet off the ground. They breathed a sigh of relief. Then they watched in horror as their aircraft ploughed into the hillside 50 yards above them. The exploding fireball engulfed their parachute and the flames swept towards their cockpit. Robbed of its retarding parachute, the capsule smacked down hard on the side of a slope below the burning F-111, then started to roll down the hill. It turned and bounced over rocks, outpacing the flames, in what the WSO described as 'like being strapped inside a metal madhouse that

A pilot from the USAF Thunderbirds display team ejects because of an emergency at very low level during an air display in 2003.

Bernard Lynch ready for one of the first airborne ejection seat tests in 1946.

Bernard Lynch strapped to the seat testing rig in the Martin-Baker factory.

Jo Lancaster, the first pilot to eject using a Martin-Baker seat in 1949.

Jo Lancaster's Armstrong Whitworth AW52.

Jo Lancaster with the author in October 2017.

Doddy Hay ejects on a rear-facing seat from a Valiant during a ground test in July 1960.

Doddy Hay prepares for the very first 'zero-zero' rocket seat ground test in April 1961.

The founders of the Martin-Baker Aircraft Company, James Martin (right) and Valentine Baker (left).

In 1962, a test pilot ejects just
before his Lightning crashes.

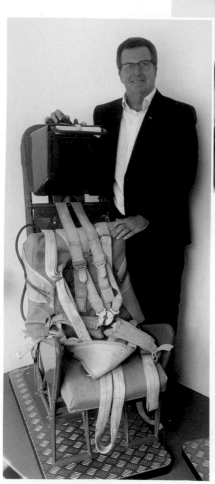

The author with a Mk1 seat similar to the
one used by Jo Lancaster in 1949.

The author with John and James Martin,
joint managing directors of Martin-Baker.
A portrait of their late father, Sir James
Martin, co-founder of the company,
hangs on the wall.

Bill and Linda Schaffner at Tan Son Nhut Air Base, Saigon, 1966.

Bill Schaffner's family in July 1970, just a few months before his death.

The wreckage of Bill Schaffner's Lightning on the dockside after recovery.

An F-100 Super Sabre similar to the one flown by Guy Gruters and Charlie Neel.

Lance Sijan, who died as a prisoner of war in Vietnam, at the USAF Academy in the early 1960s.

Guy Gruters in the cockpit.

Guy Gruters photographed during his time as a North Vietnamese prisoner of war.

Guy Gruters hugs his daughters after being reunited with the family on his release from captivity in 1973. Guy's wife Sandy looks on.

Guy Gruters (left) and Charlie Neel with their wives at a reunion event for F-100 Super Sabre Forward Air Controllers.

Meteor pilot Ron Guthrie in the cockpit prior to taking off
on the day he was shot down during the Korean War.

A recent high-speed test ejection with the Martin-Baker rocket sled.

was rolling downhill like a rollercoaster'. After six violent rotations, it eventually came to a halt on its roof, with the crew hanging upside down inside. The flames had given up the chase. Astonishingly, they were unharmed. The WSO turned to his pilot. 'Are you okay?'

'Fine,' he replied. 'But I don't want to do that again.'

Afraid that any sudden movement would start the rolling again, they gingerly unstrapped and climbed out. In spite of their brush with death, both crew expressed total confidence in Oscar Sepp's module, saying somewhat ruefully, 'It worked as advertised.' One of Sepp's colleagues admitted that he hated constructing the F-111 capsule 'because it was so damn difficult to build. But every pilot that has punched out of a wounded bird has to give thanks to Sepp. And my tube of toothpaste still has his flip cap on it!'

David Thigpen was quick to praise his lifesaving invention. 'I'll guarantee you the ride was much more comfortable and reassuring than bailing out in a regular ejection seat. The only injury incurred was when I pricked my finger crossing a barbed wire fence.'

PHÙ CÁT AIR BASE, SOUTH VIETNAM
8 NOVEMBER 1967

In South Vietnam, Guy Gruters was facing significantly greater danger. His new home was the closest airfield to the Demilitarised Zone dividing the North and South of the fractured country. He was in the middle of his tour as an airborne Fast Forward Air Controller attached to a Special Forces camp in the Central Highlands. Its task was to identify enemy activity along the Ho Chi Minh Trail, the complex military supply route running from the North to the South, through jungle and mountains. The missions were highly risky. 'We lost many friends in those early months. I didn't think much about the dangers or that we were facing death. We had a job to do. In a strange way, living on the edge amid deadly combat was exciting. Not many people get to experience living like that.'

But he was more than aware of the reality of what might happen if he had to eject over enemy territory. He vividly remembered watching the forced TV news conference in 1966, given by naval pilot Jeremiah Denton who had earlier ejected from his stricken jet during a bombing raid. As many of Denton's colleagues were discovering, successfully ejecting was often the first step on a terrifying journey. He was now incarcerated in the notorious 'Hanoi Hilton' prison complex. When his captors paraded him gleefully in front of the cameras, he refused to answer their accusations or offer any information and started blinking instead.[11] As Guy Gruters says, 'They did not realise he was blinking out a single word in Morse code':

T-O-R-T-U-R-E

Phù Cát's two parallel runways stretched towards the horizon, along a lush plain bordered by mountains. They were long enough to take military transports as well as fast jets, and the steel lattice of the control tower loomed above the trees that had survived the airfield's construction. The anti-blast walls were a constant reminder that the enemy, unheard and hard to see, was never far away.

Flying the high-speed F-100F Super Sabre, Guy's latest task was to seek out enemy targets, marking them with smoke indicators, then calling in strikes to destroy them. Their call sign, 'Misty', was chosen by the unit's first commander after his wife's favourite Johnny Mathis song. The Misty F-100F Super Sabres had a crew of two, having initially entered service as a training aircraft. All of them were pilots, who took it in turns to either fly the jet or work the back seat in charge of navigation and photography, searching out any Vietcong on the dense jungle trails and rough roads, then directing fighters to their targets.

At 2am on 8 November 1967, Guy, call sign Misty 29, was detailed for a sortie. 'We had been given the likely co-ordinates for a big ammo dump in the North and were off before dawn to find it, and then lead in F-4 Phantoms to strike it.'

Gruters was the 'guy in back' on the trip, and the aircraft commander was Misty 22, Captain Charlie Neel. Hailing from Big Stone Gap, Virginia, 29-year-old Neel had graduated from the Air

Force Academy in 1961. 'Pilots attracted to Misty were generally brave and adventurous,' he recalled. 'We enjoyed the danger and didn't just want to drop endless bombs, or "make trees look like toothpicks" as we called it. We really wanted to make a difference, doing a truly important job; helping to disrupt and destroy enemy supply routes.' But Charlie had no illusions about the dangers they faced. 'We had to spend a lot more time than most over enemy territory under intense, hostile fire, so the loss rate of Misty F-100Fs on our Fast Forward Air Control missions was four times higher than standard F-100 fighter-bombers. Our jets routinely returned to Phù Cát riddled with holes from anti-aircraft artillery shells. According to our intelligence services, there were around 180,000 anti-aircraft gun sites in North Vietnam.'

Misty pilots enjoyed their reputation for hard flying, and Charlie was no exception. 'Boy, was I punchy! You had to be to survive. I don't think I felt any fear under fire; apprehension, maybe, but as long as I was an aggressive pilot I really believed I would avoid the worst of it, and I just wanted to be where the action was.'[12] Guy was always reassured when he flew with Charlie. 'He was outstanding; a fighter pilot's fighter pilot.' On this mission into the heart of the action they both craved, they would need both courage and skill in spades.

Thousands of miles away at home in America, Sandy Gruters had become addicted to TV news. 'I could see terrible things happening over in Vietnam but it was so far out of my comfort zone that it was like watching a movie. And I had so much to do as a mother. I had to look after my family and ensure our lives continued as normally as possible. I just had to concentrate on the issues I had control over. Guy had to do the same out in Vietnam.'

After his departure, Sandy and the children had moved in with Guy's parents in Sarasota, Florida for a few months while she looked for a home to rent in the Gulf Coast city. 'I found a little wooden house and we were settling in. It was really small but it had two bedrooms and a lovely yard; perfect for us. We all settled in well, including our collie dog "Century", named after Guy's aircraft, the F-100.'

Sarasota, south of Tampa, with its white-sand beaches and enticing names like Siesta Key, was a comforting and stimulating place to raise children; a world away from napalm-torched jungles. It was where Sandy had grown up and gone to school. She still had many friends there to offer support and became close to the wife of a Marine pilot who had three children, and whose husband was also deployed to Vietnam. 'We did everything together and formed a support group of military wives. We had parties for the kids, played cards and went to dinner together. We were without husbands, and our children were without fathers.'

They also kept a close eye on news bulletins in the hope of learning more about what was happening on the front line. But the narrative was rarely positive: 11,363 servicemen lost their lives in 1967.[13] And in his State of the Union address, President Johnson told the nation: 'I wish I could report to you that the conflict is almost over. This I cannot do. We face more cost, more loss, and more agony.'[14]

Sandy tried not to dwell on the endlessly gloomy statistics. As long as Guy's letters kept arriving, she knew he was safe. And she wrote every day, as she had promised. 'General chit-chat about family news and what was going on. We also made tapes for each other so he could hear the children's voices and they could listen to him; they were a great source of comfort.'

But war was never far away. In October 1967, her friend's Marine husband was shot down and listed as missing presumed killed. 'It was truly devastating for her, and it really affected me too. It brought home the reality of what our loved ones were facing. I began to dread hearing a knock at the door, opening it to find a uniformed USAF officer standing there. The bearer of terrible news.'

As they headed out to their distinctive silver aircraft, Guy Gruters and Charlie Neel made an incongruous pair. Gruters was tall, athletic and ate up the ground in long commanding strides. Neel was almost running to keep up. His Misty colleagues joked that he was so small he had to take a run and a jump to make the first rung of the cockpit ladder. Years earlier, he had stuck layers of tape to his feet to increase his height sufficiently to make the minimum Air Force requirement.

Neel did his customary 'walk around' the aircraft, making sure everything was as it should be, that no stray screwdriver had been left lying about, ready to be sucked into the engine when he fired it up. He did a pull-up on the intake to check it was clear, then, 'just before finally climbing in, I always visually checked the ejection seat, making sure all the safety pins were in and that everything looked as though it was connected properly'. Satisfied all was in order, Charlie swung his legs over the edge of the cockpit and slid down onto his seat. He carefully fastened his shoulder and lap straps and raised the seat to its maximum height. Guy lowered his as far as it would go, checked the anti-g suit hose to his left and the oxygen hose and radio cord to his right. All were in place.

If they were badly hit by enemy fire, Charlie and Guy would grasp the lock levers alongside their thighs, pull them upwards and back to jettison the canopy, then pull the levers' integral trigger to initiate ejection.

Everything 'A-Okay', they confirmed to one another, but they still had to wriggle and stretch to get comfortable. Instead of a cush-ion, they were sitting on their survival packs, which contained an inflatable life raft, and were connected to their parachute harnesses by a 20-foot lanyard. Charlie looked out at the airfield. Apart from a few vehicle headlamps and strip-lights flickering in hangars and outbuildings, the night was inky black. Final checks done, they were good to go. The ground crew stepped away from the aircraft as it began taxiing out. Neither pilot said much as their wheels left the runway. It was no time for idle banter.

In minutes they were over North Vietnam, 4,500 feet above an apparently peaceful and serene stretch of jungle. Now they were airborne, Charlie pushed back his red headrest to improve Guy's visibility. If they had to eject, it would snap forward automatically to ensure he was in the correct position to exit the aircraft.

The sun was barely up when they received further co-ordinates from a Special Forces ground team guiding them to their objective. Charlie took the Sabre lower.

Guy suddenly shouted from the back seat. 'Big low humps, both sides of the road. Covered in some kind of camouflage.' He wasn't sure what he was looking at, but there were dozens of them. Mounds of earth covering what had to be fuel tanks or bunkers? Guy reckoned they needed to take a closer look. Charlie wheeled sharply away, and bided his time before racing back in again even lower. 'Best to pass the graveyard nonchalantly,' he thought. 'Maybe the bad guys will think we are just passing by.'

Not this time. 'I saw the muzzle fire on the ground like a giant camera flash. The large bloom of expanding light meant shells were heading upwards. Tracer glow passed either side of the jet and in front of our faces.' Charlie twisted the jet back and forth, trying to avoid it. Watching from the rear seat, Guy describes the flak as 'horrifically intense and we were right in the heart of it, surrounded by sparkling tracer and exploding shells'. Charlie was tense, 'like a banjo string, but in reality our job was to fly towards the tune those guns played. They were telling us that they were defending something important.'

Guy had seen enough. The camouflaged humps defended by multiple gun sites were so out of place in a jungle they had to have something to do with men and weapons. He called in the strike. Within minutes four F-4 Phantoms from 'Gunfighter Flight' were on their way.

Guy made contact. 'Gunfighters, we've got what appears to be large fuel storage tanks on both sides of a north–south road. Lots of guns either side. If hit, best bailout is to the east, feet wet [over the sea].' Charlie dipped his aircraft's nose towards the target. He jinked left then right to throw off the Vietnamese gunners, and fired two phosphorus rockets at the target. Thick white smoke billowed up from the jungle floor, marking its position. The Phantoms swept in to attack. Charlie pulled the Sabre into a steep climb. Their anti-g flying suits squeezed their legs, holding back the blood being forced downward and stopping them from blacking out.

Then, disaster.

'I felt a massive, solid thump in my butt, which reverberated throughout the jet,' Charlie recalled. 'The whole airframe was vibrating wildly.'

Battered by explosive anti-aircraft shells, they were still flying but, to Guy, 'it was like a giant hand had gripped the Super Sabre and was shaking us around like we were rag dolls'. Two warning lights were flashing on his instrument panel.

EMERGENCY FIRE and *ENGINE OVERHEAT*.

Charlie glanced in his rear-view mirror. The whole of the tail section was ablaze. His own warning panel showed that the aircraft's critical systems were going down fast. Electrics had gone. Hydraulics and fuel were also in trouble.

Guy now glanced in his own rear-view mirror; the sight sent him an even more devastating message. 'The entire mirror was now filled with fire. The back of our jet was obscured by flames. And they were moving rapidly towards where I was sitting.'

The pilot of the lead Gunfighter issued a staccato summation: 'Misty, get out! You're a ball of fire.' The flames had now spread to the rear of Guy's cockpit and were engulfing the wings.

'You're a ball of fire, Misty! *Get out*.'

The Super Sabre, with its large fuel load and ammunition, could explode at any moment, but Charlie still felt remarkably calm, 'even though I was sure my heart-rate was going through the roof!' He called back to Guy, 'Sit tight, man, we're still flying.' He was taking the gamble of his life. He kicked the engine into afterburner, giving them maximum thrust. Guy felt the surge of power in the small of his back.

Charlie had a simple plan: 'Get out to sea and eject in more controlled conditions and hopefully get rescued. The horrific reality of what would happen to us if we ejected over the enemy and were captured was all I could think about.' The Phantom pilot watching the terrible scene unfold was getting desperate. And angry. 'Get out at all costs,' he screamed. 'If the ejection seat is jammed, just climb out of the cockpit, but do something. NOW!'

'Negative,' Guy told him. 'We are going to ride for the coast.'

He remembered the earlier promise he had made to himself before deploying.

Eject in time.

His mantra no longer had any meaning.

'There was no way we wanted to eject where we were. As Misty Forward Air Controllers, many of my friends had conducted search-and-rescue operations to recover downed aviators. Some had seen aircrew captured by local civilians who hacked them to death with machetes. I thought there was little chance of survival if we were captured.'

'Eject! Eject!' the Gunfighter yelled. 'Before she blows.'

'Don't listen to them, Guy. Let's stay with it to the coast.'

The Sabre was now trailing a 100-foot sheet of flame. The sea was 5 miles away. Ground batteries were firing thousands of rounds to bring down the crippled fighter. The shells set off shock waves against the fuselage. Charlie kept going. 'It seemed like every man, boy, farmer and villager was shooting at us. I was beginning to doubt we were going to get out of this unscathed.'

Finally, he saw the ocean. They had made it there in just three minutes. To Guy 'it had seemed like half my life'.

'Okay, Guy,' Charlie said. 'We'll go just after we hit the beach.' He pulled his helmet visor down and prepared to punch out. Guy braced. He removed his glasses so he would not damage his eyes when he ejected. Charlie took the Sabre up to 8,000 feet. The hydraulics finally gave up. 'We started to roll left and I could no longer control it.' All the jet's systems were now dead.

'Okay, Guy. Here we go. *EJECT! EJECT!*'

Guy lifted his handgrips, which blew off the cockpit canopy. 'I then squeezed the lock levers and the seat fired me out over the tail. The chute deployed perfectly, but the seat did not separate properly and it hit my right wrist as it came free.' It shattered a bone and severed the main tendon in his thumb.

Charlie, meanwhile, had been temporarily blinded by dust and debris when the canopy released. He felt the heat and saw the flash as Guy's seat rocketed upwards. 'Then, I was on my way. I felt the blast of my own seat firing and rising out of the aircraft. That's all I remember.' During the ejection process the seat had collided with his helmet as it separated and fell away, knocking him unconscious.

Now in their parachutes, the two pilots had been spotted. A mile away, onshore enemy defences opened up with cannons and machine guns. How could they miss?

As he drifted down to the water, Guy could see Charlie 200 yards away, hanging limply from his parachute. Alarmed, he realised that Charlie's survival pack was still attached to his backside. 'The life raft was designed to automatically inflate as we released the seat pack. *C'mon, Charlie, if you don't do it before you hit the water, you're going to drown.* With so much dead weight strapped to him, he would sink like a rock.'

Guy's own life raft was inflated and dangling below him. Bullets ripped past him, reverberating through the air. The ocean was rushing up to meet him. He inflated his underarm 'water wings', which would keep him and his 90lb of kit afloat. Hitting the water, he immediately hauled the raft towards him, clambered in and started paddling for all he was worth. He had to get further away from the shore and the guns that were tearing up the waves all around him. If he could get out of range, then maybe, just maybe, he might be rescued.

Then he noticed that his heavily damaged hand had spewed a cloud of blood into the water. The ocean was full of sharks and sea snakes. Terrified, he stopped paddling and tried to kick on with his legs. More and more rounds exploded around him. He scrunched down. For a powerful and athletic man, he had never felt so vulnerable.

Charlie came to 1,000 feet above the water, his head pounding.
Holy shit. Where am I?

'The damn ejection seat had hit me with such force it split my helmet neatly into two separate pieces, which were now hanging on either side of my head.' He was dazed and disorientated. *C'mon, Charlie. Do something. Do the drills!* 'Eject and escape drills were drummed into us but I forgot to inflate my water wings and only just managed to drop my survival kit in time.'

His life raft inflated. He only came fully to his senses seconds before hitting the sea at around 30mph, and sank 20 feet below the surface until the dinghy cord stopped him falling further. 'I could see the light above me so kicked upwards until I surfaced, completely tangled in parachute lines and canopy.' He was gasping for air, coughing up what felt like litres of seawater. 'I managed to untangle myself and, as I scrambled into the dinghy, I breathed a huge sigh of relief. I had been truly lucky to survive being hit, flying a blazing jet, almost killed in the ejection, and then nearly drowning. Someone was watching over me that morning.'

Then the water began to spit around him. The shore gunners had him in their sights. His ordeal was far from over.

Charlie had landed only 200 yards from Guy but the waves were so choppy they couldn't see each other. They hugged their life rafts as the air above them filled with lead. Guy had a momentary urge to pull out his .38 service revolver and fire back, 'just to let them know they hadn't got me. But I soon realised that would be totally pointless and thought to myself, *Gee, Gruters, perhaps you are finally growing up?*' He checked his survival radio and made contact with a nearby aircraft, which came in at 50 feet and dropped a phosphorus marker to help rescue crews. It also had a less desirable effect. 'He had given the enemy gunners a perfect point to aim at and all hell broke loose.' Charlie was now talking to the Gunfighters. 'They're shooting at us a helluva lot,' he told one of the Phantom pilots. 'Do you have us in sight? Can you give us some help?' He rifled through his survival kit and launched a smoke flare to mark his position. 'We have you both in sight,' one of the pilots responded. 'Where are they shooting from?'

'Put something on the beach about 100 yards inland, right oppo-site us. I'll talk you into where I think it's coming from.'

A flight of four Gunfighters swept in and bombarded the shore-line until the whole area was smoking. By the time they left to refuel, the enemy fire began to subside. Another flight of four F-4s arrived and dropped more bombs around the beach. To Guy, what was happening was both terrifying and surreal. 'I think Charlie and I are among the few Forward Air Controllers to ever call in fighter strikes, post-ejection, from a one-man survival life raft!'

Still operating despite the onslaught, the shore battery now turned their fire on the Phantoms, which offered some respite to the two downed pilots. But not for long. About half a mile north, a dozen speedboats rammed with heavily armed North Vietnamese soldiers were heading fast towards them. This was now getting very personal and, as they closed in on Charlie, one of the Gunfighter Phantoms broke away and began a dramatic 45-degree dive towards the sea. As it streaked in at 700mph, Charlie was terrified he had become the target. He screamed into his radio: 'Hey, that's me in the raft! You're supposed to rescue me, not kill me.'

Then he saw the speedboats rocketing through the waves and realised what was going on. Risking his own life to help two fellow pilots, the Gunfighter unleashed his entire armoury at the enemy boats. As the Phantom pulled up hard to avoid impacting the sea, standing on its tail in full afterburner, the exhausts made the ocean boil. Charlie clung on for dear life as the shock waves tossed his raft around like a cork.

Half the enemy boats were sunk. The rest turned and fled, leaving their wounded comrades in the water. Guy could not believe it. 'For the next fifteen minutes we heard the horrible screams of men dying.'

It was nine in the morning, the sun was fully up, and Charlie and Guy had been in the water forty-five minutes when they spotted two 'Jolly Green Giant' combat-rescue helicopters heading for them. It was a beautiful sight.

But as the Jolly Greens rumbled in, the shore-based fire began once more. Charlie feared they were never going to get out of this alive. 'Huge geysers of water were thrown up as shells exploded. I ducked down deeper into my tiny dinghy, but it didn't offer much cover.'

Eventually, one of their rescuers managed to hover above him. A crewman's hand reached down. 'He grabbed me by the scruff of the neck and hauled me upwards. I just flew through the air and was dumped unceremoniously on the floor of the helicopter. It was one of the times that I was grateful for my small stature.'

Things were not so simple for Guy. A much bigger man, he was so waterlogged and weighed down with the extra equipment he carried that they could not lift him. Time was short. Risking everything, the pilot of the second helicopter lowered it down onto the sea. One of the rescue crew dived over Guy's head into the waves. 'The next thing I knew he came up from under me, and, treading water, managed to push me up hard. Another crewman grabbed me and pulled me over the edge of the chopper's deck. The relief was incredible. There had been so much fire, so many shells impacting within feet of the helicopter that I just could not see how we had survived. I was amazed I was alive.'

Fighter pilots called the Combat Search and Rescue (CSAR) crews 'the bravest men in the world'. Guy and Charlie, now heading away from danger to the comforting sound of rotor blades, had just seen why.

They were reunited at the nearby airbase at Đông Hà. 'Guy and I just fell into each other's arms and gave each other a huge hug, which was difficult for him with his injured arm. After our ejection I had truly wondered if I'd ever see my friend again. We had survived. It was emotional and overwhelming.'

They were flown onwards to Da Nang Air Base, a South Vietnamese facility the United States military was using as a major operational hub. It also had an excellent hospital where Guy could get his arm fixed. Before checking in for treatment, he and Charlie enjoyed a cold beer together. 'We gave each other another hug and said we'd see each other soon, once I'd recovered.

Then he walked away. I just presumed we'd be flying together again in a few weeks.'

Charlie suffered nothing more than a bad headache and would be back in a jet in a few days. 'There was a war to be fought, an enemy which still required my attention. It was my duty to get back on the horse.'

At the hospital they asked Guy if, before undergoing surgery, he would like to send Sandy a telegram telling her he was safe.

'Hell no. I'll contact her and tell her myself as soon as I can make a call.'

He couldn't wait to hear her voice.

Late that day, as the Florida sun began to go down, Sandy Gruters was sitting in her living room when the phone rang. It wasn't Guy. Her local paper, the *Sarasota Herald-Tribune*, had beaten him to it. Some reporters sniffing around Da Nang for stories had heard of his ejection. 'They asked me for a comment about my husband being shot down and rescued. I told them I knew nothing about it. I just didn't believe it. I thought it was a crank call from one of the anti-war movements, trying to upset me.' She hung up. About twenty minutes later the phone rang again. She hesitated before picking it up. 'I have a call from a Lieutenant Gruters,' the operator said. 'Will you accept it?'

'Then I heard Guy's voice.' It really was him. It was the first time they had spoken in nine months. Guy knew he was not allocated much time on the call. He also wanted to spare her the details. 'He simply told me that he'd been shot down and rescued, that he wasn't injured and there was nothing to worry about. He said he would write a letter and explain everything. That was it. He said he had to go to hospital for some checks, and rang off.'

Presuming her husband was fine, Sandy continued life as normal. 'I never heard anything from the Air Force. No one ever came to notify me officially that Guy had been shot down, that he'd ejected and then been rescued. So as far as I was concerned, it was all over.'

SARASOTA, FLORIDA
THANKSGIVING DAY, 23 NOVEMBER 1967

Guy had surgery to fix the hand injured during the ejection and
was in hospital for two weeks. Still unable to fly operationally,
and without informing his family, he had managed to hitch a lift
home on a military transport for a few weeks' leave. Arriving
in South Carolina, he took a flight to Tampa and then a car to
Sarasota.

'I didn't actually know where the house that Sandy had rented
was, so I went to my parents' home to discover my dad serving cocktails before dinner. I had completely forgotten it was
Thanksgiving.' In his nine months in Vietnam he had lost all
track of dates. 'As I strode in, my mother looked up and gasped,
"Guy?" Then everyone started hugging me and cheering. It was
good to be back.'

Unaware that her husband had returned, Sandy Gruters and
her young daughters had been to visit her own parents in
Jacksonville before driving the 250 miles back to Sarasota to spend
time with Guy's family. 'The kids were a bit of a mess, so I decided
we would all get changed and tidied up before heading over to
Guy's parents' for their family dinner.' She was rushing around
trying to get the girls ready when the phone rang. 'It was Guy's
mother, asking how long I'd be. I told her I would be there as soon
as I could.'

'Please hurry over,' her mother-in-law urged. She didn't say why.

Guy could not wait. 'All I wanted to know was where my wife
and children were now living; to go and find them. My sister drove
me over to my new house and I sneaked round to the back.'

Sandy was running late. There was a knock at the back door.
'I was really annoyed. Why would the paperboy be knocking on
Thanksgiving evening?'

Then she saw a shape through the smoked glass. 'It was too tall
for the paperboy.' She opened the door.

'Surprise!' Guy was standing there, wreathed in smiles.

Sandy was rocked backwards. 'I just couldn't believe it. It was my beloved husband. "What are you doing here?" I asked. I was in shock. It took me a few seconds to really believe he was there. I just grabbed him and held him close. Kissing him, I told him sternly, "It's a good job I have a strong heart."'

Mari was now nearly four and only knew her father from the framed photographs in the hall. 'It was very dark outside and I remember being quite scared that this huge stranger had appeared at our door with my aunt,' she remembered. 'He had a big plaster cast on his arm. I didn't really understand what was happening. They had to explain to me that this man was my daddy, home from the war to visit us. It made a profound impression on me; there was much joy and celebration in the house.'[15]

While Guy and his family rejoiced at his survival and return, Americans tuned in to the White House for the President's Thanksgiving speech. Lyndon Johnson told the nation, 'We are proud of the spirit of our men who are risking their lives on Asian soil. We pray that their sacrifice will be redeemed in an honourable peace and the restoration of a land long torn by war.'[16] So that the troops in the combat zone knew they were not forgotten, 57,000 turkeys – alongside 28 tons of cranberry sauce and 33 tons of fruit-cake – had been sent to Vietnam to provide as many as possible of the half-million Americans now serving there with the traditional holiday feast.[17]

After enjoying their own meal, Guy and Sandy spent a wonderful ten days together, seeing friends, pottering around the house, Guy getting to know his children and living a normal life. 'It was just so wonderful to have him back,' Sandy says. 'What a gift from God.'

But they both knew that the clock was ticking. Now his plaster cast was off, Guy would have to complete the final three months of his tour of combat duty. He told Sandy, 'All I have to do is get past Christmas and into the New Year, then it will be a matter of a few weeks before I return home again for good.' Sandy knew that having managed nine months apart and keeping the family together, she could easily do another three. Those thoughts offered them positivity and joy as they drove him up to Jacksonville early in December to catch his flight back to Vietnam.

At the airport Sandy saw lots of new troops preparing to deploy, about to begin their own year's tour. Family groups huddled around them. 'There were people crying, hugging. It all looked so sad. As Guy was about to board the aircraft, I said to him, "All these poor families have now got a whole *year* before they will be reunited; I just don't know how they'll cope. Thank God you and I only have three months to wait until we are back together again."'

After a final kiss, her husband walked up the steps and disappeared inside the aircraft. A few minutes later it took off, heading back to the warzone.

CHAPTER SEVEN

THE HEART OF DARKNESS

RAO NAY RIVER VALLEY, NORTH VIETNAM
21 DECEMBER 1967

Having survived the trauma of being shot down, ejecting, then being rescued under heavy fire, Guy Gruters now felt that luck must surely be with him on his return to Vietnam. 'I didn't have any lingering concerns about getting back into combat. My life had been saved by the ejection seat, I'd been given another opportunity and I just wanted to be back in the fight. We had a really important mission out there and I was desperate to play my part in it. I guess that having beaten the odds already, I had an arrogance that I would survive.'[1] Guy might have been confident of his survival, but the death toll being suffered by the USA in the region was climbing steadily. As the year drew to a close total deaths of US service personnel since 1965 had climbed to 20,000, with just over 11,000 killed in 1967 alone. In 1968 – the deadliest year of the war – another 16,899 US troops would lose their lives.[2]

Back in the rear cockpit of a Super Sabre, 1,500 feet above the patchwork of orange and green paddy fields, Guy Gruters, now promoted to the rank of captain, was on his second sortie since returning to the warzone. He had a new crewmate. Piloting the aircraft out of Phù Cát, 34-year-old Major Bob Craner had arrived in Vietnam in August 1967, a decade after winning his wings.

Experienced and self-disciplined, he calmly weaved his way over the valley floor to avoid the cherry-red tracers looping at them from the forested ridges while Guy got on with his work. Flying a fraction beneath the low cloud, he was on the lookout for artillery sites the North Vietnamese Army (NVA) had dug into camouflaged positions to protect the endless stream of trucks heading over the border for the Laos leg of the Ho Chi Minh Trail.

Craner's interest in Renaissance history and Slavic languages would count for little on this mission, but having spent six years as an instructor-pilot on the F-100, there was nothing he did not know about the Super Sabre. It was just seventy-two hours until the proposed Christmas ceasefire came into play, when the Americans would be able to do little to halt the enemy's supply convoys – but they still had ample time to take out the guns if they completed their intelligence mission and called in airstrikes.

Even Craner's cool began to slip as fresh bands of tracer from the valley floor criss-crossed the sky around his banking jet, tearing past his thin Plexiglas cockpit, missing the fuselage by inches. But the intense fire was aiding their mission – Guy had already pinpointed five gun positions. He just needed another burst of tracer from the trees to reveal the whereabouts of the sixth.

Craner broke the silence. 'Guy, you got that one yet?'

'Just give me a few more seconds, okay?'

The glow of a cannon below told them they had run out of time.

Guy tried to yell 'Break right!' but the words never made it out of his mouth. A shell smashed into the aircraft immediately behind his seat. The impact threw the Sabre onto its starboard wing. Craner fought to right it, but with both the hydraulic systems gone it was impossible. The jet rolled onto its back.

'The mountains,' Gruters urged. 'Head for the mountains.' There was no way he wanted to eject straight into the heart of hostile territory. If they reached the mountains they stood a better chance of being rescued.

'Negative,' Craner responded. Out of control, the Super Sabre was careering straight for the valley floor, headlong and upside down. There was no way that Guy Gruters was going to escape the combat zone this time around.

'Get out! Now!'

With his own advice – *Eject in time!* – still resonating after the previous incident, Guy reached for the now familiar handle for the second time in six weeks. 'I had one thought in my mind: *Are you shitting me?*' They fired their ejection seats, shooting violently downwards from the inverted jet, in plumes of flaming white smoke. They were lucky. Their parachute canopies opened just a few hundred feet above enemy territory.

Bob Craner's welcoming committee already had him covered with rifles as he floated down. Picking him up as soon as he landed in a rice paddy, they dragged him to a village, stripped him of his flight suit and boots and put him in black pyjamas. Under lock and key, he could hear trucks and tanks thundering south in readiness for the Tet Offensive. He could do nothing to stop them. Bob's war in the air was over.

With no idea what had happened to his friend and crewmate, Guy Gruters descended through the afternoon gloom. He heard gongs sounding in a village below. He knew what that meant. *Go get the pilots.* He saw the thatched roofs of a hamlet in a clearing take shape beneath him. It would be sundown in half an hour and the edge of the jungle was about a thousand yards away. If he could just make it there and hide, he could head for the ridge under cover of darkness. It was his only chance of escape. He took out his survival knife and cut through several of his parachute suspension lines to pick up speed and veer sharply towards the foliage.

Whistles were being blown below. Black-clad villagers knew precisely where he was headed. Green tracer bullets ripped through his canopy. 'I was watching the firing whipping past me and my only thought was what terrible shots they were. I sensed God must be looking after me again, and wondered if I actually had a chance of getting away.' He hit the ground, hid his parachute, made two 'Mayday' calls on his survival radio and ran. Once in the trees, he dropped down, trying to make himself as small as possible, and waited for nightfall. Around forty-five minutes later he heard footsteps crashing through the undergrowth. When he looked up, ten NVA soldiers were pointing AK-47s at him. He raised his hands and rose slowly to his feet. Guy's war as a pilot was also over and, like Bob, his post-ejection battle for survival was only just beginning. 'I'd been presented with a third chance at life, but simply had no idea what that life might look like, or how long it might last.'

His captors lashed his hands behind his back and, as they marched him along the road, they had to fix bayonets to stop furious machete-wielding villagers from hacking Guy to pieces. They locked him in a farmhouse and gave him some rice and greens. Was this a good sign? Surely the NVA would not waste precious rations on condemned men? Next day they placed him on a low stool, wrenched his arms behind his back and knotted them tightly with hemp rope which cut into his still-tender wrists. A civilian interrogator arrived, brandishing Guy's USAF ID card. He opened a large ledger and slowly removed the top of an elderly fountain pen. 'Gruters,' he began. 'What base fly from? Da Nang, Udorn? Must tell me.'

In survival school, aircrew were told never to volunteer anything beyond their name, date of birth, rank and service number. To the annoyance of his captor, Guy stuck to those rules. The interrogator raised his voice. '*You must answer all questions . . .*' Then he leaned closer, his spittle splashing Guy's cheek. 'Talk or be punished.'

'Gruters, Guy David, Captain, United States Air Force.'

In his heart, he knew what was going to happen next.

Two guards grabbed his neck and shoulders, hauled him off his stool and out of the doorway, dragging him, still bound like a pig, across a yard and into a small outbuilding. Two more guards arrived and kicked, clubbed and punched his naked chest, belly and arms. Disorientated by the pain, Guy had no idea how long the beatings went on for. Tiring from their exertions, his tormentors then moved on to the next stage. They forced him to his knees and fastened a length of knotted rope around his right bicep, coiled it under his left armpit and yanked it hard. He felt a cleated boot on the back of his neck. The soldiers wanted maximum leverage.

They're trying to rip my arm off . . .

Next, they bound both biceps so fiercely behind his back that his elbows were touching. The pain was unbelievable. He bit his lip to stop himself from howling. Massive vectors of pain surged through his torso. In minutes his arms were completely numb. They tightened the rope some more. The torture, and the questions, continued for hour after hour.

When it got too much to bear, he invented a fake cover story. He had been taught that this might stop the torture. For a while, anyway. He told his interrogator that he was an F-4 pilot from a base in Thailand. He gave them the name of his squadron and his commanding officer. The NVA officers began to smile. Their prisoner was co-operating. That evening he lay on the floor, exhausted but unable to sleep. He watched the flash and flickering shadows of the bombs that American aircraft were dropping down the valley.

The Vietnam War would grind on without him, and the increasing number of his fellow ejectees.

* * *

Some other US airmen in the warzone were having better luck than Guy. Variations on the 'traditional' ejection seat were being developed for different aircraft; some of them positively revolutionary.

The Stanley Aviation Corporation, designers and builders of the one-man escape capsule for the B-58, had also come up with the ground-breaking Yankee Extraction System. While the standard ejection seat 'pushed' a crewman out by the seat of his pants from below, the Yankee 'hauled' him from the aircraft by his harness using a 'tractor' rocket stowed flat in a compartment in the spine of the aircraft behind the seat. As the company marketing video from the early 1960s proclaimed, *'the system pulls out just the man, and his parachute, below the speeds where conventional ejection seats go sour, as straight as an arrow, regardless of the junk the pilot carries with him'.*[3]

*

Canopy is jettisoned
and tractor rocket expelled

Extraction lines begin to uncoil
and rocket ignites

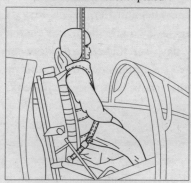

Crewman's seat folds down
as he is drawn out of cockpit

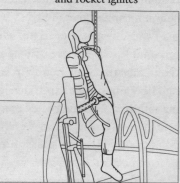

Crewman leaves cockpit
and parachute deploys

STANLEY AVIATION CORPORATION YANKEE
EXTRACTION SYSTEM

Major Bill Bagwell was carrying his usual 'junk' of combat and navigation kit. A highly experienced instructor-pilot, he was preparing for the second to last sortie of his one-year combat tour, on a routine mission in a Douglas A-1E Skyraider. The seasoned attack aircraft, first introduced in 1946, was being used extensively in Vietnam; a propeller-driven anachronism, thousands had been retrofitted with the Yankee escape system. It had already saved a whole host of lives. Including Major Bagwell's when, a few months earlier, he had been forced to use it to escape a flaming Skyraider after being hit by North

Vietnamese anti-aircraft fire. Luckily, like Guy Gruters' first shoot-down, he too had been rescued from enemy territory.[4]

Bagwell and his co-pilot took off on a mission to attack NVA units that had crossed into northern Laos and surrounded a Central Intelligence Agency (CIA) base. 'Few of us had ever been up there,' he recalled. 'We grabbed some extra maps to help us get familiar on the way.' As they flew towards their target, the clouds were building but in the distance they could still make out some of the higher peaks of the Himalayas.

As the CIA operatives retreated, the pilots' task was to destroy any remaining American equipment before the 'bad guys' could get hold of it. 'We tuned our radios to the combat frequency. It was easy to tell things were pretty tense.' As they dropped their napalm on the targets, following it up with 20mm cannon rounds, the enemy on the ground began returning fire. Bagwell could clearly see the muzzle flashes on the ground and had a grandstand view from the right-hand seat of what happened next. 'Suddenly our right wing exploded. Fire spread rapidly over it and the skin began to peel off.' The Skyraider's ammunition began to ignite and the aircraft began a fatal roll. As the control flaps broke away, his co-pilot shouted, 'I can't hold it.'

It was time to go.

Bagwell was first man out. He pulled the D-ring between his thighs, activating the fully automatic Yankee system so his cockpit canopy should immediately have jettisoned. Unbeknown to Bill it didn't, but it made no difference as he was already on his way.

He was wearing two harnesses, the first holding him to the seat, the second one, worn underneath, was similar to a standard parachute harness. This one was connected to the large 'tractor' rocket stored horizontally behind his headrest. As soon as Bill had pulled the D-ring, the rocket had raised itself to adopt a near-vertical launch position. A catapult then propelled the whole rocket unit upwards on two 12-foot lines which were connected to Bagwell's harness. They began to haul him, and his seat, upwards in the cockpit.

The back of Bagwell's seat was connected to the sitting section itself by a hinge at the rear. So, as it began to rise up a set of rails, the base automatically folded under and downwards, meaning he was now in a near-standing position in the cockpit. The straps holding him into the seat then automatically disconnected. As the rocket reached the 12-foot extent of its lines under the force of the catapult, it ignited and blasted upwards, spinning above Bagwell like an enormous, bright white firework, trailing thick black smoke and dragging him from his cockpit, *like a giant hand had reached down and yanked me upwards by the scruff of my neck.*[5]

When the rocket extractor had expended 90 per cent of its energy, it severed the towlines connected to the harness, and utilised its final 10 per cent of thrust to fly safely clear of the tumbling Bagwell whose parachute, assisted by a smaller, rocket-powered drogue, was then opened to its full extent. It had taken about six seconds from pulling the handle.

The Yankee was lighter, more compact and cheaper than the conventional ejection seat used by Guy Gruters. As the metal seat remained in the aircraft, it only needed a fraction of the power to drag the pilot clear. And it provided safe descent from an aircraft stationary on the ground – offering a true 'zero-zero' ejection system, well before those particular ejection seats were in service. Though not suitable for high-performance jets, it was now standard in over 3,000 Skyraiders and other, slower, aircraft originally designed without ejection seats.

These stats were of little interest to Bill Bagwell, who was now 'hanging out in the parachute straps over some very unfriendly country'. Apart from the challenge of landing on a hillside now covered in hazardous tree stumps, two thoughts hammered through his brain: *Where's my co-pilot?* and *Why me . . . AGAIN?*

For a second time – unlike Guy Gruters – he was rescued from enemy territory by helicopter.

'Climbing into the chopper, I realised my face was covered with blood and I had to puke.' A flap of skin was hanging from his forehead; he had a thumping headache and was suffering from concussion. It later transpired that the cockpit canopy had failed to blow clear so, as the extraction rocket fired, it simply dragged him through the shattered Plexiglas. A plastic surgeon would later repair the section of torn skin on his forehead, creating a few more wrinkles to hide the scar.

Shortly after Bagwell arrived at the main clandestine CIA base in Laos, his co-pilot showed up. He too had been rescued around twenty minutes after hitting the ground. 'While part of my extraction system failed, his didn't work at all,' Bagwell said. When his co-pilot pulled the D-ring it had simply come away in his hand and nothing triggered. 'When that happened, he just unhooked from the seat and the rolling airplane tossed him out. He had gotten a load of gunpowder from my extraction rocket in his eyes and someone was trying to rinse them out with water.'

'Who cares about the friggin' seat?' his friend declared. 'It's my ass that matters, and it's here.'

Back at their home base in eastern Thailand, they were met 'by a lot of happy guys and a lot of excitement'. Someone handed Bagwell a beer. 'I asked where the champagne was. They said we got only beer the second time we're shot down!'

* * *

There would be neither beer nor champagne for Guy Gruters and Bob Craner after their own second shoot-down. The day after his ejection, Bob was paraded barefoot, in shackles, around the six-gun batteries he and Guy had been hoping to destroy. His tour took a whole day. The villagers yelled at him, punched him and beat him with sticks along the way.

Badly bruised, he was still able to make a mental note of everything he saw. Each 37mm gun crew consisted of eight men. Ten for the 57mms. Roads crammed, nose to tail, with military trucks. US Intelligence had no idea of the scale of this preparation for the Tet Offensive. If only he could get the information home. On the way back to the village where he was being held, his guard was playing with Bob's survival radio and accidentally switched it to transmit. Two Misty pilots out searching for their missing friends picked up the signal.

'I got you, Misty. Where are you?'

The soldier stared at the piece of equipment, puzzled. Cool as anything, Craner motioned to him. 'I can fix it.'

The guard thrust the radio at him and Craner was able to tell his fellow pilots where he was, that he loved his wife and would see her at the end of the war. To offer comfort to Guy's wife Sandy, he also said he knew he had ejected and he, too, was almost certainly captured. Finally, he detailed the gun-battery manning levels and the amount of traffic on the roads. The Misty pilot then started talking to the guard in French, promising, 'If you release your prisoner, he will take you to Saigon and give you much gold.'

Gold! The guard got the gist of that. He grinned and pointed at the radio, not knowing the voices were coming from enemy aircraft. Then he saw the F-100F sweep around the bend of the river and realised he had been taken for a ride. He became very angry and raised his weapon, threatening to shoot Craner as the jet was beaten back by enemy flak. The rescue was off.

Eight thousand miles away, in Florida, Sandy Gruters was in her living room putting up Christmas decorations with Guy's sister Mary Anne. She glanced out of the window and saw a military vehicle pull up in the car port. A man in uniform got out.

'Here they come,' she thought. 'Weeks after it happened, to finally tell me officially that Guy had been shot down and rescued.' She had just received an old letter from her husband telling her that he had to go back into hospital. 'I opened the door to a lovely Air Force captain, with a huge smile on my face. "You've come to tell me that my husband was shot down, haven't you?"'

He looked puzzled. Why was she so happy? She invited the officer inside. 'Mrs Gruters, I'm afraid it's very bad news. Your husband has been shot down over Vietnam. He is currently listed as missing in action.'

Sandy cheerily looked him straight in the eye. 'I'm sorry, Captain, but you've made a mistake. My husband is back in hospital after his ejection.' Now the Captain was really confused. 'The Air Force doesn't make mistakes like this, Mrs Gruters.' He handed her an official letter.

There it was in black and white. Sandy suddenly realised the enormity of what she was being told. 'I was just stunned; simply shattered. How on earth could Guy be shot down *again*?'

The next few hours were a blur.

'I was in shock. It was as though God had lifted me out of my situation so I was actually floating, looking down on my life unfurling. I was watching myself in the living room, talking to the Captain, seeing Mary Anne on the phone telling Guy's parents he was missing. I watched the Captain telling me that the Air Force would be in touch with more information. But he would only come back in person for two reasons: to either tell me that Guy was a confirmed prisoner of war. Or that he was dead.'

As the news spread, visitors arrived. Sandy prepared food and drinks for them but her mind was elsewhere. *Guy, where are you?* She sat alone to write a letter to her beloved husband. 'I told him that I believed he had ejected and was alive, and that if he had died I would know it. I wrote that I would look after his daughters and continue as normal and wait for him to return safely, which I *knew* he would. I told him that we loved him and that we would see him again.' She had little choice but to believe her own words.

The next morning Guy's father came over to check how she was. Sandy had just given the girls their breakfast and they were sitting at the table. She was in a rocking chair next to them when she saw the Air Force vehicle pull into the driveway again. The Captain stepped out. She felt her pulse race. 'I knew I was about to find out if Guy was alive or dead. There was no in-between.'

She walked to the door, filled with dread, and opened it. 'This time it was the Captain who had a huge grin on his face. "Mrs Gruters," he said, "I have a much better Christmas present for you. Your husband is confirmed as a prisoner of war." I was overjoyed. Many people wouldn't have thought that was much of a Christmas gift, but I did. My spirits soared with relief.'

Christmas 1967 would be a difficult occasion for the whole

Gruters family. In those early years of the war, few people truly understood what being an American POW in Vietnam really meant, or just how long their shared ordeals would last. 'Maybe if we had known what lay ahead, we would have been overwhelmed,' Sandy says.

THE BAMBOO PRISON
26 DECEMBER 1967

As Sandy Gruters had been putting on a brave face for her family and doing her best to enjoy Christmas Day, Guy was taken out of the village house where he was being held and shoved onto a waiting truck. Minutes later he was joined by Bob Craner. It was the first time they had seen each other since they had ejected five days earlier. They exchanged grins but the dark bruises around their eyes now matched their black pyjama suits. They were ordered not to talk to each other, but as the wagon rolled out Bob managed to whisper that he too had endured the rope torture. He had also invented a cover story, but had not been able to lie about his aircraft because his captors had seen and identified the wreckage. With no idea where they were being taken and what lay in store, Guy hoped he could convincingly amend his cover story to match.

After an all-night journey, bouncing along roads ruptured with bomb craters, Guy Gruters and Bob Craner were bundled at dawn into a building constructed from sturdy bamboo poles thicker than a man's arm and a rice-straw roof. POWs called it the 'Bamboo Prison'. The complex of single-storey cell blocks was a holding jail and interrogation centre on an army base near Vinh, the biggest city in North Central Vietnam, 180 miles south of Hanoi.

The two pilots adjusted their eyes to the dim light. The cells, either side of a central hallway, were 6 feet by 6 with dirt floors. A guard ordered them to stop outside an empty cell while he went inside. Craner saw piles of rusty handcuffs and implements of torture. Spiked boards, wire whips, ropes and pulleys. The guard returned with cuffs and chain-linked ankle irons; they were restrained and thrown into separate cells. Craner was the first to be questioned. His interrogator was an English-speaking, rat-faced officer known to inmates as 'the Rodent'.

Because he spoke in a very loud voice and the interrogation booth was close to his own cell, Guy could hear all the questions and answers. If he could memorise them, he could match his cover story to Bob's.

'Who is your Wing Commander?' the Rodent demanded.

'Colonel Robert E. Lee.'

'Commander of squadron?'

'Lieutenant Colonel Stoney Jackson.'

Craner was pretty sure that the Rodent would not have heard of either of these American Civil War Confederate generals, but he now wanted the details of other key squadron members. Craner gave him the names of football players. Guy Gruters could not believe that the Rodent was swallowing so much bullshit.

After two hours, Craner was taken back to his cell. It was Gruters' turn. He stumbled along in his ankle irons. The Rodent was not happy. From questioning Craner he knew that Guy had lied about being an F-4 pilot. What else had he lied about?

'You say, Takhli base in Thailand. Craner say Phù Cát. You tell us many lies.' He accused him of criminal conduct, the price for which was severe punishment.

Gruters had to think fast. His mind was seething with images of wire whips and spiked boards. 'I said that I was supporting F-4s, not flying them. You knew I was flying an F-100 because it crashed in the valley.' His first interrogator had misunderstood. Takhli was just a temporary staging base. Of course Phù Cát was his home base. The Rodent nodded and wrote it all down. 'Good. Now you tell the truth. Commanding Officer name?'

'Colonel Robert E. Lee.'

'Squadron Commander?'

'Lieutenant Colonel Jackson. First name Stoney.'

'You spell.'

Gruters obliged. The interrogator compared his notes. Guy

Gruters was off the hook. He and Bob Craner had avoided further torture. For the moment.

Their cramped cells swarmed with flies. It was difficult to brush them away with cuffed hands. The heat and monsoon humidity made breathing difficult. Their unwashed bodies stank. The two pilots clung to Article 3 of the Code of the US Fighting Force: *If I am captured I will continue to resist by all means available. I will make every effort to escape.*[6]

Ejection seats may have prolonged Gruters' and Craner's lives, but they may not have saved them, and as 1968 loomed into view, promising yet more repetitive interrogations and threats, Guy began to realise that they might face years of the cruelty.

On New Year's Day, the distant thud of anti-aircraft batteries echoed late into the night. Through the wall of the Bamboo Prison, Guy Gruters heard a truck pulling up. Then shouts of command. A tailgate clattered. A soldier jogged past to the doorway of the cell block. A new prisoner was arriving.

'You awake, Bob?' Guy whispered to his friend in the nearby cell.

'I'm awake.'

'You hanging in there?'

'Damn right. Room service is lousy, but I can't complain about the price.'

Torch beams suddenly snaked along the floor. Through a narrow gap in his cell partition Guy saw two guards dragging a tall, barely conscious, skeletal figure down the corridor. A plaster cast on his left leg scraped along the ground. He was not wearing a uniform, just filthy black pyjamas like theirs. They threw him into the cell opposite Bob Craner's. The guards disappeared. The block returned to darkness, the only sound the scrabbling of rats in the straw on which they slept and mosquitos coming out to feed. 'What's your name?' Craner whispered across the hall. All that came back was a muffled moan. The new prisoner slept until noon the next day when the Rodent arrived. Bob Craner listened to him working through his usual list of questions.

'What airplane you fly? Which squadron?'

'I'm ... not ... going ... to ... tell ... you ... anything ...' The man gasped for air between each word.

The Rodent started yelling in rage. 'Name of your commander!'

'Can't talk to you,' he stuttered. 'It's against the code.'

It was difficult to hear everything, but Craner recognised the accent. He was one of theirs.

'What base you fly from?' Their fellow airman said nothing.

The Rodent called a guard and Craner now heard the sound of a bamboo club smacking into bone. The prisoner shrieked in agony. Military boots then administered a kicking to what sounded like every part of his body; until the sharper screams told him they were now targeting the airman's injured leg. He could not imagine what a broken bone must feel like when it was clubbed and kicked.

The ordeal lasted an hour and a half. From somewhere, somehow, this brave aviator still managed to dredge up defiance. 'All right, you son of a bitch, wait till I get better, you're really going to get it.' The interrogations and the beatings intensified. The new arrival roared in torment. But he would not be beaten. 'I'll get you, you fucker. I'll kick your ass.' Beyond spluttering his date of birth, rank and service number, he would not give them anything.

Guy could almost feel the pain himself. 'Hearing a man being tortured is possibly worse than being tortured yourself. Their suffering strikes through your body. It is impossible to block out the terrible sound and fear of what they are doing to him. And to know that it could be you next is simply overwhelming.' Bob and Guy shouted at the guards to stop. Guy caught sight of their comrade as he staggered past, en route to the stinking latrine. 'He looked like a living skeleton, and was a mass of raging sores. He had a mangled hand, and I could now see the cast on his left leg ran from his waist to his ankle.'

Conversation between prisoners was forbidden in the POW camps. Those who broke the rules were whipped with rubber hoses and car fan belts. But when the guards changed shift, Bob risked whispering across the hallway to tell the pilot he needed to invent a cover story, as they had done, to stop the torture, at least for a while. He was in no shape to keep taking the beatings.

The man replied hoarsely that he could handle the punishment.

Early next morning the new prisoner started calling out. 'Hey, somebody! Take me to the latrine. I got to take care of myself.' The guard shouted at him to shut up but the cries for help continued. Craner and Gruters were then wrenched from their cells and told to take the man and clean him up. They flinched when they saw his skeletal body curled up on the fouled straw. He had been beaten and kicked so severely he was now unable to stand, let alone walk. It was only when they took an arm each to carry him outside that Guy realised he was as tall as they were.

'This guy is really pretty big,' he said to Bob. Hearing Guy's voice, their fellow prisoner turned. 'Aren't you Guy Gruters?'

'Yes.' Guy did a double-take. 'Who are you?'

'Lance,' the man replied.

'Lance who?'

'Sijan. Lance Sijan.'

'No . . . oh no. Not Lance . . .'

When he had last seen Sijan at the Air Force Academy four years earlier, he had been a super-fit athlete and skydiver. Now he was a tangled, frail, tortured mess. 'I was rocked to my very soul. It was such an awful shock to realise this was the man so many of us admired back at the academy.'

Guy looked at the battered 25-year-old with tears in his eyes. 'I just couldn't believe it. It froze my heart.'

They carried Sijan to the latrine and gently washed the faeces, blood, urine and mud from his body. As his torture and interrogation continued, the guards allowed them to repeat the process several times over the following days, during which, bit by bit, he managed to tell them his story.[7]

He had been flying F-4C Phantoms out of Da Nang, the northernmost of the Air Force's major bases, since July 1967. Sixty-six combat sorties in Vietnam so far. When not in the air he had spent a lot of time in the gym lifting weights, strengthening his legs and arms. If he was forced to eject and survive on the ground, he was determined to be prepared. He had been one of the star performers during jungle survival training in the Philippines. Now he was no longer capable of even washing himself.

On 9 November 1967, the day after Guy Gruters' first ejection, he was briefed for a night mission to destroy a river crossing just inside Laos on the Ho Chi Minh Trail. As Sijan's Phantom released its payload, it suddenly turned into a massive ball of fire. All six bombs had detonated prematurely, less than 50 feet from the aircraft, which then tumbled down, hitting the jungle canopy like a burst of napalm. The flames burned out quickly, leaving impenetrable darkness. Those still circling overhead listened for a signal from the downed Phantom's crew.

Nothing.

The pilot was lost presumed dead, but Lance Sijan had survived. He had escaped in the nick of time, ejecting up through a blast of orange flame. Miraculously he survived, but Martin-Baker ejectee number 1,886 was in a bad way. The explosion, the ejection and crashing down into forested mountainside had given him a compound fracture of his left leg. Two pieces of bone several inches long had broken through the swelling. He also had a fractured skull, concussion and a mangled hand.

That night and all of the next day, Sijan lay on the ground, veering between unconsciousness and delirium. He managed to dig out his survival kit and patch up his wounds well enough to drag himself along the forest floor, but, being almost 200 feet below the canopy, he had no way of spotting or signalling to potential rescuers.

For an astonishing forty-six days he somehow evaded North Vietnamese patrols, surviving on rainwater, bugs, leaves and ferns. He did establish radio contact with searching aircraft but they couldn't get to him before his survival radio batteries were exhausted. Shortly after dawn on Christmas Day, a North

Vietnamese convoy found him lying in the road, just 3 miles from where he had ejected on 9 November. Sijan told Guy and Bob that at one point he was left alone and managed to overpower a guard, grab his carbine and crawl off into the jungle again. It had taken an entire village most of a day to find him. He was proud of that. The Vietnamese guards had then fitted the crude cast to his leg, to immobilise him and prevent him from escaping again, as much as to mend his broken bones.

Even though he now frequently lapsed into delirium, Sijan was still capable of lucid moments. He talked urgently to Guy and Bob about escaping. 'How secure is this place? What are the chances of getting out of here?'

Craner was astounded by his courage. What nurtured his belief that he could spring free, even though he could barely move one foot in front of the other? When they carried him, they felt the tremors of pain that were coursing through his feeble body. 'He even tried doing arm exercises so he would be ready to take part in any escape.' There was no escape plan, but Craner played along to keep Sijan's spirits up. 'We're doing a recce, Lance. We'll keep you posted, buddy, don't worry.'

'Count me in. Okay?' Sijan smiled faintly. 'I can handle my end of it. We're going to get out of here, aren't we?'

'Damn right, Lance. We're all in this together.'

They tried yet again to persuade him to invent a cover story to bring his torture to an end. But Sijan wasn't interested. 'They're not really hurting me too bad. You guys just figure some way that we can get out of here.'

The questions continued and the torture and beatings began again. Sijan was now near unconscious and his disturbed muttering made no sense. His leg was gangrenous and starting to rot. When they could, Bob and Guy tried to feed him slowly so he did not choke. Even in his stupor he tried to crawl to the cell door and freedom. A guard rushed in wielding a bamboo club and beat his injured leg like he was playing baseball. Heavy rain now fell constantly. The temperature in the Bamboo Prison plummeted. Lance Sijan began to cough. He was fading badly.

A battered old Soviet 2½-ton truck pulled up at the jail on 10 January. The Rodent told Gruters and Craner they would be leaving for Hanoi that evening and would be responsible for caring for the 'criminal' Sijan. They would all be loosely roped so Guy and Bob could feed and wash him.

As soon as it was dark, Gruters and Craner carried Sijan out and laid him on the straw on the truck bed. They sat on wooden benches either side of him. Four armed soldiers bustled in and took up position by the tailgate. They shared the space with two 55-gallon fuel drums standing in pools of water. The truck lurched off into the night, along a bombed-out road. It was only doing 10mph, but Gruters and Craner felt every jarring rut and pothole. The next crater launched one of the fuel drums into the air. Craner leapt across to shield Sijan's prone body. Then, the second drum rolled onto his mangled right arm. Gruters manhandled it off him as Sijan's screams were drowned by the soldiers' laughter.

The truck accelerated like a demented rollercoaster. 'Bob and I took turns cradling Lance between our legs with his head and shoulders on our bellies and chests, holding him tightly to our bodies while trying to prevent the drums from crushing us.'

Although now cushioned by his compatriots, Sijan's body was still taking a beating and the rest of the journey, much of it in cold, slashing rain, was just as punishing. Taking shelter under a tarpaulin, they massaged Sijan's face, soothing him, encouraging him. But he was not responding. Craner put his ear to Sijan's mouth. There was no sign of breathing. His flesh was cold.

'Guy, he's dead.'

'No! Not Sijan. He can't be. Not after all this!'

Both men were sobbing. They massaged his face again and made him as comfortable as they could.

Then Lance stirred and he was briefly lucid again. Craner couldn't believe it. 'He still wanted to be caught up with what our escape plan was, and how we were going to get out of the situation.' Despite his desperate condition, Sijan was still full of fight.

Guy Gruters dried his eyes. Lance had come back from the dead. 'He's going to make it, Bob. He's really going to make it.'

In the middle of the third night, 13 January, the truck stopped at a roadblock. Peering out from under the tarpaulin, Guy Gruters saw the twinkling lights of a city. They heard orders being barked. Large metal gates swinging open on rusty hinges. They peered through the darkness and sheets of rain at a forbidding nineteenth-century prison. Its high concrete walls were topped with shards of glass and rolls of electrified barbed wire.

They had arrived at Hỏa Lò.

American POWs called it the 'Hanoi Hilton'.

Guy now remembered Jeremiah Denton's blinked message: T-O-R-T-U-R-E.

Back home in Florida, the January weather was typically warm but not humid. The Christmas tourists had gone, leaving the golf courses and beaches to the residents. The orchids were out and the sky was full of birds escaping heavy snowfalls in the north. Sandy Gruters was enjoying the sunshine and the blooms as much as anybody, but she was living one day at a time. She was still in shock. She also knew that she had no choice but to just keep going. 'I had children that needed to be cared for, bills to pay, a house to keep neat and tidy. Our lives in America had to continue, even though our loved ones were in mortal danger 9,000 miles away.'

Every time Sandy saw TV coverage of US airmen who had ejected only to be captured and paraded in front of the cameras, she flinched. She cast her mind back to the hopeful words and feelings she and Guy had shared at Jacksonville Air Base only weeks before. *We'll soon be back together again.* She prayed that Guy's ordeal would soon be over. But in her heart she knew it would not and, out of sight of the children, she sometimes buckled. 'I always tried to protect the girls but I still occasionally felt overwhelmed by my situation. So I would go for a drive and sit in a parking lot where I could break down in tears alone, and think, *Why me? What have I done to deserve this?*' Then one day she noticed a family with a severely disabled child. Seeing their obvious challenges put her own into perspective. 'I was healthy, had two wonderful, healthy children, and a good family. Yes, I had a cross to bear, but others had heavier crosses than mine.'

Like all wives whose husbands were prisoners of war or listed as missing in action (MIA), Sandy had received the instructions from the US State Department telling them to keep quiet and not to talk about their husbands' treatment or status in Vietnam. They should never speak to the press, give interviews or show any personal mail. And they should never intercede in any way on behalf of a prisoner without State Department approval, as *independent actions could seriously damage negotiations being conducted on behalf of the prisoner.*

In short, wives and families should carry on as if nothing had happened and let the good folk in Washington follow their policy of 'quiet diplomacy'. Sandy was desperate to find others in her situation to talk to. Only they could really understand what she was going through.

Guy and Bob had carried Lance Sijan into the Hanoi Hilton on a wooden pallet. Nobody else knew he was alive. At least their own wives knew what had happened to them, even if they had no idea where they were being kept or how they were being treated. They were allowed to wash Lance at a cistern in the courtyard. They carefully removed his soiled pyjamas but some of the fabric stuck to his weeping sores causing further pain. Once inside the jail, the three were taken through a series of clanking steel doors to the central interrogation centre where they were shoved into separate cells. Gruters' was just 7 feet long and a few feet wide. There was no window and the walls dripped with water which formed puddles on the floor. He began to shiver violently, so squatted against the driest wall and hugged his knees for warmth. The inspection hatch in the door was thrown open and the guard bellowed at him to stand up.

Sleep deprivation was one of their key weapons. The lights were left on all night and at regular intervals propaganda broadcasts were relayed over Tannoys at ear-piercing volume. 'Confess your crimes! Thank the Vietnamese people for saving your life.'

Guy was very concerned about how Lance would cope without them.

Early the next morning the loudspeakers began bellowing their next twelve hours of propaganda. Bob Craner was marched from his cell across a courtyard to one of four 'quiz rooms'. The commanding officer of the Hỏa Lò Prison was waiting for him. The diminutive captain was also in charge of interrogation – a responsibility he took very seriously indeed. In his early forties, his moon face and dark eyes led the prisoners to nickname him 'the Bug'.

Unlike the Rodent, the Bug was educated, spoke more English and was painstaking. He also had access to much better military intelligence against which to measure their confessions. Craner remained standing. The Bug told him that if he failed to co-operate he would have no choice but to send him to a torture room. Room 18 was the worst. Its stucco-covered, soundproofed walls were spattered with the blood of men who had been beaten for hours with whips and clubs.

Thanks to the exhausting journey to Hanoi and the lack of sleep, Craner was able to give a convincing impression of a man whose resistance had already been broken. Slowly and methodically, he repeated the story he had invented in the Bamboo Prison, detailing the command structure and missions of the fictitious F-100 squadron at Phù Cát.

The Bug seemed satisfied. For the moment.

Guy was next and on the way to the 'quiz room' he saw guards kicking other prisoners in corridors, booting them towards another interrogation or dragging them with nooses around their necks like sick dogs, willing them to retaliate so they could make them pay the ultimate price. Guy knew he was completely expendable and resisted the temptation to fight back. 'I wanted more than anything to get home and rejoin my wife and children.'

After three days of aggressive questioning, where their fictitious story seemed to hold up, Craner and Gruters were reunited with Lance Sijan in a three-man cell. He was lying on a wet cot in his own waste and his lungs crackled as he tried to form a few painful words. 'Check how ... secure ... the gate is ...'

Although still hellbent on escape, Sijan was now almost constantly delirious, so his friends took turns to watch over him through the night, two hours on, two off. In spite of all their efforts to keep him clean and calm, Lance now looked so awful the guards were no longer coming into the cell because they found the sight of him so repellent. On 18 January he developed pneumonia. When he breathed, his lungs rattled. A doctor visited with some guards. Two of them lifted Sijan onto the pallet and began carrying him away.

Suddenly Lance Sijan came around; there was terror in his eyes.

Guy tried to reassure him. 'You're going to get some real treatment now. In the hospital.'

Sijan managed to dredge up his old voice. 'Oh, my God,' he shouted. 'It's over.' Then he cried out again. 'Dad! Dad, help me. I need you.'

The steel door slammed shut. His two cellmates slumped on the edge of a bunk and wept. They had come to love Lance, his values and his courage.

Two days later Gruters and Craner were moved from the Hanoi Hilton to 'The Plantation', another prison camp on the edge of the city, once the home of the French colonial mayor. Shortly after their arrival, Craner spotted the Bug in the courtyard. Summoning his courage he beckoned the interrogator over. 'Captain, can you tell me the condition of my friend Lieutenant Sijan?'

The Bug permitted himself a thin smile. 'Sijan die.'

Guy and Bob could barely believe the news. After everything he'd endured, Lance had finally been forced to surrender. They were heartbroken.

'To hear Lance had died was simply overwhelming, I was utterly crestfallen,' Guy says. 'He had survived his jet exploding, ejecting, evading for days on end. And, in reality, they had simply tortured him to death over a period of weeks. Even today, over fifty years on, it's still painful for me to imagine what horrors he suffered after they took him from us. The pain, the hatred, the disbelief began to take me over. How could they do this? What kind of people could torture a man to death? I was beside myself with grief. But he really kept the faith under appalling punishment. It was the most incredible thing I have ever seen in my life and his example inspired us to try to do the best we could. He lived and died upholding the American fighting man's code of conduct to never surrender, and if captured to resist by all means possible. He was simply a sincere, honest soldier doing the best he could, doing what he understood to be right. He was a hero.'

Lance Sijan's post-ejection ordeal was mercifully over, but Guy Gruters, Bob Craner and hundreds of other prisoners were looking at a very uncertain future as the conflict ground inexorably on.

CHAPTER EIGHT

THERE BUT FOR THE
GRACE OF GOD

RAF BINBROOK, LINCOLNSHIRE
SEPTEMBER 1969

It was nearly two years since Charlie Neel's great friend had been shot down for a second time, and subsequently captured. But Guy Gruters was never far from Charlie's thoughts. 'After we'd ejected and been rescued that first time, I had not seen or spoken to him,' Charlie remembered. 'I had been told Guy had been shot down a second time and possibly captured, but had no other information. Letters could take weeks or even months to move across continents, international phone calls were almost unheard of. And, to be honest, most of us just moved on to the next task. So, in reality, I had no idea if Guy was alive or dead. At that point, little was known about the POWs, and it was quite a burden to bear. Some of my other friends had not been as lucky as me and were also missing in action. But you can't let it affect you. What good would dwelling on it do? I just had to hope that Guy had survived ejecting that second time, had been captured, and that I would see him again at some point. Like many of us who survived, I often reflected, *there but for the grace of God go I*.'[1]

He owed it to Guy to stay positive, which was his default setting.

Since Vietnam, his career had taken an unexpected path. Major Charlie Neel had been invited to Great Britain to fly the English Electric Lightning fighter as part of the prestigious USAF/RAF pilot exchange programme. He couldn't wait to get started as he parked the VW Valiant estate car he had bought on his arrival in the UK six months earlier outside the guardroom of 5 Squadron's base at RAF Binbrook.

His new home was never going to feature on many postcards. Located on the Lincolnshire Wolds, a range of low hills, one squadron commander reckoned RAF Binbrook was best known 'for its miserable winter weather and its semi-isolation'.[2] In late September, as Neel's wife and their three young children scanned their new surroundings, heavy cloud, driving rain and 50-yard visibility were far from unusual. But the airfield's close proximity to the east coast made it the perfect base for air combat training exercises over the North Sea at the height of the Cold War.

In the aftermath of the Second World War, the United States, the Soviet Union and their linked allies had been locked in a long, tense stand-off. Though the parties were technically at 'peace', the period was characterised by an aggressive arms race, proxy wars, and ideological bids for world dominance.[3]

Formed shortly after the First World War, 5 Squadron had already served with distinction in outposts of the British Empire and was the first to be equipped with the Mk6 Lightning. Charlie caught a glimpse of the gleaming silver machines on the airfield as he drove his family to their new quarters. Just over 55 feet long, with 60-degree swept-back V-shaped wings, the supersonic fighter's two powerful Rolls-Royce turbojets, housed one above the other beneath the statuesque tailplane, gave it a full-bellied look. And ensured that what the aircraft lacked in grace and beauty, it more than made up for in raw power. It was capable of over 1,500mph, climbing at 20,000 feet a minute to a ceiling in excess of around 60,000. Although, at the time, the exact figures were a closely guarded secret.[4]

As the USAF exchange pilot, Charlie wasted no time introducing himself to his new colleagues in the Officers' Mess, and he had already revised his wardrobe in preparation for his 'tour with the Brits'.

'I had been warned that the RAF had quite different dress codes to what I had been used to in the USAF, and that I would need something called a *lounge suit*, which I learned meant that the jacket and trousers had to be made of matching fabric. I also discovered I'd need better shirts, shoes and even a smoking jacket. Luckily, I'd had everything tailor-made in Hong Kong for just a few hundred bucks while I was on R&R from Vietnam.' Ever the maverick, Charlie had also made sure to pack his favourite US Air Force flying boots and helmet. Not exactly RAF-approved equipment.

One of Charlie's new friends was Chris Coville, later Air Marshal Sir Christopher Coville, then on his first squadron posting. 'The Lightning was the best aircraft in the world,' Coville said. 'We couldn't wait to get our hands on it. It looked and felt big. It was extremely challenging to fly and you were always operating at the edge of your abilities, but it was exciting and the camaraderie on the Squadron was fantastic.'[5] A fighter pilot to his core, Charlie Neel was delighted with the prospect of getting to grips with the aircraft. This was what fast-jet flying was all about. There was a downside. Design flaws meant that internal fires had sent some jets spinning into the ground. There had been many successful ejections, alongside some fatal crashes, in those early years. But all that mattered to Charlie Neel and his fellow 5 Squadron pilots was that they were on the RAF's front line, at the controls of the best fighter in the world.

And its role as an interceptor was clear. If Soviet intruders even looked like threatening the airfields housing the Vulcan, Victor and Valiant, Britain's V-force bombers, the Lightning and her pilots were tasked with seeking out and destroying hostile aircraft with the jet's armoury of air-to-air missiles and cannons.[6] Chris Coville and his fellow pilots had no illusions about the very real threat they faced from the Soviet Union. 'We knew that if war broke out and we took off, there would be very little to go home to because of the reality of nuclear weapons,' Coville remembered. 'In fact, the chances were that we probably wouldn't get home, as our instructions were quite clear. After we had engaged the formations of Russian nuclear bombers heading towards our coast, and we had used up our limited armoury of missiles and guns, we were to use our jets as weapons and ram any remaining enemy aircraft to bring them down. So, in any scenario, the chance of personal survival was rather limited!'

As a highly experienced combat veteran at the age of just thirty, Charlie Neel had supervised some of Chris Coville's ongoing training. Their first two sorties had focused on getting the best out of the jet and its weapons system. The third was to be more fun – full-on air combat. Dogfighting.

In the late afternoon on 22 September 1969 they changed into their flying suits and boots, and grabbed their gloves and helmets. As usual, Charlie was wearing his USAF helmet and boots. In the briefing room, they checked the weather reports and had a final run-through of their objectives. Charlie Neel walked out to the airfield and climbed into the cockpit of aircraft XS926, reassured as always to see the Martin-Baker Mk4B ejection seat waiting for him. Although the Mk4 was still a relatively early 'bang seat' (the new 'rocket' seats being developed had not been incorporated into older aircraft) this model had already saved the lives of nineteen Lightning pilots alone.[7]

He sat down and adjusted the integral safety harness, which would hold him firmly into the seat against the lumbar cushion and internal horseshoe-shaped parachute pack that provided extra cushioning behind his shoulder blades. This rig, designed for the new generation of fighters, was now fitted to more than thirty different aircraft, including in America, India, Italy and France. It was certainly more comfortable than the old seat in his F-100.

Charlie removed the dust cover from the personal equipment connector (PEC) unit on the right of his seat and snapped in his own connector. Another new development in the ejection programme, the PEC was a system for connecting the vital personal services to the aircraft itself. The 'man portion' of the kit hung from the life jacket and consisted of a small metal 'plate', around 2 inches wide and 8 inches long, into which was connected the pilot's anti-g hose, telecommunications cable and oxygen supply. The 'seat portion' of the system had a matching 'receiver' unit so the pilot could simply clip all his connectors directly onto the seat – and hence the aircraft systems – with one simple movement. On ejection, the seat portion would detach from the aircraft systems as the seat rose up the rails, and then when the pilot separated from the seat, the man portion would disconnect from it. It was a remarkably simple invention but saved considerable time when strapping in, and made the ejection process much faster, and more reliable.

Charlie then reached down to attach another clip on his life jacket to the lanyard of his personal survival pack (PSP), and sat down on the cushion-topped solid fibreglass box which contained his personal life raft and other survival equipment like flares, emergency water and signalling devices. If he ejected, after separating from the seat he could reach behind and unclip the 10-kilo PSP so it was no longer attached to his backside, where it might cause injury on landing.[8] Still fastened to him by the lanyard, it would hang beneath him, ready for use if he landed on water. He leaned forward to loop the leg-restraint cords through the metal D-rings in the garters above his calves.

Charlie then put on his helmet and pulled the harness straps over his shoulders and the lap straps across his upper thighs, inserting all four metal connectors into the single quick-release harness fitting. He was now strapped firmly to his ejection seat. Finally, he ensured he could reach up to the face-blind ejection handle. He also had a second firing handle in the front of the seat, between his legs, which he could use if excessive g-force or injury prevented him from reaching up. Having ejected once already, he was fastidious in making sure he could easily get to the handles, 'even though I never really thought I'd ever need to use them again'. The last thing he did was to remove both safety pins from the two firing handles. His seat was now live. He would reinsert them after landing, to make the jet 'safe for parking', before climbing out. He was ready to go.

Taxiing out to the runway, both Lightnings blasted off the tarmac and headed out to the North Sea and the exercise area. At 25,000 feet, 60 miles east of Flamborough Head on the Yorkshire coast, Charlie keyed the radio button. 'Chris, you're ready to graduate to real aerial combat manoeuvres.' It was dogfighting time.

The two Lightnings started with a 'visual split' – initially flying parallel, each turned 45 degrees outwards, heading away from one another. Neel banked hard right, Coville went to his left. After a set period of time, when enough distance had built up, they turned back towards each other, squaring up like boxers, primed and ready to charge in for the fight. Who would get into position to 'fire' the first shot? Who would blink? Coville reckoned he was 'in for a real walloping. After everything he had accomplished in Vietnam, Charlie was clear favourite to win.'

But Neel, uncharacteristically, had made a mistake. He suddenly found himself dazzled by the setting sun and was having real problems maintaining visual contact with his opponent. 'Sensing this, I pulled high,' Coville recalls. 'It was obvious that he had lost contact. I gradually worked my way round towards a firing position in his six o'clock. Boy, this would keep me in beer for a long time!'

Coville was jubilant, he had Charlie Neel, a combat-proven fighter pilot, at his mercy.

'Are you still visual?' Charlie drawled.

'Affirmative.'

Then Coville did a foolish thing. He gave away his own position. 'I'm in your six o'clock.'

Neel laughed. 'Got you, you little bastard!'

He hauled his aircraft into an incredibly tight, aggressive turn to the right in an attempt to counter his younger opponent's tactics.

Coville was shocked. 'I thought Charlie's manoeuvre had to be beyond the aircraft's safe flying capabilities.'

This turned out to be the case. 'Almost immediately I saw the nose of Charlie's Lightning pitching up violently and going into a spin.'

Neel's manoeuvre had been so aggressive that he had lost control and the aircraft was no longer flying. 'My jet was now just a large lump of metal plummeting towards the sea.' Coville tried to keep Neel in sight. 'Then he went into layered cloud and disappeared from view. I tried a number of radio calls but all I heard was a single strangled gasp. And then nothing. Had he recovered from the spin? Had he ejected or had he stayed too long and crashed into the sea? Was Charlie alive or dead?'

Neel wrestled to regain control, but knew it was hopeless. His aircraft was now spinning upside down, and he was hanging in his straps. 'I couldn't believe I was going down again. I simply thought, *Charlie, there's no more point trying to save one of the Queen's aircraft. You've been shot down before and nearly died.* As I spun downwards through 10,000 feet it became apparent I'd need to take the ejection ride. Again.'

Charlie Neel reached for the ejection handle for the second time in less than two years. 'I was still inverted, hanging with my butt a few inches off the seat. I just knew when it went off it was going to hit me hard. And that it would hurt. But I still pulled that darn yellow and black handle as hard as I could.'

A charge immediately blasted the cockpit canopy clear of the aircraft. If it failed to jettison, the seat gun cartridges would not fire in order to prevent catastrophic, possibly fatal, head injuries. In that event, he would have had to pull the canopy lock to jettison it manually, and then fire the seat again.

But everything was working perfectly. As the canopy vanished, the slipstream slammed into the cockpit. Just 0.6 seconds after pulling the handle, the main ejection charge launched Charlie's seat up its guide rails at around 80 feet per second. His leg-restraint cords tightened as the seat rose, holding them firmly in place, and the seat's PEC unit detached from the aircraft, activating his own emergency oxygen supply. Almost immediately, the two secondary charges fired, propelling him, at around twenty times the force of gravity, out of his doomed Lightning. Half a second later he was safely clear, and the drogue gun, mounted on the seat's left side, fired out two small parachutes to slow and stabilise its descent, reducing his forward speed until the main parachute could open without risk of bursting.

After around 1.25 seconds, the final sequence began. A plunger released the scissor shackle holding the drogues which then played out the main parachute lines from the back of the seat. At the same time, Charlie's own harness and leg restraint connections were opened, his PEC was disconnected from the seat and he was pulled free. He felt a heavy upward jerk as his rapid descent was halted and he began to float down under the blossoming parachute canopy. The whole sequence had taken around three seconds, but all Charlie recalled of the event was that 'it hurt like hell, as though I'd been hit by a giant locomotive!'

In some ways, that's exactly what had happened. Because Charlie had been inverted and hanging in his straps, he was not in physical contact with the seat when he pulled the handle. Even though he was separated by mere inches, as the seat rose at 80 feet per second it had hammered into his backside and – as he would later discover – the upward force had caused compression fractures at T11 and T12 in his upper spine.

Descending under the parachute and in considerable pain, he reached behind his lower back and squeezed the clips to release his survival pack. It fell around 15 feet, attached to his life jacket by its lanyard. With memories of his fall into the Gulf of Tonkin after his first ejection a couple of years earlier, Charlie inflated his life jacket so that the yellow stoles surrounded his neck, then hauled his survival pack back up, inflated the life raft and dropped the fully deployed orange dinghy, allowing it to swing beneath his feet. It was not the *precise* procedure taught by the RAF, but this time Charlie was determined to be prepared. 'I knew that the icy waters of the North Sea were going to offer a very different proposition for survival to the hostile waters off the Vietnam coast.' Five minutes later, he splashed into the waves at a sedate 22mph, fully aware that, like his friend Guy Gruters, he'd been lucky to escape death a second time.

It was now almost total darkness, not ideal conditions for a North Sea rescue. Charlie pulled over his already inflated life raft and scrambled aboard, covering his shoulders with its day-glo canopy to take the edge off the cold. He reached into his life jacket and fished out his locator beacon to ensure it was transmitting its homing signal. Rising and falling in the heavy swell, he waited. Alerted by Chris Coville's radio calls, a Search and Rescue helicopter had already been launched. Around thirty minutes later, Charlie heard the beat of rotor blades in the distance and ignited a flare from his survival kit. The water around him glowed red, then began to froth and swirl. The rescue helicopter winched him onboard as it hovered overhead. Soaked and shivering, his back aching badly from the kick up the pants delivered by his ejection seat, he was dropped back at Binbrook. The medical centre referred him to a specialist for further scans, which revealed the extent of his injuries.

A Board of Inquiry was set up by the RAF Directorate of Flight Safety to establish the cause of Charlie Neel's accident. Convened after every such mishap, they had sweeping investigative powers, and could take months, in some cases years, to reach a potentially career-defining conclusion, which was thus keenly awaited by all involved. The board judged that although Charlie had mishandled his aircraft, the error had been compounded by turbulence caused by Coville's Lightning, leading to him spinning out of control with no choice but to eject. Neel joined 2,562 others who owed their lives to a Martin-Baker ejection seat, and was the twentieth Lightning pilot to use the Mk4B – writing off a £1 million (£20.6 million today)[9] aircraft in the process. To add insult to the back injuries which would ground him for nearly two months, the board also took issue with Charlie's use of his USAF flying helmet and boots. A very senior RAF officer made his displeasure extremely clear; flying kit in Her Majesty's Royal Air Force was *not* a matter of individual taste.

Charlie took it all in his stride. 'Because of the fractures, my back still looks like a bullfrog today. But I had survived again and I was truly grateful for my ejection seat. I'd been given a third chance at life.'

* * *

In Vietnam, Guy Gruters was fighting a very personal battle to establish if *he* was still grateful for *his* ejection and survival. After Lance Sijan's brutal death, he had begun to truly loathe his captors. The strength of this emotion surprised him. The hatred was consuming him. He wanted them dead. 'All the POWs in the Hanoi Hilton were hanging onto life by their fingernails. It felt like we'd be there a lifetime. That I would never get out alive. I presumed we would die in prison, beaten and tortured to death like Sijan. I had begun hearing voices in my head telling me to take my own life. I realised that I was disappearing down a dark hole.'

Then came a turning point. A few months into his ordeal he got down on his knees in his cell and prayed to God for help.

'Although I had been born and brought up as a Catholic, my faith was not particularly strong.' Guy had never been a regular churchgoer. He liked a drink and a smoke, as did most of his fellow pilots. 'I realised that if I wanted to get back to Sandy and the kids, I needed to stay strong. It wouldn't be fair on them for me to quit and take my own life. I had to fight that hate. That's when things began to turn around for me. I stopped hating my torturers and began praying for them. Life as a POW became somewhat more survivable.'

He and his fellow prisoners still suffered terribly. 'We were regularly beaten and put in stocks and manacles in solitary confinement. The living conditions were horrendous.' Bread and water twice a day. The bread was full of rat droppings. 'If you dug out all the rat excrement you'd have nothing left to eat.' The loaves also contained worms, some of which were alive. 'When we bit into it, the worms would bite us. We killed the worms and ate them. They were the only nutrients we had.' At least the worms in what little water they received were dead because it had been boiled.

The 'toilet' or the 'Hanoi Honey Bucket' sat in the corner of the cell. 'That bucket would not be large enough for two men so we had raw sewage in our cell all the time.' There was precious little toilet paper, Guy says, 'so we had to use our hands, which we then tried to clean by rubbing them up and down the rough walls'. The stench in those airless, windowless concrete boxes, which were freezing in winter and stifling in summer, was unbearable. Guy would lie flat and attempt to breathe the fresher air that came in through the tiny gap at the bottom of the door, which was also the only source of natural light.

Not surprisingly, disease was rife. Guy suffered dysentery for nearly two and a half years and also caught dengue fever, a mosquito-borne tropical disease, which he thought was going to kill him. Maltreatment and torture were a regular part of life for the prisoners. Meat hooks on the ceiling were utilised as a favourite form of torment. Prisoners were hoisted up on their arms until their limbs turned purple and swelled to twice their normal size. Others were strung up with their hands bound to their ankles and the ropes were tightened until they could not breathe. Many begged to die. Others were forced to kneel on gravel until the skin was ripped from their knees. The deep cuts from the leg irons and the stocks caused infections that went untreated. Often unable to use a latrine or bucket their soiled bodies were a magnet for rats and cockroaches. Waterboarding and prolonged solitary confinement were also part of the repertoire.

To go from a busy military life, where every minute was accounted for, to the relentless monotony of incarceration was yet another form of torture. Guy was allowed to receive or write only a handful of letters. At their home in Florida – a place he often dreamed of – Sandy was writing to him regularly, hoping to keep him up to date with family news and major events like the assassinations of Martin Luther King Jr and Robert Kennedy in 1968, or Neil Armstrong's 'one small step for man, one giant leap for mankind' on the surface of the moon the following year.

The prisoners had found a way of communicating with each other by tapping messages on the walls, using a form of Morse code, not least to broadcast the latest results of the 'how long is your tapeworm?' competition. Guy's cellmate proudly held the record after pulling a 32-footer out of his backside. It helped maintain morale. But it was his rediscovered and deepening faith that was really getting Guy through. 'It gave me great comfort over the next few years.

The cruelty and brutality were always there but I had no doubt that I was in God's hands.' Another thought made him smile. Before he had been shot down the second time, he'd been told that Charlie Neel had been chosen for the RAF/USAF exchange programme. In his darkest days as a captive, he imagined how Charlie must be having the time of his life flying a Lightning. How he envied him.

Little did he know.

* * *

As the Vietnam War wore on, development of ejection systems designed for combat situations continued rapidly. For good reason. Increasing numbers of American aviators were being shot down over enemy territory. Most of them, like Guy Gruters and Bob Craner, were captured. Studies of a one-year period of air action showed that 70 per cent of those pilots and crewmen who ejected within 5 miles of being hit were captured. One hundred per cent of those ejecting within 3 miles of being shot down ended up in enemy hands.[10]

Navy and Air Force experts were working desperately with industry to develop an ejection seat that could be 'flown' by the pilot after ejection, long enough to cross into friendly territory, or to find a safer landing site from which he could be rescued. They had come up with some truly astonishing adaptions to the 'standard' seat.

The simplest was the Bell Aerosystems 'Rogallo Wing'. After the pilot ejected, instead of pulling out the main parachute, the drogue parachute would deploy a triangular-shaped, canvas-covered 'wing' stowed in the rear of the ejection seat. It extended and snapped rigid with an action similar to opening an umbrella. Around 7 feet wide at the rear and 13 feet long, once the wing was fully deployed a small engine mounted in the back of the seat started automatically and could be controlled by a throttle on the armrest. The Rogallo Wing looked not unlike a motorised hang-glider and was fully operational in six seconds. A simple control next to the pilot allowed him to make the basic movements of turning, climbing and descending. But with the wing fixed to the back of the now-horizontal seat, and the airman still strapped in as normal, it meant he would be flying the device facing the ground, held in by the harness. Not a particularly comfortable sensation, and early tests demonstrated considerable 'unease' about the positioning. Though, clearly, any discomfort or fear would be countered by the realisation that the pilot was hopefully avoiding capture. Of course, they could not land the contraption while still fixed to a heavy seat; once clear of enemy territory, they would still have to release the harness, fall away from the seat and then parachute down.

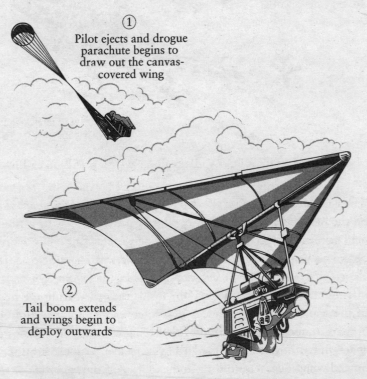

① Pilot ejects and drogue parachute begins to draw out the canvas-covered wing

② Tail boom extends and wings begin to deploy outwards

BELL AEROSYSTEMS ROGALLO WING

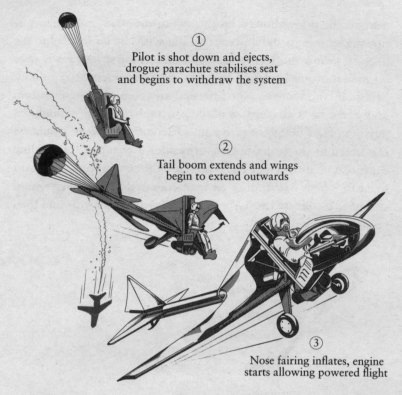

① Pilot is shot down and ejects, drogue parachute stabilises seat and begins to withdraw the system

② Tail boom extends and wings begin to extend outwards

③ Nose fairing inflates, engine starts allowing powered flight

FAIRCHILD STRATOS CORPORATION ESCAPE SYSTEM

The Fairchild Stratos Corporation came up with another variation on the 'wing' concept, but it looked very different – a tiny aircraft held within the ejection seat itself. The drogue parachute would deploy as usual after ejecting and once the seat had slowed sufficiently it would deploy a telescopic 14-foot tail boom, consisting of a number of sections nested one inside the other, from the back of the seat. The spring-loaded, plastic-covered tailplanes and fin, housed in notches in the boom, then popped out, and a series of spars, hinges and cables pulled all these elements taut. Aluminium hinged wings began unfolding and extending from each side of the ejection seat and locked into place, and the engine beneath it started to idle. Two small wheels beneath the seat snapped outwards to form landing gear. A rubberised nose fairing then inflated around the pilot's legs to complete the 'fuselage'. The ejection seat transformed into a small, fully aerodynamic 'aircraft' with a wingspan of 16 feet in eight seconds. Using an integral armrest control stick, the pilot could manoeuvre the craft and control the engine, flying away from the combat zone. Although it was technically feasible to land the machine, it was envisaged that most pilots would simply unstrap and parachute to safety.

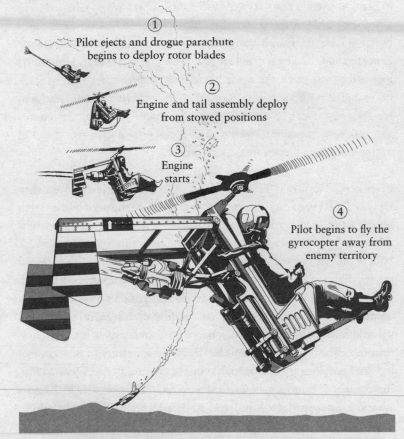

① Pilot ejects and drogue parachute begins to deploy rotor blades

② Engine and tail assembly deploy from stowed positions

③ Engine starts

④ Pilot begins to fly the gyrocopter away from enemy territory

KAMAN AEROSPACE CORPORATION GYROCOPTER

Designed for the US Navy, the Kaman Aerospace Corporation's gyrocopter was the most James Bond-like of the current crop of inventions. Christened the SAVER – Stowable Aircrew Vehicle Escape Rotoseat – it was the only system designed to take a pilot all the way to the ground without having to separate from it, although he still had the parachute option.

Four seconds after the SAVER's ejection seat had fired out the pilot and the drogue had stabilised it, a 15-foot, two-bladed helicopter-like rotor stowed in the back of the seat was pulled out and locked in place above his head. Two parallel, fin-tipped arms – shaped to fit along each side of the seat – then hinged themselves upwards until horizontal behind the airman at shoulder height. The turbofan engine beneath them, fed by jet fuel from a tank under the seat, ignited. Its forward thrust made the rotor blades spin (autorotate) and the stick between the pilot's legs gave him full flying control. Airman and seat weighed just over 700 pounds, and now resembling a modern auto-gyro aircraft, it gave an ejected pilot a machine with a top speed of 115mph and an escape range of 57 miles.

Elsewhere, Goodyear was developing the Pilot Airborne Recovery Device (PARD), to carry aircrew to safety. Once the pilot had ejected, his parachute deployed and seat jettisoned in the usual way, a 37-foot-diameter balloon then inflated above the conventional parachute. A butane burner housed in its rigging then ignited to heat the enclosed air, just like a hot air balloon. The occupant wore the butane fuel pack on his back, which fed the burner through a hose in one of his para-chute lines. He could ascend or descend by operating a fuel control valve, but any direction of movement was obviously at the mercy of the prevailing wind. Testing of the device demonstrated a float time of half an hour, keeping the occupant out of the range of small arms fire. Rescue would be effected by specialised retrieval aircraft which had a system to fly over the balloon, snagging it up and hauling in the pilot.[11]

All these machines were as remarkable as the first ejection seats and offered ever-increasing means of survival and escape from both stricken aircraft and hostile territory. Drawing board to live test-ing was one thing, but would they make the grade in time to save aircrew lives in Vietnam? As far as the politicians and the military were concerned, peace talks were now a possibility. Maybe the end was in sight.

CHRISTMAS EVE, 1969

Two years after Guy Gruters' second ejection, Sandy left their young daughters Mari and Sheri with Guy's parents and boarded a flight to Paris. Sandy was now a member of the National League of POW/MIA Families, formed in 1967, and it was the first time she had ever left the country. Service wives had suffered enough from government indifference and secrecy and had created the group to ensure their loved ones were not forgotten. Now talking openly and stridently to newspapers, TV and radio – and politicians – about the prisoners and those missing in action, they were carrying their mes-sage across the nation, sharing their stories. Many families, Lance Sijan's included, still had no idea if their loved ones were alive or, in his case, dead. Few knew of the brutality the men were enduring and the campaign for openness was gaining traction.[12]

Sandy and 150 family members of American POWs now hoped to lobby the North Vietnamese delegation at the Paris peace talks. Their trip was financed by the Texan tech billionaire Ross Perot, who would make an unsuccessful bid for the White House in 1992. Sandy and the other wives were to be disappointed. 'We tried to meet the North Vietnamese, but they refused to see us. So we went to Notre Dame Cathedral instead and prayed for our loved ones. Then simply returned home on Christmas Day evening. We had been on the ground in France for just six hours.'

Back in New York there was a hotel reception for the Paris group and a presentation about the campaign. Santa Claus arrived to give presents to some of the children. 'It was a picture of normality amid all the chaos and hurt, and I was just overcome with grief and loss. I burst into floods of tears. Guy was a prisoner, I was in New York, the children were in Florida, the trip to Paris had been fruitless, I'd been on a plane for twenty-four hours and it was Christmas Day.' Santa came over, held her hands and asked if she was okay. 'Then he lifted up his beard and said, "Honey, you need a drink." Which I certainly did. It was a wonderful moment and he really lifted my spirits.'

Even though Sandy did her best to protect her daughters, Mari, who turned six in 1970, was beginning to understand that her father was a prisoner of war and what that meant. 'I knew that Dad was in a bad place and I was being shielded from the reality, but the sense of the anguish the adults clearly felt filtered through to me. As the years dragged on, we had no idea when or if he would come home. An older child had warned me that he might be dead. They were not being cruel, just realistic.'

Veteran journalist and TV broadcaster Walter Cronkite was a nightly visitor to the Gruters household, as he was to living rooms across America throughout the war. 'We saw the conflict playing out on our TV screen,' Mari says. 'We were watching it unfold as distant images of death, destruction and suffering, but were also living a personal, horrific reality.' Some mail from Guy had at last been allowed through. 'It made me cry to see the tiny letters my mother received. She would sometimes cry and explain that our daddy loved us but that he couldn't really tell us what was happening out there.'[13]

School, piano and dance lessons, trips to the beach and family parties at weekends brought a semblance of normality for the girls, but the war was never far away. Many POW/MIA families saw each other regularly. There was a lot of fun and games for the kids but Mari always sensed the sadness, especially around those who had lost fathers and husbands. 'I remember being with John McCain's then wife Carol and her sons at one of these get-togethers.' McCain (who would later make a run for US President) had been shot down in October 1967, and on ejection he fractured both arms and a leg, and nearly drowned when he parachuted into a lake. His captors refused to treat his injuries and subjected him to the same brutality as Guy Gruters and Bob Craner. 'Nothing much was said about the prisoners, or our personal situations. We all just knew we had a shared suffering. And Carol McCain was a wonderful lady. I really looked up to her. She would never quit when fighting for the POWs.'

Sandy Gruters was equally determined to see it out. There was not a moment when Guy was not central to her thoughts. She was convinced he was still alive. 'I always felt in my heart that if he was dead, I would know it.'

* * *

Charlie Neel was not so sure. He still had almost no information about his friend in Vietnam, and his own time in England was coming to an end. He had written to the USAF pilot who would take his place in the exchange programme.

> *Dear Bill,*
> *Welcome to the Royal Air Force. Come prepared to enjoy yourself, to take it easy and to fly a really hot ship.*[14]

Captain William Olin Schaffner was an instructor with a combat training squadron in Texas. Born in Ohio in 1941, his flying pedigree was impressive. Early in his career his boss, Colonel 'Chuck' Yeager, the legendary test pilot and first man to break the speed of sound, was unstinting in his praise. 'This ambitious and highly capable young officer shows outstanding career potential. I recommend his promotion well ahead of his contemporaries.'

In 1967 Schaffner completed a two-year tour of Vietnam, having flown 290 combat sorties on F-102 Delta Daggers. Like Guy Gruters' wife Sandy, Bill's wife Linda had become an avid TV news watcher during her husband's absences. 'Seeing reports of casualties I began to understand the dangers he faced. Bill was an intensely focused man, proud to be a fighter pilot. He was thrilled when he got the news that he had been selected for the USAF/RAF exchange programme. He was always striving to be the best aviator possible. He was military through and through.'[15]

In his letter, Charlie promised to add Bill Schaffner's name to the housing list and gave him a thumbnail of the training he would go through prior to joining 5 Squadron, most of it good.

> The bad thing about the tour is that you won't get much flying time. Generally the blokes don't get in a big sweat about anything. Your biggest problem will be slowing down.

He warned Schaffner that he would lose his suntan, but

> the good things are innumerable. Great airplane, interesting and fantastic people and places. Lots of good parties and formal dinners; an all-round enjoyable experience.

Charlie also had some sartorial tips.

> The general standard of dress is a bit more formal. You will need a suit or two, as opposed to a sport coat. And bring your mess dress and a black dinner jacket. And don't bother to bring a bunch of USAF flying clothing.

He made no mention of his recent ejection faux pas before adding, 'I wear my USAF boots, helmet and flying jacket quite a bit around the Squadron.'

He signed off:

> Congratulations on your assignment. We are looking forward to seeing you 'on the town' in London.
> Cheers, Charlie.

CHAPTER NINE

THE 'KNOCK AT THE DOOR'

RAF Binbrook, Lincolnshire
8 September 1970

With his reddish hair and healthy tan, 5 Squadron's powerfully built, though diminutive, new USAF exchange pilot stood out even before his smile revealed a mouthful of brilliant white teeth. Bill Schaffner was less ebullient than his predecessor but could hold his own with the best of them, both on the ground and in the air. Welcomed to Binbrook by the Squadron Boss Alan 'Chalky' White, Bill and Linda Schaffner moved into a house on the base with their three small boys. David was six, Glennon five and baby Michael, born just seven months earlier.[1]

The Schaffners settled into Binbrook quickly and happily. 'We both loved being in the UK,' says Linda. 'Everything about it, the villages, and the countryside, and especially the people who were so wonderful and friendly. Early on, the whole Squadron came over to our house for a spaghetti dinner; there was so much laughter and fun. It was incredible to be part of this RAF family.' The Schaffners' next-door neighbours, Chris and Irene Coville, were already firm friends. Chris, who had been on the sortie where Charlie Neel had ejected for the second time, could tell that this new USAF exchange officer would fit in well. 'Bill adapted to squadron life incredibly quickly,' says Chris. 'He was clearly a very experienced and capable

pilot, although new to the Lightning. He was also extremely generous with his duty-free booze.'

'We were such a close group on 5 Squadron at that time,' remembers Irene. 'We were in our twenties, deep in the countryside,
socialising together regularly. The Schaffners were charming people
and loved by everyone. Bill himself was tanned, good-looking and
always smiling. We were all living life to the full.' But alongside the
parties, their life on a remote Cold War airbase revolved around the
ever-present threat from Soviet attack. All personnel, on the ground
and in the air, were required to be fully prepared for any eventuality,
and that level of preparedness was regularly tested to breaking point.

On 8 September 1970 the day began routinely enough for Linda.
'It was three days before Bill's twenty-ninth birthday. After he left
for work in the morning, I saw the two older boys off to school
then got on with my chores around the house, while looking after
baby Michael.'

Then, everything changed. An echoing siren earlier that day
had announced the arrival of a NATO Tactical Evaluation team
and warned the Squadron to be at 'readiness for war'. Tactical
Evaluation exercises, 'TACEVALs', could be called without notice
to test a squadron's readiness in the face of a whole range of eventu-
alities, designed to represent the first days of an escalating conflict
with the Soviet Union. Alan White briefed his pilots and ground
crew. The first challenge was to prepare for the predicted onslaught
from Soviet bombers attacking the UK. 'There was a rush of activity to get the maximum number of Lightnings serviceable and live
missiles out of the bomb dump and onto the aircraft.' Across the station, personnel were readying its defences and every available, fully
armed and fuelled jet in preparation for any incoming assault. When
that first ground phase of the exercise was completed satisfactorily,
the weapons were removed. Phase two would take place in the air.[2]

Pilots and ground crew waited for the evaluators to feed in war
scenarios designed to seriously challenge their capabilities. The
atmosphere was tense, but they didn't have to wait long. British

aircraft posing as Soviet bombers were out over the North Sea, threatening to stray into territorial waters. The Squadron's challenge was to find them, shadow them and warn them off, or, if necessary, practise shooting them down. 'Enemy' aircraft would suddenly change course, ignoring all instructions to keep away. This was cat-and-mouse at speed.

The demand on pilots was relentless as the evaluators kept tweaking the scripts and ramping up the pressure to wrong-foot them. Lightnings roared airborne in rotation to challenge 'the invaders' then headed back into Binbrook, tyres squealing, to refuel. They had to be 'turned around' – readied for flight – at full operational speed and sent straight back out again. Timings were critical. Stopwatch stuff. Slackness was marked down. The ground crew needed to perform like a Formula 1 pit team if the Squadron was to make the grade.

Bill Schaffner had only been at Binbrook eight weeks and was rated 'Limited Combat Ready' even though he had notched up 121 flying hours. This meant that he could intercept the enemy, but had yet to complete training for the more challenging task of shadowing hostiles that had strayed into British territorial waters or shepherding them in for questioning. Shadowing and shepherding: flying a few hundred yards from a suspicious aircraft at any altitude, in all weathers, day and night, in or out of cloud, and with lights on or off, taxed even the most experienced pilots. Those deemed not fully combat ready usually had to make do with support duties during evaluation. But it was all hands on deck for this exercise, so as a highly experienced fighter pilot, Schaffner, alongside his friend Chris Coville, was stood down after the initial call to readiness and told to reappear for the night shift.

Back home, Linda had made Bill his favourite lunch of fried chicken and mashed potato. 'We ate together at the dining table. Afterwards, Bill relaxed in the living room, still in his flight suit, lying on the couch and holding baby Mike on his chest. He held him up and stared intently at him for a long time. I watched them both, wondering what he was thinking. It was unusual, because Bill had always assumed a more "traditional" role with the children and, like many men of his generation, was not so open with his affection and feelings.' Bill had lost his own father when he was only twelve years old.

Late afternoon, a car drew up outside their house. It was time to go back to work. As he left, Linda said, 'Bill, I do love you, and I'm sorry.' They'd argued a few days before and were still making up. He turned back at the open door. 'I'm sorry too. I don't know why you love me.'

'I watched Bill as he walked down the path to where Chris was waiting to take him over to the Squadron. He got in and they drove off.'

'It was a pleasant evening when we reported back,' Coville remembers. He and Schaffner walked into the sparsely furnished crew room and bagged a couple of ageing chairs. There was a bar, which dispensed nothing stronger than black Nescafé. The walls were adorned with the usual array of plaques and photographs presented by visiting RAF and overseas air force units. Through the window they could see the apron where a line of silver Lightnings glinted in the high-intensity sodium lighting, waiting to meet the next round of challenges.

The pressure had been mounting during the day. The strain on both men and machines was painfully obvious. Schaffner and Coville were chewing the fat when Alan White gave them news they had not been expecting. They were to prepare to scramble. After the way the exercise had already progressed, White believed that Schaffner would not be involved in shadowing or shepherding enemy aircraft at low level. 'Bill still needed to complete a few sorties before being formally declared operational, but I assessed him to be competent to take part in the flying phase.'[3]

At 7.34pm Schaffner was ordered to his fighter, XS894, as the weather began to deteriorate. He strapped into his ejection seat and checked the lanyard to his personal survival pack was fastened in place. Sitting stationary on a taxiway and having to wait was what pilots hated most. He was finally scrambled at 9.25 and allocated the call sign 'Charlie 45'.

He double-checked his ejection seat was live and the safety pins were correctly stowed, then swung his Lightning out of the dispersal area and blasted off the runway in a sheet of flame from the twin tail pipes. Three minutes and five seconds later he checked in with the Master Radar Station at RAF Patrington, who would be controlling the interceptions.[4] He climbed to 10,000 feet, still unaware of the type or altitude of the simulated attacker he was about to target.

The tactical evaluators suddenly changed the script. They now wanted a shadowing and shepherding operation – something Schaffner had not been cleared to perform. It was now dark, and he had only notched up eighteen hours of night flying on the Lightning. A four-engine Avro Shackleton, posing as a Soviet bomber, was heading towards the Yorkshire coast. Speed: 185 mph. Altitude: 1,500 feet. The slowest a Lightning could be flown in this type of operation was around 230mph.

Bill radioed his Radar Controller at RAF Patrington at 9.33pm. The radio transmissions were all being recorded.[5]

'Patrington, Mission 45 airborne.'

'I have you contact,' Patrington responded as he saw Bill appear on his radar screen. 'Target is north-west of you, range 35 miles.'

At 9.33 and fifty-five seconds Patrington called, 'Buster, Buster! Target range 28.'

Buster was the command to increase to maximum speed towards the 'target', which was now just 28 nautical miles away. Bill was tasked to replace another Lightning that had to break off and return to base. Then the NATO evaluators changed the challenge again. The enemy was showing signs of defecting. Schaffner's job would be to intercept and, if necessary, escort it to Binbrook. This was a serious technical challenge for an inexperienced Lightning pilot.

He dropped to 5,000 feet.

Two minutes later, Patrington came back on. '45, make speed decimal 95, over.'

He was being told to increase to 0.95 Mach, 728mph. The

Lightning had phenomenal acceleration. In seconds it could be travelling at well over 12 miles a minute. Schaffner couldn't conceal his surprise.

'Roger? That's pretty fast.'

The phoney Soviet bomber was now 12 nautical miles off Flamborough Head.

At 9.37.30 Schaffner reported, 'Visual contact with a set of lights in that area.'

'The target speed is 185mph.'

Schaffner would be on top of it in under a minute. He could see the Shackleton's navigation lights. Over the radio his voice sounded distorted and strained, the effect of g-force on his body and the challenges he was facing. Two minutes later he reported 'contact with two aircraft'. The other was the Lightning he was due to relieve, which had been on station for an hour.

At 9.39.14 he radioed its pilot. 'I'm going to have to do some manoeuvring to get this speed burned off.'

Bill needed to perform some high-g S-turns to lose momentum while staying in visual contact and not losing altitude. The other Lightning pilot reminded him to lower his flaps for greater control at very low speed.

'I got 'em down, babe.'

'Roger, 45.' His colleague pulled his Lightning away. 'Turning port at this time.'

The Shackleton crew were unaware of the radio calls as they were separated from the mock intercept on a different radio frequency. They saw the first Lightning's flashing white lights turn away, to be replaced by Schaffner's own.

Patrington reminded Schaffner to keep a sharp lookout.

'Roger, I'm watching 'em,' replied Bill.

The night was pitch-black, the weather had worsened, and Schaffner probably could no longer see the horizon. The workload in the cockpit was reaching breaking point as he concentrated on shadowing the slow-moving aircraft.

At 9.41.27, Patrington updated him.

'*Estimated range from the coast now 5 miles. If he comes within 3 miles he is to be escorted to Binbrook.*'

Bill responded a few seconds later, at 9.41.35.

'*Roger.*'

It was just eleven minutes and thirty-five seconds since he had blasted off from RAF Binbrook. The Shackleton saw the lights of Bill's Lightning flash once and then disappear.

At 9.42.20, the instruction changed again.

'*If target aircraft approaches within 3 miles of the UK coastline, he is to be directed to land at Waddington. I say again, directed to land at Waddington.*'

There was no response.

A few more seconds passed. Bill Schaffner still did not reply.

'*45, be advised you are dark to me at this time.*'

The controller could no longer see Schaffner on his radar screen and was growing increasingly frantic. A succession of urgent radio calls rang out.

'*45, you are dark to me at this time, check target's heading and your own . . .*'

'*45, this is Patrington, nothing heard . . .*'

The Chief Radar Controller came on the radio.

'*Charlie 45, Charlie 45, this is Patrington, do you read? Over.*'

'*Charlie 45, Charlie 45, do you read? OVER.*'

'*Charlie 45, Charlie 45, this is Patrington . . . over.*'

Only the hiss and crackle of static filled the radio.

Still operating on a different radio frequency, the Shackleton pilot was unaware of the unfolding drama and simply presumed that the second Lightning had also been ordered to return to base. Then his own Radar Controller called him. '*The Lightning aircraft exercising with you has disappeared from radar. Request you carry out a search in your immediate area.*'[6]

A few minutes later, Chris Coville was scrambled from Binbrook with the simple order: '*Search for any sign of Charlie 45 visually, or on the emergency radio frequency.*'

A lifeboat and an RAF Marine Branch rescue vessel were also alerted. The Shackleton was leading the search and dropping flares but Coville was unable to see anything in the inky water. 'The penny began to drop. Something terrible had happened to Bill. It was obvious that he had crashed into the sea, but there was still hope. If he had ejected and triggered his survival beacon – like Charlie Neel before him – it would transmit a signal to allow rescuers to home in on him.'

There were no bleeps from Schaffner's survival radio. Coville returned to Binbrook with a heavy heart. He had come to regard Bill as a close friend and 'the chances that he had survived any crash were quickly reducing'. With every passing minute, his feeling of dread increased. 'I landed to see the padre in the crew room, never a welcome sight on a fighter squadron. My boss Chalky White and the Station Commander both turned to me. "Nothing?" they asked. I shook my head. The Station Commander looked at Chalky grimly, saying, "Let's go." They collected the padre. The trio that no wife ever wants to see on the doorstep set off to where Linda Schaffner was totally unaware that her life was about to change for ever. It was time for the *knock at the door*.'

Nearby, on the 'married patch' of military quarters, Chris's wife Irene was enjoying a drink with a friend at the home of Esme White, the Squadron Commander's wife. A regular gathering, the mood was light and happy as they laughed and swapped stories of their day. 'I heard the sound of someone coming into the house but I didn't see who it was,' Irene recalls. It was Alan White, who had stopped by to inform his wife what had happened before heading to Schaffner's home. 'Esme went out of the room and I could hear anxious talk. A few minutes later, she called our friend to join her. A sense of dread began to rise in me. They returned looking really distressed and told me that an aircraft had gone missing during the exercise. Instinctively I put my hands to my face in shock. *Was it Chris?*'

'"No, no," they said. "It's not Chris."'

Esme told Irene they had to go out immediately, but urged her to stay and finish her drink. 'Although I was truly relieved it wasn't Chris, they didn't stop to tell me who it was. All I knew was that a fellow wife would soon be receiving the dreaded "knock at the door", and some truly terrible news.' A few streets away, the Station Commander and the padre were already arriving at the Schaffner house.

Irene waited for ten minutes. When they did not return she decided to go home. As she moved towards the exit, the Whites' bad-tempered basset hound gave a low growl and barred her way. 'He obviously thought I was an intruder and would not let me by.' It was a strangely comical experience amid the tragedy of the night. 'I tried talking nicely to him but he growled even more fiercely. In the end I had to climb, like the burglar he obviously thought I was, through the lounge window to escape. Outside, the station was deathly quiet. No aircraft noise, no movement, just silence. It felt very peculiar.' Irene sent her babysitter home and waited for Chris to return. 'I can't remember how long I sat there wondering what had happened. And who was dead.'

* * *

Having put the children to bed, Linda Schaffner had settled down in front of the TV in her nightclothes, watching *A Summer Place*. The 1959 romantic drama had a memorable instrumental theme tune by Percy Faith and his orchestra which had spent many weeks topping the US charts. She was about half an hour into the movie when she heard a car draw up outside. 'I went to the window and saw a man in RAF uniform and a priest getting out. They were coming to the house.'

Oh no, not me.

There was a firm knock at the front door.

Linda turned on the hall light and opened it gingerly. The Station Commander and the padre stepped inside, their faces etched in sadness. Five-year-old Glennon, who had not been sleeping because of the wind in the eaves, heard male voices and came down from his bedroom to see what was going on. 'I can still see this image in my mind today,' he says. 'Two dark figures backlit by the outside light. My mom told me to go back up but I sat on the stairs looking through the banisters.'

Linda was not taking in her visitors' message. It was all a blur. They kept repeating the word 'missing'.

'What did it mean? Bill couldn't be missing. I had only seen him off a few hours earlier.' Their tone became more urgent and, suddenly, Linda understood all too well.

Glennon saw his mother fall backwards against the hall cabinet and the two men grabbing to catch her. 'As she collapsed to the floor they knelt next to her, trying to help her up. She seemed distraught. I was so scared. What had happened to my mom? I simply didn't understand what it all meant. I ran back upstairs to my bed.'

Linda remembers little of the hours that followed. Alan White and his wife Esme arrived. 'They tried to explain what was happening but I couldn't take it in. It was as if it was all happening to someone else and I was a spectator. I was stunned and distraught. The RAF doctor arrived with a box of Valium to calm me down.' But nothing could take the edge off her mounting grief. 'I was in total shock. I felt numb, wounded. When was this all going to be over? I had been told Bill was *probably* dead. But what did *probably* mean? How could they know? I stayed up most of the night, hoping, praying, that they would find him. Every time I heard a noise outside, I rushed to the window to see if it was him.'

Chris Coville headed home later that night, despondent at the loss of his friend. 'I tried to walk past the Schaffners' house undetected, so Linda didn't think I might be Bill returning. To avoid any sound of my boots on the pavement, a sure sign of returning aircrew, I walked on the grass verge on the other side of the road.' Nonetheless, their lounge curtain was pulled to one side. 'I saw Linda's frightened, tearful face staring out in forlorn hope that some miracle had happened, and it was Bill returning. It was a poignant image. One I can still clearly see in my mind today;

it was an image that echoed around many RAF bases in those years.'

Chris's wife Irene was waiting anxiously. 'Have they found him?'

'I shook my head. "Bill's dead." I then poured myself a very large whisky.'

Discussing the events of that day over fifty years ago, Chris retrieves his battered RAF logbook for the author and thumbs through it. He reads the entry aloud: *'8th September 1970 – TACEVAL* ... Then there is an asterisk, and it simply says, *Captain Bill Schaffner lost.'*

The search was called off at midnight.

Linda slept only a couple of hours. 'I woke and stared at myself in the bathroom mirror. It was as though I was looking at a different person. Had the previous night's events all been a dream? Surely this couldn't be happening?'

Some of the Squadron wives, including Irene Coville, came to cook, do the laundry and keep her company. 'There was little to be said,' Irene says. 'What does one say to a new young widow with three children? I don't think that Linda really believed Bill could be dead. I'm not sure I did. It was so painful, and we felt so useless to Linda and the boys amid their terrible loss, so far from their family, thousands of miles away in America.'

Chris Coville was back flying with the rest of the Squadron when the search for Bill Schaffner resumed that morning. There was no sign of his aircraft and it was still not known if he had ejected. 'We got airborne with a little trepidation at first, but were soon looking ahead into the blue sky as our beautiful jets flew in a last salute to another lost friend.' Life in the military moved on quickly. It had to. Aircraft crashes were commonplace. But the individual scars remained. 'We lost a number of colleagues in those years. I never got over it.'

Chris was right to remember the sheer numbers of aircrew being killed in aviation incidents at that time. Between 1960 and 1970, the British military *alone* lost 630 aircraft, resulting in 479 deaths.[7] In the same period, Martin-Baker registered 2,611 successful ejections worldwide, on its own seats.[8] With many nations using their own ejection seat manufacturers, it would not be unreasonable to suppose that these figures could be increased perhaps tenfold, possibly more, to indicate worldwide ejections in that decade.[9] Although the conflict in Vietnam raged through much of this period, most accidents were a result of intense peacetime training. It was a risk all aviators accepted.

That same day the US Air Force sent a liaison officer to Binbrook. Linda was told that she should pack and leave the UK immediately. Even for the 1970s Cold War machine, it seemed a remarkably hardhearted reaction. 'I think they thought that as Bill was presumed dead there was no point in having his family staying on the base. They thought we would quickly become a burden to "the system". They probably didn't mean to be so cruel; that's just the way it was back then. I was in despair and confused. I think I accepted that Bill was gone but I still held out a thin thread of hope he might be found alive. I was determined not to give in to them.'

But there was still no sign of Bill Schaffner or his Lightning a week later, so, once the immediate searches had been called off, she did decide it was best to leave. Irene Coville went to Lincoln to buy new jackets for David and Glennon. Linda wanted them to look smart for their trip home. 'I felt I had to go back to America, to acknowledge that Bill was dead. I had three young children to care for. I couldn't wallow in self-pity in England. I had to put my pain to one side; my life, and theirs, had to continue.' Chris and Irene Coville called round to say goodbye the day before their departure. 'She came to the door desperately trying to smile and keep up a brave face. It was heartbreaking.'

'I was just twenty-four and Chris was twenty-five,' remembers Irene. 'Being so young, I suspect that we found words difficult, so a hug had to do. I think the Squadron was deeply affected by Bill's death. I certainly was and promised that, in the future, I would always give Chris a warm send-off each time he went out to fly. Just in case. But, to be honest, after the initial shock we just got on with life. Losing friends was nothing new in those days.'

On Linda's flight home, one thought refused to disappear.

Bill's body had still not been recovered. Had he managed to eject? Was there perhaps a scintilla of hope?

They arrived in Columbus, Ohio, and stayed with a relative. Her husband was a helicopter pilot who'd been killed in Vietnam, and her brother, a pilot, had also perished there. Linda felt she was part of 'a family hit by multiple tragedies'.

On 12 October, a Royal Navy minesweeper picked up a bulky object on its sonar, 5 miles north-east of Flamborough Head, lying in mud on the seabed, about 100 feet from the surface,[10] A diver went down. 'It's an aircraft with the letter F on its tail.'[11] It was Bill Schaffner's Lightning, still largely intact. Those listening in from the search vessel, including the members of the RAF's official accident investigation team, could hear the diver's heavy breathing as he clambered over the wing. They couldn't quite comprehend what he said next.

'Cockpit closed. Looking in now. It's empty. No sign of the pilot.'

Bill's Lightning was winched back to the surface. Although it had broken its back during the crash – or subsequent salvage – and had a few missing fuselage panels, it was remarkably undamaged. Photographs taken at the time show the distinctive swept-back wings, with the aircraft's registration, XS894, emblazoned on the underside. The 5 Squadron maple leaf insignia was clearly visible on the tail.

At this point, the confusion began to set in. The cockpit was empty, the canopy was closed, but Bill's seat harness had been disconnected. The PSP lanyard which had once attached him to his survival pack and dinghy was now dangling freely from the closed cockpit, beside the inverted red triangle symbols on the fuselage beneath the canopy, warning of the dangers of explosive ejection seats. The mystery deepened – what had happened that night? And where was Bill?

His jet was housed in a hangar at Binbrook, hidden from view by a large tarpaulin, awaiting the accident inspectors. Chris Coville could not resist the temptation to peek through the giant steel doors. 'The last of the seawater was oozing out of Bill's aircraft onto the hangar floor. I was filled with a sense of horror at how quickly the life of somebody I had known had been snuffed out; yet another reminder that life as an aviator was a dangerous business.'

The official Board of Inquiry, set up the day after the accident by the RAF Directorate of Flight Safety, went to work. A host of questions needed answering to establish what had happened to Lightning XS894. Chief among them was the whereabouts of the pilot. His ejection seat was still in the aircraft with the harness undone. Had it failed, or had he not pulled the handle? And if not, why not? Where was the body? If he had managed to unstrap and climb out, why was the cockpit canopy closed?

So far, only one thing was for sure. The exact point of impact had frozen on the altimeter and the air speed indicator. Bill Schaffner had hit the water at 180mph.

* * *

Sandy Gruters knew just how Linda Schaffner felt. She still had little news of her own husband Guy, who had now been a POW in Vietnam for over three years. 'It was the not knowing that was so hard to deal with,' she says. 'As the military wife of someone who is missing, you almost have to put them behind you, and try to live one day at a time. How long could this go on for? You can't predict the future, so you deal with what you have in front of you at that moment. What else could we do? I always hoped and prayed that the ordeal would be over soon, but I knew in my heart that it would not. I just kept asking myself, *Why me?*' Revisiting this period of her life – some fifty years behind her – with the author, Sandy begins to cry.

'I wouldn't wish that ordeal on anyone,' she says after composing herself. 'But you just have to cope as best you can and my faith played a huge part in sustaining me. There was a lot of sadness, fear and trauma to deal with, but I had to get through it.' Like Linda

Schaffner, she had a young family to care for and bills to pay. There was no time for self-pity.[12]

* * *

Linda Schaffner had found herself and the boys a house in Ohio and their furniture had arrived from the UK. She received a letter from the chief of the exchange programme which looked after all USAF personnel serving with the RAF in the UK. The letter she had been dreading. It was coldly brief and to the point.

'This is to advise you that the aircraft which your husband Captain William Schaffner was flying on 8 September 1970 was recovered from the North Sea on 12 October. However, his body was not inside the aircraft. The whereabouts is undetermined at this time. I will advise you immediately of any new developments. I extend my deepest sympathy.'

That was it?

'It was inexplicable. Where was Bill's body? I had no husband to bury and no information about what had happened to him. How was he *not* in the aircraft? Had he ejected? I just couldn't understand it.' That rather blunt letter was quickly followed by an altogether warmer and more personal one from Lieutenant Colonel William Heckendorn, who was also part of the exchange programme and had escorted the family back to the USA. 'I just wish that there were more that my wife and I could have done for you while you were here. You can be sure you were in our prayers and will continue to be so.'

Heckendorn was part of the investigation team and had witnessed the recovery of her husband's aircraft. He had been keeping Linda up to date with developments as much as was allowed. 'As you know, I have been on the Board of Inquiry, and we're still in session.' The board had interviewed forty-four witnesses about the events leading up to the accident and its aftermath, but not one of them had seen the Lightning ditch into the sea. 'The whys and wherefores are still somewhat of a mystery,' Heckendorn admitted. 'The fact that Bill was not in the aircraft when found is also puzzling as his ejection seat was still in it.' This revelation – almost mentioned in passing – was the first time that Linda had been told that Bill's ejection seat was still in the jet, but that he was not. She struggled to take in the information.

Heckendorn went on to explain that, 'this is an extremely thorough investigation, and *nothing* is being overlooked. But I am afraid it will be lengthy, and it may be some time, a month or two at the earliest, before we are informed of their findings. I would truly like to be able to tell you exactly what happened but, at this point, I can't. Our heartfelt sympathy to you and the children. May God bless you.'

If the finality of losing her husband was indescribably painful, so was the mundanity of tying up loose ends. In early November, Linda received another letter from the USAF exchange programme. 'Enclosed is a cheque refund from your TV rental and a letter from your bank. RAF Binbrook has compiled a bill for telephone charges, which I will pay and deduct when I sell the car. About the car, I have heard from the lender that they need a death certificate to settle the loan. Please could you send me a copy? I hope you are getting settled and are fairly comfortable.'[13]

Towards the end of 1970 a memorial service was held for Bill Schaffner at the Lockbourne Air Force Base, Ohio. After the speeches and tributes, Linda, dressed in black, gazed up at a flypast in 'missing man' formation; an aerial salute during which one fighter engages its afterburners over the crowd and soars upwards, out of the formation, leaving a gap in memory of a fallen pilot. A bugler played the twenty-four notes of 'Taps', the American equivalent of the 'Last Post'. 'Seeing those planes going over with one jet missing was really hard to watch. It's the only thing I can remember about the ceremony . . .' As she describes the occasion to the author, Linda again breaks down in tears. 'I felt so weak, and nearly collapsed in the same way I had the night they told me Bill was missing. A family member stepped forward to support me. It finally came home to me. I was alone.'

At the end of the solemn service, an airman, resplendent in full dress uniform, stepped forward to present Linda with the Stars and Stripes, folded into a triangle; the iconic symbol of a lost loved one. She refused to take it. With no body to bury, it had not shrouded Bill's coffin. Standing beside her, second son Glennon was confused. 'As a 5-year-old I thought to myself, *That flag looks pretty cool. I would really like that.* So why didn't my mom take it? I presumed it was because she didn't want to carry it. Maybe it was too heavy?'

The neatly folded United States flag would never be collected by the Schaffner family.

CHAPTER TEN

'My Daddy's Coming Home!'

Vietnam, 1972

After the death of Captain Bill Schaffner, his widow Linda and their three sons had eventually settled in Chicago, the legendary 'Windy City', on the shores of Lake Michigan. And although the RAF's Board of Inquiry – the long-running official investigation into the accident – was eventually completed in June 1972, nearly two years after the incident, the family were never told the conclusions. It may seem inexplicable today, but the reality of the time meant that there was little official communication with the families of the deceased after an accident. Some information might be provided by the deceased's superiors or colleagues on base, but as Bill Schaffner's family was now on another continent, they had been told nothing since those letters in the immediate aftermath of the accident. 'As far as we knew the incident was closed, and everyone else had moved on with their lives,' says Linda. 'We were in the dark about what had happened that night, why Bill was not in his cockpit, that the ejection seat was, and that his body was never found. We certainly never saw the accident report or even heard anything about it being finalised.' Indeed, as was the custom of the time, it had been given a 'Restricted' security tag and the full report was only seen by a select few. The RAF had moved on; it had a Cold War to fight.

That same year, Linda married a schoolteacher, a widowed Korean War vet. 'Our stepfather was a wonderful, loving man,' Bill Schaffner's son Glennon says. 'He had three children of his own. There was a sense, though, that talking about our deceased dad might damage the new family dynamic.'

As far as the family was concerned, there were no further reports about Bill Schaffner's disappearance, and no one was looking into it. They were wrong, but it would be many years before they received further, truly dramatic information.[1]

* * *

Guy Gruters had now been languishing in cramped, sordid and brutal jails in Vietnam for five years. He had no idea that his younger brother, Terry, was now climbing aboard a giant B-52 Stratofortress on an airfield in Thailand for a night bombing raid. Its target was a surface-to-air missile (SAM) site close to Hanoi. As co-pilot in the six-man crew, Terry was 'very excited' at the prospect of dropping bombs near the North Vietnamese capital. 'That was where my brother was, and I had great hopes of seeing him get out of there.' Terry had graduated from the Air Force Academy a year after Guy's capture in 1967. But because his brother was now a POW, he was actually exempt from service in the combat zone. There was no way that he was going to take that option. Instead, he had completed his pilot training in September 1969 and then took 'the assignment that would get me to Vietnam the quickest'. He wanted to be near the action. And his beloved elder brother.

Guy's wife Sandy had pleaded with him to reconsider. 'What happens if you get shot down like Guy? How will the family cope?' She couldn't bear the thought. But Terry Gruters' mind was set. 'Forget it, Sandy. I'm staying out there until either Guy is freed. Or I'm dead.' He was now on his third tour in the region and part of Operation Linebacker: the campaign to bomb the regime back to the negotiating table as Christmas 1972 neared. The Americans were determined to force an end to what was now a very unpopular war and, in the first eight days of the operation, the B-52s had hit a string of targets in North Vietnam.

Terry could not wait. Missile strikes had dogged his most recent Linebacker missions. 'I remember seeing the holes in our aircraft after we had returned to our base and thanked God for what a great machine the B-52 bomber was.' The Boeing was a black and menacing 216-ton monster, 159 feet long and almost 41 feet high. Powered by eight thunderous Pratt & Whitney engines on a 185-foot wingspan, it carried a huge load of bombs, mines and missiles.[2]

Approaching their target, the pilot and co-pilot were up front. Side by side behind them, in rear-facing seats, came the gunner, who could remotely fire thousands of rounds a minute from the aircraft's tail-mounted guns, and the Electronic Warfare Officer (EWO), a specialist trained to identify and counter enemy air defence systems.[3] All had upward-firing ejection seats. On the deck below them, in the belly of the beast, the 'route' navigator and the radar navigator were strapped into downward-firing ejection seats.

As their briefing had predicted, Hanoi was completely blacked out. They were flying at 50,000 feet, out of the range of anti-aircraft guns, but as Terry and his crew neared the target he saw countless SAMs launching ahead, rocket plumes blazing. 'So many that the windscreen was full of them. Like a giant firework display.' The bay beneath them opened and the payload released. Time to head home. Suddenly the Stratofortress rocked like a rollercoaster. A missile had exploded beneath Terry's wing. 'Three of the engines on the port side went out. We could not restart any, so had to take emergency measures to shut them down safely. The fourth on that side was still running but the aircraft began to lose altitude.' The pilot's legs had been lacerated by shrapnel. The remaining port engine caught fire and, as they descended, they became an inviting target for the SAMs.

Terry tried every trick in the book to keep them in the air. None of the crew wanted to go the same route as Guy. 'What a strange feeling it is to realise you might soon be dead,' Terry remembers. 'That touches your spirit to its depths. There I was, looking out over the darkness, hearing all the noise of the plane, watching the fire, and thinking life might end soon.'

Just like his brother, he clung to his faith, and eventually they limped over the border to Thailand where they could safely eject without fear of capture. Now in contact with rescue helicopters and flying over dense forest at just 3,500 feet, 'we followed the recommended order of bailout. Those in the most precarious positions, the two navigators sitting beneath me on the lower deck of the plane and who had downward ejection seats, would get out first.'

Terry and his pilot would be the last to go. The radar navigator, the gunner and the EWO all ejected safely. The second navigator was less fortunate. 'The hatch beneath his ejection seat had not blown. He said he would have to jump out of the hole where the radar navigator's ejection seat had been and deploy his parachute once out of the aircraft.' The pilot told him to go ahead. 'We waited a little while and heard nothing else on the intercom.' Assuming he had made it, the pilot told Terry to get out.

'I pulled my roof ejection handle. The roof blew off, just like in the movies.' He felt a rush of relief. If the Perspex canopy over his seat had stayed in place, he would have had to clamber down the ladder to the lower deck and throw himself out of the open hatch. He then told the captain he was ready to go. 'I pulled my trigger. I was out into the night. I remember laughing as the seat fell away from me and my chute opened. I was quite surprised to be alive. What a beautiful night it was, as I drifted down. It was now very likely that I would live. I was no longer sitting in a plane waiting for it to explode. Tears of joy were in my eyes.' Just before he landed, Terry saw the B-52 hit the ground and explode in a massive fireball. He hoped his pilot had made it. 'I prayed for him and all the crew.' Crashing through the trees and landing unscathed, he could now hear the beat of helicopter rotors above the branches.

All six crew had made it and were reunited the next day to share their tales of escape and survival. Terry felt it was like listening to one miracle after another. The two Gruters brothers now had three ejections between them.[4]

* * *

On 8 January 1973, North Vietnam returned to the Paris conference table. Rumours of America's withdrawal began circulating around the POW camps. The peace agreement was signed on 27 January. Operation Homecoming, a massive airlift to repatriate the POWs, would launch on 12 February.[5]

Still unaware of Terry's participation in the conflict, Guy Gruters was both elated and dumbfounded. There had been so many times when he thought he would spend the rest of his life in a stinking cell, or have it cut short by the kind of treatment suffered by Lance Sijan. That he would never again see his wife, the daughters he barely knew, the rest of his family, or his country. But in early February, daring to believe that freedom was in sight, he sat down and wrote a short letter to Sandy. His excitement shone through every line. He could not wait to get home. 'The next few days,' he told her, 'will seem much longer than all the last five years.'[6]

At 7.30 on the morning of Monday, 12 March 1973, Sandy Gruters was woken by the phone in her bedroom. The caller was a US Air Force Casualty Officer from Randolph Air Force Base in Texas. 'He told me that Guy was being freed that week. He was coming home nearly five and a half years after his ejection, five and a half years of hell. It was the happiest awakening of my life.' She still could not quite take it all in, even though Guy's February letter had arrived the previous Friday. It had only been the seven lines permitted, but she had read it over and over. Her daughters, Mari, now eight, and Sheri, seven, sensed something very exciting was going on when they heard Sandy calling Guy's parents to tell them the news. They ran into the bedroom. As soon as she told them their father was being freed 'they jumped up and down on the bed like a couple of frogs'.

'Is he really coming home?' they asked.

Sandy's face was all smiles. 'He really is.'

'I can't wait until my daddy gets home,' Sheri shouted, over and over. Mari could not wait either. 'He had said in a letter that maybe I could get a horse when he got back.'

The phone in the Gruters household that morning was red hot as Sandy called Guy's five siblings and close friends. Terry was in South Dakota, just home from his last Vietnam tour, and reliving every moment of his ejection and what he called his and his crewmates' 'miracle escape'.

On Wednesday, 14 March, Guy and his fellow POWs – almost all of them ejectees – were loaded into nineteen trucks. Three US Air Force transport jets with distinctive red crosses on their tails were lined up at Hanoi's Gia Lam Airport, ready for take-off. Viewers at home watched their televisions spellbound as POWs of all ranks were carried on stretchers or shuffled uncertainly towards the aircraft.

Guy remembers the take-off vividly. 'Once airborne, there was intense cheering and clenched fists raised. We were overjoyed.' They landed on friendly soil, Clark Air Force Base in the Philippines, the next day. As he showered, the sensation of warm, comforting water cascading over him was intensely emotional. 'At times it had been hard to be positive over those long hard years as a POW.' He wondered how much America had changed in the years he had been away. How much had *he* changed? Like all the POWs, he was medically and psychologically assessed. After a haircut and a shave, Guy put on a crisp white shirt and a new, pressed US Air Force uniform. Outwardly, all the trappings were back in place, but how much physical and mental damage had he and the others suffered? And would they ever truly recover?

The US Air Force confirmed that Guy Gruters was now safely out of Vietnam and journalists had been to interview and photograph Sandy and her daughters. Guy's story and news of his release was all over the *Sarasota Herald-Tribune*, the paper that had brought her the news of his first ejection all those years ago. She rushed around the house, getting it ready for his homecoming, applying finishing touches, making sure that everything was in place. Then there were errands to do. She took the girls shopping to buy new dresses. Neighbours cried out happy greetings as they passed. Sheri told one shopkeeper, 'My daddy's coming home. He hasn't been home for nearly six years.'

In the midst of her joy, Sandy was still contemplating what sort of 'daddy' would return. And, still thinking of others less fortunate, she told a reporter, 'I'm elated for those who are coming home, but I'm very sad for those who aren't.'

On Friday, 16 March, Guy's flight left the Philippines. 'We simply couldn't believe it was all over and that we were out.' Some of the POWs were shouting with joy; others were in tears. 'We were coming back to the United States and real freedom.' During the trip, the prisoners, several smoking American cigarettes for the first time in years, caught up on some of what they had missed while they had been incarcerated, discussing the latest football results or lamenting the assassinations of Robert Kennedy and Dr King. None of them knew about the women's liberation movement, or that the Beatles had split up.

MAXWELL AIR FORCE BASE, ALABAMA
SATURDAY, 17 MARCH 1973

Sandy Gruters had flown to Alabama from Florida the previous day. She was now sitting in a queue of cars on the USAF base, radiant in a pink dress and white jacket, waiting for the return of the POWs. Outwardly calm, her heart was pounding as the bulky transport aircraft taxied slowly towards the red carpet. A crowd of 400 waited expectantly. Television crews were poised to roll their cameras. 'I was walking on the moon,' Sandy says. 'I was overjoyed but nervous too. Five and a half years is a long time to be separated from your loved one. How would Guy be? Would he be how I remembered him? Had I changed as well? Were we now different people? What did our future look like? There were so many questions, so many unknowns.'

Guy's old friend and compatriot Charlie Neel was in the driving seat. They had not seen each other since being rescued together after their first ejection back in 1967. Charlie had been chosen as the Gruters' family liaison officer, to be on hand to help Guy readjust to 'normality'. Whatever that might mean. 'I had heard a few weeks earlier that Guy was coming out of "pokey", and that I was nominated to be his escort when he got home. I was just delighted beyond belief that my friend was alive . . .' Speaking to the author by video call from his home in Colorado, Charlie pauses to gather his feelings, turning away to gaze out of the window at the blue winter sky. 'I had often wondered if I'd ever see him again. It was one of the few times in my life I remember weeping.'

The big transport jet turned slowly until its exit door was facing the crowd. A helmeted six-man honour guard bearing an Air Force flag and the Stars and Stripes took up position either side of a red carpet, to welcome the next batch of the 591 prisoners repatriated by Operation Homecoming.[7]

After what seemed like an age the aircraft door opened. First off was the senior-ranking man on board, and future presidential candidate, Colonel John McCain, who had been shot down and badly injured in 1967. There was no brass band or speeches, simply applause, relief and a few banners. One by one eighteen others followed, in date order of their capture. As each of them reached the end of the carpet, a car holding a loved one pulled up to take them to another part of the base for their family reunions. Then Sandy saw her husband. Nearly six years after they had parted at Jacksonville, Guy was standing at the top of the steps, ramrod straight and as smart as ever. His uniform looked several sizes too big for him and the collar of his shirt was loose around his neck. As his feet touched American soil, a beaming smile lit up his face. He marched as smartly as his emaciated legs would allow, saluting the flags and the assembled generals.

Charlie Neel bit his lip. His own vehicle was a two-seater, so he had borrowed an air-force blue four-door Dodge Aries staff car for the occasion, and the big sedan had been stalling all morning. He prayed it would not do so again and mess up all the other reunions. He feathered the accelerator and eased it towards the end of the carpet. Sandy was transfixed. Her husband was now just feet away. 'It was such a wonderful, indescribable feeling to see him in person. I had waited so long. It was a gift from God I didn't feel I deserved, but was very happy to have.'

Guy opened the rear door and climbed in. And there she was, nearly two thousand days after he had ejected over enemy territory. No words were spoken as he fell into the arms of his wife. 'Then, we were hugging, kissing, laughing. It was just so incredibly emotional to be out of the dirt and degradation of a prison camp. I simply could not believe I was home. It was like coming from hell, back to heaven.'

Suddenly Guy recognised the driver. 'Charlie Neel! What are you doing here?'

Charlie had wept when he saw his old friend walk down the aircraft steps. 'To see him again after so long, alive, about to be reunited with his family, was an overwhelming experience.' Charlie's wife, sitting in the front passenger seat, handed him a tissue. She knew he was on the verge of tears again. 'Put your sunglasses on, Charlie,' she murmured.

Neel was being motioned to move on. He accelerated hard, wheels spinning, trying to ensure the engine did not stutter in front of the dignitaries. His charges in the back seat were catapulted around the compartment. 'We were both thrown around as Charlie sped away,' Sandy says. 'We were trying to hug and kiss but we ended up rolling around and everyone was in fits of laughter.' While the other vehicles ferried happily reunited couples away like the final scene in a movie, Charlie's car was bunny-hopping to the base Visiting Officers' Quarters. Guy didn't mind, he was concentrating on his beloved wife. 'Me and Sandy played "kissy face" as we drove to meet the rest of my family.'

The Gruters clan were excited and nervous in equal measure as they peered out of the doorway of the bungalow where they waited. The men in jackets and ties and the women in their pretty dresses all looked as though there were attending a wedding. Then they saw the Dodge coming slowly around the corner. As it pulled up, Guy's brother Peter, twelve years his junior, rushed forward and opened the rear passenger door. Guy stepped out and the two brothers embraced tightly, swaying slightly, overcome with emotion.

His father, Bertram, tall and imposing in his black suit, took Guy's hand and held him in his arms. Neither wanted to relinquish their hug. Then Guy's mother stepped forward, in tears. There was a chorus of cheers as she reached up and threw her arms around her long-lost son, holding his face in her hands, staring into his eyes, making extra sure he really was standing there.

Mari and Sheri were desperate to touch their dad, but Sandy Gruters gently held her daughters' shoulders. She did not want Guy overwhelmed as he clasped and hugged his sisters. He would be reunited the following day with his fellow ejectee and brother, Terry. His girls could wait no longer. They rushed over to their father and clambered into his arms. Guy found the strength to gather them up and clutch them to his chest. The rest of the family looked on as father and children were reunited, tears rolling down their faces. 'The father I never really knew was back in my life,' Mari says. 'It was almost too much to take in, for an 8-year-old. There was so much joy and celebration around his homecoming that it was rather overwhelming.'

After medical checks Guy was free to resume his 'normal' life, but he had returned to a very different nation. From TV shows and new car models, to the world's first mobile cell phone and a movie called *The Godfather*, it was hard to take it all in. The World Trade Center's iconic twin towers were a month away from opening in Lower Manhattan. Political corruption was in the air as the Watergate scandal was exposed. Most challenging of all was the realisation that America, now tired of a war in which 58,220 US servicemen had been killed and nearly 1,600 were still missing, had turned its back on Vietnam. Around a quarter of a million South Vietnamese soldiers and some 1.1 million North Vietnamese and

Vietcong fighters lost their lives, along with as many as 2 million civilians.[8]

By the time of Guy's release, the US Air Force had flown close to 352,000 sorties in North Vietnam, losing more than 3,700 aircraft and 5,600 helicopters. Rescue crews are credited with saving some 1,200 aviators who had ejected in what is said to have been 'the greatest combat aircrew recovery force in the history of aerial warfare'.[9]

Guy was in no fit state to compute those statistics. All he knew for sure was that it seemed few people were acknowledging the sacrifice. His aunt lent him and Sandy her house on Ponte Vedra Beach in Jacksonville for a month. Instead of heading straight home to Sarasota, they could start to piece their lives back together beside the ocean. The luxury low-rise resort boasted golf courses and tennis courts in a sculpted landscape of tall, elegant palms and vivid floral displays. The place almost made Guy's eyes smart after the drab, colourless twilight world he had so recently left behind.

They spent the first two weeks walking along the beach, hand in hand, just talking and re-establishing who they were, and where they were headed. Although Sandy's love and a strong family had sustained Guy through the tough years, he was a stranger to them. 'For much of that time Sandy didn't know I was alive. We had been apart so long, and so many things had happened to us individually, it took a long time to build that connection again.'

'It was difficult for each of us to truly understand what the other had endured,' Sandy says. 'Guy had gone through torture, brutality, isolation and fear. I had suffered fear of the unknown, isolation and not knowing if or when my husband might return to me. I did not suffer the physical ordeal Guy had, but his time as a POW had been truly challenging for me.'

Mari and Sheri joined them at the beach house for the final fortnight. Sandy recognised the challenges that obviously lay ahead. 'I had been in sole control. We had our own routines, our own way of working, and now this "stranger" was part of our little group. We had to learn how to live together. To be a family again.'

'In those early days, I sensed that my dad had suffered some truly horrific experiences,' Mari says. 'But no one really knew quite how bad, and, initially, little was discussed. Although I think my grandmother and my mother must have had some understanding; I would regularly hear them talking in whispered voices about what he had endured. And while we had our own joy at our dad's survival and return to us, some of the friends we had spent the previous five and a half years with did not. There would be no joyful homecoming for them. Fifty years on I often think about those kids whose dads did not make it back. It still makes me cry to this day.'

Sandy admits that 'it took many years to make it all work, but that month in Ponte Vedra Beach was the most important gift we could have been given. It was time getting to know each other again. The start of our new life together as a family.'

For the Gruters family, Guy's experiences, and those of his fellow ejectees Bob Craner, Lance Sijan and Charlie Neel, would resonate for ever. Sijan would eventually be awarded the Medal of Honor – the USA's highest military award for valour – after Craner and Gruters revealed his astonishing story of courage.[10]

Guy still reflects on his own survival. 'I often think that I survived those two ejections to give me a second, then third chance at life. I am still living my best life because of the ejection seat. Yes, I had to survive cruelty and depravity for over five years, but that's when I rediscovered my faith in God, which sustained me then, and still does today. I'm alive, I'm happy, I have my family. What more could you ask?'

Perhaps Sandy Gruters' thoughts reflect those of many whose loved ones' lives were saved by ejection seats. 'We were *all* given another chance. We had five more children and adopted another, so now have eight children and twenty-one grandchildren. That extended family is only around me because Guy survived two ejections. And their aftermath. The wonderful life we have today is a direct result of that survival. He was so lucky, and we were blessed by God to be given this life. At a recent party where we were all gathered together, one of our friends looked around and said to my son, "Just think, if your dad had not survived those two ejections, you, and most of these people, would not be here today." That's quite strange to think about, isn't it?'

* * *

For Bill Schaffner's family, there had been no closure. Or further information about the Lightning crash which had seemingly killed him. But the past had a habit of resurfacing in unexpected ways. In 1976, Bill's third son Mike, then aged five, watched his mother drag a red wooden trunk, about 4 feet by 5, out of a closet. He had never seen it before. On the front were printed the words: '*Captain William Schaffner USAF (Deceased)*'. She proceeded to take out the contents. 'There was memorabilia from his time in Vietnam, photos, his name patches, insignia and his captain's bars. Then she showed us his logbook, with a picture of a Lightning stuck inside it. I couldn't believe it. "That's my dad's plane?"'

Linda nodded.

Oh wow!

'In the years after the accident, there was little conversation about my dad,' remembers Glennon, his second son. 'We didn't mark the anniversary of his death. There was a sense of needing to move on. I think Mom was trying to shield us from any disruption or sadness. But I still remember today my recurring dreams about that dark night when the RAF men came to our door.'

Mike Schaffner, only a baby at the time, might not have known his father, 'but he was my hero. I had a thousand questions. I wanted to know everything and hoped to go into the Air Force and be a pilot, just like him.' As he went through his teenage years, Mike became more affected by his father's absence. 'I cried in bed that I would never know him through adulthood. He would not be there for me, to offer advice, whatever direction my life took. All I was going to have was that red box.' The mystery of the unused ejection seat remained unsolved. Other stories filled the newspapers. Linda rarely talked to her sons about their father. 'Without a body it was difficult to draw a line,' she remembers. 'For many years I dreamed that there would be a knock at the door and Bill would be standing there. Maybe he was returning from a secret mission and hadn't been able to contact anyone. It was silly, but without his body, there was always a glimmer of hope.'

As the years rolled on, wherever she was, Linda Schaffner was always drawn up short when she heard the theme song to *A Summer Place*, the movie she was watching on that dark September night in 1970. It took her straight back to the moment the uniformed RAF officers knocked at her door to tell her that her husband was never coming home.

PART IV
1981–1987

THE FREEZING WATERS
OF THE SOUTH ATLANTIC

CHAPTER ELEVEN

'IT COULD HAVE BEEN ME . . .'

COMANDANTE ESPORA NAVAL AIR BASE, ARGENTINA
1 AUGUST 1981

Lieutenant Commander Rodolfo Castro Fox, of the Argentinian 3rd Air Naval Fighter and Attack Squadron, allowed himself a smile as he hauled himself into the cockpit of his ageing fighter-bomber. This was the big league. He was leading five other American-made A-4 Skyhawks in the South Atlantic leg of one of the most ambitious Cold War exercises ever. United States-led 'Ocean Venture '81' would run for ten weeks and involve 120,000 service personnel from fourteen nations. Its goal: protecting sea lanes in the Caribbean, Atlantic and Baltic from Soviet aggression, and practising maritime communications in the event of war. Castro Fox and his fellow pilots took off to rendezvous with the Argentinian aircraft carrier ARA *Veinticinco de Mayo*. Named after his country's 25 May 1810 revolution, she had originally been built for the Royal Navy as HMS *Venerable*, and would be the base from which they would hone their skills, alongside their friends and allies, including the British.[1]

The flying bug had bitten Castro Fox the moment he saw a Korean War F9F Panther, one of the US Navy's first carrier-based jet fighters, on an advertising poster. Beneath it was the legend, 'You can fly this plane.' Convinced by the slogan, he joined the Argentinian Navy in 1961. 'I had good physical co-ordination,' he says. 'And I had been driving motorcycles and cars since I was twelve years old.' Pictured in his late teens coolly astride his motorbike, he looked like a cross between James Dean and Elvis Presley.[2]

He got his wings at twenty and rose rapidly, with a later stint as a flight instructor. Now thirty-nine, he had more than 250 'hitches' – hooking the high-tension arrestor cable on the flight deck that stopped his aircraft careering off the bow – to his credit. 'Every time I took the cable there was no room for carelessness; landing safely was an intense experience loved by all naval pilots.' Having already completed two successful take-offs and returns in the morning, he was on his third sortie of the day, preparing for his final approach back to the ship. 'I finished my turn to face the carrier at 500 feet. Below, to my left, was the white wake of the ship that contrasted with a calm, greenish-blue sea. Ahead was the inverted yellow "T" marking the beginning of the runway.'

As Castro Fox descended, his thoughts turned to his upcoming vacation with his wife and four children. To his right he could see the carrier's control 'island' and antennae, and a lazy plume of smoke from its funnel. On the port side, a yellow, ball-shaped indi-cator helped him line up his approach. The high stern of the ship swayed slowly. The Landing Officer's instructions kept coming over the radio. Castro Fox cut his speed and, as he descended, the vessel loomed larger. He crossed the stern at just under 150mph, heading for the arrestor cables.

Six tons of aircraft shuddered as it slowed to less than 100mph in 65 yards. He had hooked cable three. Secure in his ejection seat, Castro Fox throttled his power down to minimum. 'Suddenly my head was rocked back and hit the ejection seat hard.' The Skyhawk's nose lurched forward. The cable had snapped and was snaking wildly around the deck. Now unrestrained, the jet picked up speed. The front of the flight deck was coming up fast. Castro Fox didn't panic. He'd been here before. He knew what to do. He thrust the throttle forward, powering his engine to 100 per cent so he could simply take off, come around, and attempt another landing.

Too late. He was running out of deck. 'I now did not have enough

Royal Navy gunner Neil Wilkinson before the Falklands War. Neil would shoot down Argentinian pilot Mariano Velasco on 27 May 1982.

HMS *Coventry* sinking after being hit by the bombs dropped by Mariano Velasco on 25 May 1982. An emergency life raft carrying survivors can be seen in the water on the left.

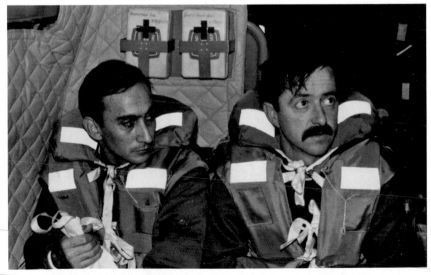

Ricardo Lucero and Mariano Velasco, who both ejected after being shot down during the Falklands War, are repatriated after the end of the conflict.

Rodolfo Castro Fox, who was seriously injured when his ejection seat failed to fire properly in August 1981.

Neil Wilkinson at Mariano Velasco's crash site in the Falklands. Aircraft wreckage is still clearly visible.

Wilkinson and Velasco reunited in Argentina in 2011. Parts of Velasco's aircraft wreckage sit on the table.

The Jaguar flown by Steve Griggs, pictured before he was mistakenly shot down by an RAF colleague.

Griggs holding a mounted ejection handle presented by the RAF Armourers to commemorate his survival.

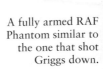

A fully armed RAF Phantom similar to the one that shot Griggs down.

Kate Saunders with a Harrier before the crash.

A seriously injured Saunders in hospital following the crash.

Ash Stevenson and Kate Saunders are reunited. Kate is wearing gloves and long sleeves to hide her burns.

Ian McLean lies on the ground near an ambulance following his first ejection from a Jaguar in 1983. His orange life jacket is just visible, as is his yellow survival pack near his parachute.

Ian McLean points to his name on the wall while visiting the Martin-Baker factory after his second ejection from a Tornado in 1990. Countless ejectees visit the company to thank the staff for their life-saving seats.

Pictured here in close formation, a Tornado and Jaguar similar to the aircraft involved in Neil Johnston and Ian McLean's mid-air collision in January 1990.

The author with Neil Johnston after his ejection. The pioneering repairs to his broken legs are clearly visible.

Tom Moloney climbing through the shattered cockpit of his Jet Provost on landing after his brother Des fell out.

Tom and Des airborne again in 2022.

The Tornado test flight carried out in 1988 with the cockpit canopy missing.

The author and pilot John Peters in 1991, wearing full
flying kit for the Tornado's Mk10 ejection seat.

1 Life jacket: the inflatable
stoles are held in the bulky
section around the neck
2 Arm restraint clips
3 Personal Equipment
Connector (PEC) which
connects oxygen and g-suit
hoses, and communications
cable, to the ejection seat

4 Communications connector
from the flying helmet
5 G-suit which fits tightly
around the waist, thighs
and calves
6 Lower leg restraint garters
7 Upper leg restraint garters
8 Flying helmet and oxygen
mask

9 Arm restraint clip at the
top of the Velcro seam it
pulls down
10 Oxygen hose and
communications
connectors into which the
crew plug their helmet
connections

speed to take off again.' Fifty-four yards from the edge. He throttled back hard and stamped on the brakes. The Skyhawk skidded left. He was still travelling too fast.

The Landing Officer was yelling over the radio. 'Eject! Eject!'

Castro Fox reached down between his legs with his right hand and pulled hard on the yellow and black handle. 'I felt a dull explosion behind me. The canopy was jettisoned. Now, I thought, I will be blasted out by the seat rocket.'[3]

But nothing else happened.

His ejection seat stayed rooted to the cockpit floor.

The failure of an Argentinian ejection seat was not a one-off. With the country in the grip of a military dictatorship, the United States had refused to supply replacement explosive cartridges for the ejection guns. Many aircraft now had cartridges that were fast approaching, or even past, their end-of-life dates. All the crews understood that they could not be assumed to work properly; it would be a rather brutal lottery as to who might be sitting on a suspect seat. Castro Fox's ejection cartridges had expired in terms of reliability the previous December; he was out of luck.[4]

His Skyhawk kept slewing left. The wheel on that side was the first to lose contact with the deck. Trapped in his cockpit, still strapped to his failed ejection seat, he plunged over the edge and down into the freezing waters of the South Atlantic. 'I remember vividly the events and images leading up to that point, almost in slow motion,' he says. Hitting the sea was the last thing he recalls as he was knocked unconscious.

It was just five seconds since the cable had snapped, and three since he had pulled his ejection handle. He should already have been under his parachute, floating down to safety. Instead, he was sinking into the depths, still out cold. Dying.

Suddenly, the seat cartridges now ignited and fired properly, propelling him up the rails, out of the cockpit and through the water like a torpedo. Then, miraculously, the ejection sequence continued underwater; the mechanism automatically cutting his straps to release him from the seat and firing the drogue gun to deploy his parachute. The seat, along with his jet, was plummeting to the ocean floor. Still unconscious, Castro Fox was helpless and alone, trapped in darkness, inhaling water through his nose and mouth, a dead weight drowning in claustrophobic free-fall. For those watching the horror unfold from the flight deck of *Veinticinco de Mayo*, things did not look good.

Then, Castro Fox began to slowly rise. The buoyant air trapped between his body and his thick rubber immersion suit was coming to his rescue. Inch by inch, foot by foot, he ascended, his sodden parachute lines dangling beneath him like limp spaghetti. Two minutes later Rodolfo Castro Fox bobbed back to the surface.

Rescue swimmers dived in. Spotters on the deck noticed that Castro Fox was attempting to swim, but was only using his right arm. When the seat had fired he was still holding onto the throttle with his left hand. 'It was a serious error,' he says. His arm had smashed into the side of the cockpit as he and the damaged seat rocketed out. He had broken the radius and ulna, the bones linking his elbow and wrist. He had also fractured some ribs and seriously dislocated his shoulder.

The rescuers freed him from his parachute and passed a sling under his shoulders. A helicopter frothed the waves as it lowered a line. The swimmers motioned to the winch-man that Castro Fox was ready to go. As he rose from the water, he was so badly injured he slipped out of the sling and fell back into the sea. At the second attempt they landed him on the ship's flight deck and forced him to vomit the water from his lungs. They rushed him on a stretcher to the bow aircraft elevator, the quickest way to the ship's operating room, where he suffered his first heart attack. The medics managed to revive him. When they got him down below, he had a second. Again, they saved his life. Racing against the clock, the doctors stabilised him and catalogued his post-ejection injuries. Apart from the broken bones, there were enough traumas and wounds to fill a medical dictionary.

His close friend and fellow Skyhawk pilot, Lieutenant Commander Alberto Jorge Philippi, telephoned Castro's wife,

Stella, at the family apartment south-west of Buenos Aires, telling her that her husband had survived a crash and subsequent ejection, but not revealing the full details. It was not the last telephone call of that nature the friends would make.

Castro Fox was transferred to a mainland intensive care unit. His facial injuries, caused as he impacted the sea, were so bad that the nurses refused his request for a mirror. Stella and their children were able to visit once he had been moved to a ward. If any of his kids had aspirations of pursuing a medical career, the sight of their father was enough to make them aim elsewhere. Rodolfo Castro Fox's illustrious flying days were surely over? As he began a lengthy convalescence, bandaged and stitched, with his arm encased in a heavy cast, the irony of an ejection seat that had both saved his life and nearly killed him did not escape him.

* * *

Although the problems with the Argentinian seat cartridges were well known within their own military establishment, ejection seats were still saving lives around the world. The ever-complex designs trialled during the Vietnam War, allowing crews to 'fly' the seat after ejection, had been put on the back burner, and 'traditional' ejection seats, using explosive cartridges or, increasingly now, rocket packs pioneered by Martin-Baker, were the method of choice. James Martin's seats alone had saved the lives of close to 5,000 airmen across the globe.

He died, aged eighty-seven, a few months before Castro Fox's ejection, having dedicated most of his life to saving the lives of stricken aircrew. Thousands of messages of condolence flooded into the factory from grateful ejectees. An obituary in *The Times* hailed him as 'one of the great engineers of his generation. He did not have to learn the laws of motion, he knew them intuitively.' A packed memorial service was held at St Clement Danes, the Royal Air Force church in London. Former test pilot Bill Bedford gave the address. He had become Martin-Baker ejectee number 511 in

1961, when he had escaped from a prototype vertical take-off and landing (VTOL) aircraft which would pave the way for the Harrier, the so-called 'jump jet'. He told the congregation: 'As one of those whose lives have been saved by his genius, I can say from my very heart: never in the history of aviation has the safety of aircrew and the happiness of their families owed so much to one man. Thank you, Jimmy.'[5]

Thank you, Jimmy would be the words on many aviators' lips in the coming months.

* * *

On 19 March 1982, his eight-month convalescence after his near-fatal ejection almost complete, Rodolfo Castro Fox was watching a TV news report of a group of Argentinian scrap metal merchants who had landed from a supply ship onto an inhospitable, treeless ridge in the South Atlantic. Of little interest except to visiting scientists, the island of South Georgia had no permanent population. The party had a contract to dismantle a derelict whaling station. They duly raised an Argentinian flag and sang their national anthem. Problem: South Georgia, 870 miles east of the Falkland Islands, was a British Overseas Territory.[6]

Able Seaman Neil Wilkinson had joined up in 1979 to 'see the world and get paid for it'. It was also the fresh-faced, fair-haired Royal Navy gunner's escape from a humdrum job in a Leeds printworks, and his mum's violent boyfriend. His first ship, HMS *Intrepid*, a Landing Platform Dock (LPD) designed for delivering troops, tanks and equipment, had taken him onto all manner of Caribbean, Egyptian and Mediterranean beaches. 'The trips abroad were endless, and for a young man of twenty-two it was a great adventure.' But now, thanks to radical defence cuts, *Intrepid* was languishing in a dry dock in Portsmouth, waiting to be sold or scrapped. Her captain and many of the crew had already deployed elsewhere, and he was desperate to join them. He dreaded going home. He got his wish. 'My next draft came through. I was to join HMS *Sheffield*, a Type 42 guided missile destroyer.'[7]

The Royal Navy had broadened Neil Wilkinson's horizons, but he had never heard of South Georgia; nor the Falkland Islands, a rugged and remote archipelago sitting 300 miles east of the southern tip of Argentina. Rodolfo Castro Fox, now nearing a return to military service, knew exactly where they were. And, like his fellow countrymen, he referred to them as the *Islas Malvinas*.

Two-thirds the size of Wales, East and West Falkland, and hundreds of small islands and islets, were variously claimed in the sixteenth century by Britain, Spain and Portugal, and had been the subject of much conflict and many changes of ownership ever since. But the Union Jack had flown over the capital, Port Stanley, since 1840.[8]

Argentina, however, had never renounced its claims to this harsh, unforgiving landscape, strewn with boulders and pummelled by wind and rain, and never more so than in 1982. The dictator, General Leopoldo Galtieri, an impulsive man with a liking for Scotch, found himself grappling with 600 per cent inflation and sweeping unemployment, and allegations of appalling human rights abuses and the disappearance of opponents.[9] What better than a patriotic diversion to unite the nation and save his political neck?

Early on the morning of Friday, 2 April, a spearhead of 100 Argentinian marines landed on East Falkland, met no resistance, and headed for Stanley to reclaim the Malvinas. Another 2,000 troops followed in their wake. All that stood between them and the capital were sixty-nine Royal Marines, and, after a heroic but brisk defence, the Argentinian flag was flying over the red post and telephone boxes of Port Stanley, now renamed *Puerto Argentino*. The following day, British Prime Minister Margaret Thatcher told a packed House of Commons that a Task Force of Royal Navy and civilian ships was already being assembled to ferry troops, aircraft, fuel and weapons to the South Atlantic.[10]

The naval dockyards at Portsmouth and Plymouth Devonport hummed with activity. Neil Wilkinson would not now be boarding

HMS *Sheffield*. He was to remain on HMS *Intrepid*, which was being reactivated to join the Task Force. On 3 April he sat on the edge of his bunk and wrote a letter home.

> *They are calling everyone back, so bang goes my leave. The dockyard is getting ships ready. Aircraft carriers HMS Invincible and Hermes are loaded up with Sea Harriers and there should be a squadron of helicopters arriving soon. Five destroyers are leaving within the next two days and eighteen frigates have been turned around in the Med. I bet you're biting your nails at the moment but don't worry. We may not even get there, but this is what I joined for.*

Rodolfo Castro Fox was now back on board aircraft carrier *Veinticinco de Mayo*, leading a unit of Skyhawks. Amazingly, he had returned to duty even though he needed a mechanic to help him open and close his cockpit canopy. His arm had never fully recovered after his failed ejection. He could not operate the landing gear lever with his left hand so had to cross his right arm in front of his body to reach it. 'When my wife said goodbye to me, she thought we were going on another routine sea exercise,' he said. 'Only the commanders knew our destination.'

The mission was soon clear. 'We were supporting operations to recover the Malvinas.' He and his fellow pilots followed the invasion minute by minute on a portable radio high up in the carrier's control tower. 'Gentlemen,' he told his team, 'the time has come to start working.' The mood was euphoric. 'My men could not hide their joy upon learning of the success of the recovery of the islands.'

* * *

On Monday, 5 April, a vanguard of Task Force warships and aircraft led by HMS *Invincible* sailed from Portsmouth to Ascension Island, a small volcanic outcrop and military staging post south of the equator, 4,300 miles from the UK. A 4,500-man force had boarded requisitioned ocean liners and countless other vessels in a show of strength to encourage the enemy to go home.

Ten days later, Neil Wilkinson and HMS *Intrepid* joined them. Launched in 1964, *Intrepid* would have a vital role to play in the Falklands, hence her reprieve. She could transport 2,000 assault troops, six 30-ton tanks and assorted other vehicles. In the stern, she carried four large landing craft for amphibious assault. Four smaller craft were suspended on cranes. As an LPD, she could offload the force at any location.

Crowds ten deep, waving Union Jacks, gave her a rousing Portsmouth send-off. Tugs shooting jets of water into the air and small civilian boats, red ensigns flapping, accompanied her into the Solent. 'Leaving harbour was amazing,' Neil Wilkinson says. 'Standing on the deck and seeing all those people cheering for us, wishing us well, the balloons and banners; it was such a lovely feeling. War was very far from our thoughts; we just presumed that we would sail south, rattle the sabre, the Argentinians would leave and we would come home without a shot being fired.'

The climate got warmer as *Intrepid* headed south. Weapon drills against the clock dominated every day. Neil was part of a trio of gunners manning one of the Bofors 40mm anti-aircraft guns. Designed in the 1930s, it was now a relic, but it had a fine record and could still blast 120 shells a minute.[11] When a live test-firing was ordered, a line of 50-gallon oil drums was dropped into the ocean. Neil climbed into the aimer's seat, rotated the gun's motorised platform into position and lined up the first drum through its circular sight, a spider's web mounted to the side of the barrel. He opened fire, and, one by one, blew them out of the water. Recalling events forty years on, he still smiles with pride at the memory. 'We were certainly ready for action, but still not really expecting it.'

After passing Ascension Island the mood changed and so did the weather. The sea swelled and the winds became keener. *Intrepid* ploughed on through massive waves. There was less bullish banter about teaching 'the Argies' a lesson. The occupying troops showed no signs of turning tail.

The profusion of briefings on enemy forces, aircraft types and weapons brought home the very real possibility that Neil would soon be shooting down Argentinian fighters. 'I knew that if any aircraft were destroyed, the crew might eject and we could end up capturing some. I had seen films from Vietnam of crews being blasted out of the cockpits in their ejection seats. It looked incredible, almost unreal. It was difficult to imagine how anyone could survive something like that. But, in reality, I didn't give the Argentinian pilots – or their ejection seats – much thought. My job was simple. Kill them, before they killed me and my friends.'

With every nautical mile the conflict was getting closer. On Sunday, 2 May, the 12,000-ton Argentinian Navy light cruiser, *General Belgrano,* was spotted in the 200-mile maritime exclusion zone around the Falklands. The British nuclear hunter-killer submarine HMS *Conqueror* had her in his sights, and, just before 4pm Falklands time (7pm in the UK), unleashed a fan of torpedoes. The explosions were so severe the captain was unable to send out a distress call. At 4.24 he ordered the crew to abandon ship. Minutes later, 770 men in orange high-vis life rafts huddled together in the cold and darkness as the *Belgrano* went down. Some 323 had been killed in the attack. In the aftermath, the vast majority of the fleet scrambled back to the safety of port. The Argentinian naval threat had effectively been eliminated. Now all her aircraft would have to fly from the mainland.[12]

Any British elation at the sinking of the *Belgrano* was short-lived. At 11.04 on the morning of 4 May a trail of fire, exhaust and white smoke announced an Argentinian Super Étendard strike fighter's launch of an Exocet anti-ship missile. Around 30 miles away, HMS *Sheffield* had been policing the Falklands Exclusion Zone since 1 May. With almost no warning, the Exocet smashed into her starboard side with the force of a runaway train, tearing a 9-foot by 4-foot hole 8 feet above the water. *Sheffield* was quickly engulfed in flames. For four hours her crew fought valiantly – and fruitlessly – to save her. Twenty sailors lost their lives, and she became the first British warship to be lost in thirty-seven years.[13]

The conflict over ownership of the Falkland Islands – or Islas Malvinas – had begun.

Neil Wilkinson was shattered when he heard the news. 'The loss of *Sheffield* was a complete reality check, a real shock. She was going to be my next ship; I could have been on board when she went down. I had no idea if any, or how many, of my friends had been killed. I just kept thinking, *It could have been me ...*'

There was a frenzy of activity on the lower decks. 'It was like an ants' nest, people moving about with weapons, making ready,' Neil remembers. 'We were handed flak jackets and white anti-flash hoods and gloves to protect us from shrapnel and fire. It was all getting very real. We weren't going home; we were going all the way.' As the leader of his gun crew he checked the ammunition by the Bofors. Each case held four clips of shells, greased up to protect them from saltwater and weather and to keep the barrel lubricated and ready to load. But could his old-fashioned weapon and its Second World War ordnance take on modern jets?

He took a moment to write home. The fate of *Sheffield* and those who lost their lives hung heavy.

> *I just hope I make it through all this. Sod the medal, there's a lot of fighting going to happen. Everybody's writing, because it may be our last letters. I'm sorry to sound so blunt but it's pointless saying it any other way. I can't say I'll see you in the near future because fate hasn't decided that yet, but don't forget, I love you all.*

As the Argentinian aircrews trained to take on the approaching Task Force, Rodolfo Castro Fox's superiors told him that his physical handicap disqualified him from flying combat missions. He informed them – politely – that he was obliged to disobey their order. He would not send his pilots into aerial combat if he did not go with them.[14]

In spite of their success against *Sheffield* with the French-built

Super Étendards, the omens for Argentina were not good. Many of
their aircraft, including the Skyhawks, were in poor condition, with
faulty ejection seats, cracked wings and a shortage of basic parts.
Early confrontations had cost the lives of twelve inexperienced aircrew.[15] By 15 May, they had already lost sixteen jets and helicopters.
Only one pilot had ejected to safety. The two-man crew of an ageing
Canberra bomber were also presumed dead. They had ejected but
were never rescued.

Although heavily outnumbered, the twenty-eight Royal Navy
Sea Harriers they faced were regarded as superior jets, and their
pilots, despite a similar lack of combat experience, were some of the
world's best trained. Of their three early losses, one had been shot
down by anti-aircraft fire, and two had collided in bad weather.
None of the British pilots had ejected and had already been posted
as killed-in-action.

As the clock ticked over to 21 May, the day selected for the British
landings, *Intrepid* broke away from the aircraft carriers and battle-
ships of the main Task Force and began a steady run under leaden
skies along Falkland Sound, the stretch of water separating East
and West. Support ships fully loaded with Rapier missile batteries,
battlefield helicopters, engineers, artillery and medical facilities
accompanied her, escorted by the frigate HMS *Ardent* and the
destroyer HMS *Antrim*. They took up position in San Carlos Water,
a deep inlet just off the beaches of East Falkland. 'Then everything
went quiet,' Neil Wilkinson recalls. 'There was nothing but the hum
of our own ship and the dozens of others around us, all hidden by
darkness.'

Intrepid's crew had been at action stations since late the night
before. Her stern doors clattered open and landing craft poured
out, packed with elite troops heading for the beachhead. HMS
Antrim's twin 4.5-inch guns delivered a protective barrage of 250
high-explosive rounds over the next half-hour to give them cover.
'Looking out over my gunsights as dawn broke clear and cold over
the Falklands, I could see the lines of the mountains silhouetted

against the sky,' Neil remembered. They reminded him of the Scottish Highlands.

So far, there had been no response from the enemy. Where were they? Task Force Commander Admiral Sandy Woodward wrote in his diary: '07:40. Still a deathly hush. Extraordinary.'[16]

The calm would not last.

Reports of the British incursions began to filter through on the mainland. The Brigadier General commanding the Argentinian Southern Air Force (Fuerza Aérea Sur) took his time to establish that they were genuine. He had sixty-two serviceable strike fighters, thirty-nine Skyhawks and twenty-three Daggers (an Israeli derivative of a supersonic French Mirage) at his disposal, reinforced by eight Skyhawks from Castro Fox's squadron taken off the carrier *Veinticinco de Mayo*. Only once he was sure that the British had actually landed did he order almost all of them into action.

The Argentinian pilots would have to operate under highly unfavourable conditions, carrying out low-level attacks over water with only minimal targeting intelligence, without escort, and at maximum range from their bases. But they were all up for the fight.

The pilots of two Daggers fired up their engines shortly before 10am. Flying just under 100 feet apart and hugging the contours of the landscape, they arrived at the hills to the west of Falkland Sound. 'When we crossed them we had an unforgettable sight, no less than eight British ships,' one remembered. 'We approached diagonally from the stern so they would not have time to use their bow gun or missiles.'[17]

HMS *Antrim* was their closest target. The Battle of San Carlos was underway.

'Without warning the Daggers were on us, coming straight down the Sound,' one British sailor recalls. The destroyer was turning at full power as the jets raked her with cannon fire. 'I was standing by a door and the flight commander shoved me down into the fuel store where we made a big bundle on the floor. They screamed over the top of us with a tremendous roar, leaving the back of the ship enveloped in steam. The locker where we kept flares, phosphorus and other nasty things was pouring smoke.' Four bombs hammered into the water, exploding over *Antrim*'s stern and hammering her sideways. Minutes later, more jets streaked in to attack. A Sea Wolf anti-aircraft missile brought down one of them. Its pilot failed to eject.

As more attackers descended, it felt like a shooting gallery.

Amid the maelstrom, *Intrepid* continued to offload men and machines. Neil Wilkinson had been at his post on the starboard battery all night when he suddenly noticed the sea begin to boil. 'I didn't understand what I was looking at. Then I saw a jet coming very fast over a ridge and realised it was bullets hitting the water.'

Others followed. Wave after wave.

For a second, Neil wondered if he could pull the trigger. He had trained, but without the fear or urgency he now felt. He lined up an aircraft in his sights. 'When it's for real,' he says, 'you just open fire.' He lost count of the rounds. 'My crewmates were working their socks off to keep me loaded.' The noise of the battle surrounding him made him shudder. As the jets screamed towards *Intrepid* he was sure he had hit one. 'You don't think of the man flying it. It's just a war machine. You need to destroy it before it can kill your colleagues.'

Royal Navy Sea Harriers were soon on the scene. The cacophony of the jets, ships' guns and the blast of missiles was mesmerising. Up on the Bofors platform, Neil was in his own adrenaline-fuelled world. He couldn't hear the ship's loudspeakers, so he asked the radio operators to alert him to incoming aircraft. 'They held up scrawled notes on the bridge windows which said "Aircraft taken off" or "Inbound – ten minutes out".'

Argentinian jets ducked and dived so low he found it difficult to get them in his sights. 'Their pilots were very brave. We had been told that their bases were so far away they wouldn't be able to reach us. Yet here they were, giving us a welcome none of us were truly prepared for. One of them even waved at us as he sped by. There was more than an element of cheek in that. We waved back.'

Río Grande Naval Air Base, Argentina
21 May, early afternoon

Rodolfo Castro Fox had been on the first attacks but returned – frustrated – to base with his formation of six Skyhawks. Technical errors in their navigation and attack systems meant they could not find their designated target. He was determined to be at the heart of the action, but while he waited for the problem to be fixed, a new attack mission was ordered. His old friend Lieutenant Commander Alberto Jorge Philippi – who had called Castro Fox's wife to tell her of his near-fatal ejection the previous year – could not believe his luck. He was going to lead it. Supremely focused, he was joining his brothers-in-arms, defending the Malvinas.

He took off with two other jets, and they climbed to 27,000 feet to join up in formation. Each aircraft was armed with four 500lb Snake Eye low-level bombs, plus 200 rounds for their two 20mm cannons. Their objective: a frigate 385 miles away. Time to target: fifty-eight minutes.

Philippi and his fellow pilots flew the last 50 miles 100 feet above the water.

The weather at low level was appalling. Sheets of rain reduced visibility to less than a mile. Philippi called his two wingmen to tighten formation. Marcelo Gustavo Márquez tucked in behind him and José César Arca took up position to his right. Philippi's Skyhawks screamed in low over the beach at 500mph and into Falkland Sound, heading for the masts and antennae of the frigate 5 miles ahead. HMS *Ardent* was pounding the airstrip at Goose Green on East Falkland, where a large Argentinian ground force was stationed. The words of Psalm 23 echoed in Philippi's mind. *Yea, though I walk through the valley of the shadow of death, I will fear no evil.*

He gave the order to attack. *Ardent* was now moving surprisingly quickly into the middle of the Sound, so the Skyhawks had to adjust their approach. 'I assumed they had detected us,' Philippi says, 'so I turned to the right, then smoothly to the left, trying to hide my aircraft from their radar.'

Ardent was steaming at maximum speed across his nose. He saw her gun turret turn to face him, and her shells ripping up the water ahead of him as he began his attack run. Philippi returned fire with the two cannons mounted beneath his cockpit. 'I pressed the trigger. Both jammed after firing only a few shots.' He pulled back his control column. Climbing to 300 feet above the steel-grey water, he felt horribly exposed, but if he went any lower, his bomb fuses would not have time to arm once they had left the aircraft.

Wingman José Arca was in similar peril. 'There were explosions very close to my aircraft. Then I saw a missile leaving the ship. I broke hard to the right to avoid.'

Philippi dropped his four bombs right over *Ardent*. Arca came on the radio as the last of them hit. 'Very good, sir!'

Arca was too close to Philippi for comfort and as he dropped his own bombs he was rocked by the explosions of his leader's. 'It looked like Dante's *Inferno* as I flew through a thick ball of flame and smoke.' He called Philippi. 'One bomb on the stern.' Marcelo Márquez's voice broke through the static. 'Another one on the stern.' More orange-tinged smoke spewed skywards from HMS *Ardent*. Its Lynx helicopter had been destroyed, and the Sea Cat missile launcher was blown 80 feet into the air before crashing back down onto the flight deck. The water supply had been breached, flooding the stricken frigate. Communications between the bridge and control centre had been severed. Casualties were mounting.

At least Alberto Philippi and his friends had survived the onslaught; he now led the trio low and fast in their escape to the south-west. Using the same route to return home, he reasoned, was the best way to avoid the enemy.

* * *

The Argentinians were not having it all their own way. Later that day, three Daggers were homing in on the assembled ships when two Sea Harrier pilots spotted them 500 yards apart and barely 100 feet above the waves. Three minutes from the target and travelling at 575mph, the Argentinian pilots saw their foes just 160 feet above them. As the Harriers engaged, a call went out over the radio from the Daggers. 'Damn! Break right, drop the ordnance.' They jettisoned their bombs to lose weight and gunned their engines, ready for the fight.

As they climbed to try to engage the Harriers, the skies became a tangled mass of aircraft vying for position. A heat-seeking Sidewinder missile left the rails of a Harrier at 1,900mph in a stream of white smoke and flame, and harpooned into the spine of the first Dagger, between its cockpit and its tail. The jet buffeted up and down like a porpoise, burst into flames and went into a flat spin, the pilot ejecting successfully. Unlike some others, his seat had saved his life.

Another Sidewinder streaked away towards the remaining Argentinians. The warning call went out too late. The second missile found its target. Another pilot ejected successfully and floated down beneath his parachute.

A third British missile locked on to the last remaining Dagger, spearing into the fleeing jet. 'My God, what's happening?' screamed its pilot as the Sidewinder blasted his aircraft apart. Travelling at a little over 500mph, he lost all control and reached for the yellow and black handle between his legs, pulling it sharply upwards.

Three Argentinian jets downed in a matter of seconds, with three successful ejections. The pilots spent the night out in the open on West Falkland, but were eventually rescued and returned home. They were the lucky ones.[18]

* * *

The sight of thick smoke billowing from *Ardent* was vivid in Alberto Philippi's mind as he led his trio of Skyhawks away from Falkland Sound, job done. Streaking to safety, he was beginning to think that he and his compatriots were in the clear. There was thicker cloud ahead to cover their tracks.

Marcelo Márquez, bringing up the rear of the Argentinian trio, saw things differently. 'Harrier! Sea Harrier on the left.'

Philippi's blood ran cold. He ordered his formation to drop their tanks and bomb racks to lose weight and squeeze out more speed.

In horror, Márquez watched a Harrier streak through the clouds ahead of him, making straight for the rear of Philippi's aircraft.

José Arca was a thousand yards behind his flight leader. 'I saw a Sidewinder being fired and then flying into Philippi's exhaust nozzle.'

'I felt a violent explosion on the tail,' Philippi recalls. 'The whole plane shuddered and started to climb.' He pushed his stick forward with all his strength but there was no response. His hydraulics had gone. Then he saw the Harrier to his right, just 150 yards away. 'He was coming in for the kill.' He was out of options.

Except one.

'They've got me, I'm going to eject,' he told his wingmen. 'I'm okay.'

Or was he? The recommended speed for ejecting was 155mph, and no more than 400. Philippi was travelling 150mph faster than that. He tried throttling back and opening his airbrakes. There was little change. Keeping his wings level, with his right hand on the stick, he reached down for the handle with his left. An image flashed through his mind: a friend who had broken his neck when ejecting from a Skyhawk. Another victim of a malfunctioning seat cartridge.

Philippi yanked hard. There was a huge explosion as the seat fired. He felt a stab of pain in his neck as he was hurled out of his aircraft. Then he blacked out. When he came to, his ejection seat had gone, his parachute was open and he was dropping towards Falkland Sound.

Seeing the missile taking Alberto Philippi out of the picture stunned José Arca. His shock intensified when he looked behind and to his right. Instead of Márquez on station, he saw the Harrier that had just shot down his leader turning towards him. Its pilot locked his heat-seeking missile onto Arca's jet pipe, but it misfired and fell into the sea. The two jets shot across Falkland Sound no more than 50 feet above the waves. There was no way a Skyhawk could outrun a Harrier. The Harrier pilot lined up his cannon and launched a two and a half-second blizzard of 30mm shells into Arca's aircraft. 'The hit almost caused me to crash,' Arca remembers. 'I was just 10 feet above the water. I lost all hydraulic pressure, my oxygen supply and some electrical power. I prepared to eject.'

Then he changed his mind.

He wasn't going down without a fight.

Gripped by a new resolve, he hauled his beleaguered jet around to face the Harrier. 'The combat lasted for about forty seconds and then he left me,' Arca says. 'Maybe he was short of fuel, or had run out of ammunition.' The Argentinian pilot knew he would never make it back to the mainland. The stick was heavy in his hand. He was low on fuel. He had six gaping holes in his left wing and four in the right. But he was determined to nurse his aircraft east, to the runway at nearby Puerto Argentino – the renamed Port Stanley.

Meanwhile, Marcelo Márquez was heading straight and level to the south-west, seemingly unaware of the second incoming Harrier. The British pilot tried to launch a missile at his fleeing enemy. Nothing happened. He tried again, but with no success. He reached forward, selected 'guns' and squeezed the trigger. A couple of short volleys to find his range. Closer and closer. Blazing cannon shells hit the fuselage just behind Márquez's cockpit.

The Harrier was close to colliding with the Skyhawk when the Argentinian jet disintegrated in a ball of flames. With no time to pull away, he closed his eyes and sank into his ejection seat as he flew through the debris, convinced it would take him down too. As Márquez's right wing broke away from the burning fuselage, the Harrier somehow made it through the tiny gap and headed home to HMS *Hermes*.

Alberto Philippi's ejection blast had ripped off his helmet and oxygen mask. When he came down some 200 yards from the shore, his dinghy failed to inflate, and he had to cut himself free from parachute lines that were snared in thick weed. Drained by the effort of dragging himself out of the freezing water and onto the East

Falkland beach, he triggered his emergency beacon and lay prone until he could summon enough energy to stagger to his feet.

As sunset loomed, there was no shelter from the wind. Soaking wet and in shock, he started to shiver. He pulled out the 8-inch German hunting knife he always carried on missions and began to scrape out a foxhole. The physical effort warmed him a little, and two thoughts now occupied his mind. Had Márquez and Arca survived? And would God prevent him from being taken prisoner?

Trapped in his aircraft, Marcelo Márquez had plunged to his death. It is not known if he tried to eject. But José Arca was still airborne. Just.

Fuel continued to drain from his punctured wings. He wrestled with the controls of his Skyhawk as he flew low over undulating ground at 575mph. He tried to make contact with the Puerto Argentino tower so he would not be downed by the airfield's massed anti-aircraft defences. He couldn't get through. 'After a lot of calls I contacted an army helicopter and asked them to act as radio relay to the ground.' He reached forward for his landing gear lever as he approached the runway and felt a satisfying clunk as the nose wheel locked down, followed by the right main wheel. Then he saw an ominous red cockpit warning light telling him his left wheel had not followed suit.

The airfield controller confirmed it. 'The landing gear has gone. I can see the sky through the holes in your plane. Go and eject over the bay.'

All Arca had wanted to do was save his aircraft. This was a major disappointment. 'I had no choice. I climbed to 2,500 feet over the water and cut my speed to 195mph, ready to eject. It's a very hard moment, as you don't know what to expect, but you just do it automatically. I pulled the upper handle.'

After the sudden explosion, he remembered tumbling through the air 'like a stuntman' until the seat fell away. His parachute blossomed, hauling him sharply upright. After the tumult of battle, 'I was surrounded by total silence.' He inflated his life jacket and watched the aircraft drift away. Suddenly his pilotless Skyhawk performed a smooth turn. 'It didn't want to abandon me. It was descending and coming back to hit me!' He had survived astonishing aerial combat and ejection; would he now be wiped out by his own aircraft? It just missed him, then circled and came at him a second time.

The airfield control officer watched the whole scene unfold. 'We had ordered him to eject over the sea, not on the land where there might be some minefields. We saw him go out, tumbling around in the air like a little puppet until his parachute opened. To our surprise, the plane seemed to become alive and ready to play a nasty joke on us. For some moments it flew towards the pilot, as if to crash into him, then towards the town, then to the airfield in a lively and playful flight. It looked as if it were happy to be free from its master. Considering the damage it might cause, we ordered our guns to destroy it, but surprisingly enough and in spite of the aim of our gunners, it went on flying without being touched, as if the shells refused to hit a friendly plane. Finally it crashed on the beach and was smashed against the rocks.'[19]

Arca released his parachute as soon as he hit the waves. He had not deployed his dinghy because he could see an army helicopter racing towards him. But the crew was not trained in rescue techniques. 'They didn't have a crane to hoist me. After half an hour of failed attempts and only surviving thanks to my immersion suit, the pilot put the skis into the water.' Arca grabbed one of them and wrapped his legs around it, clinging on for dear life for the 500-yard ride to the airfield. A few feet above the ground, Arca jumped clear and the helicopter landed. He then clambered inside and was flown to hospital. Once again, in a matter of minutes, another three Argentinian jets had gone down. Two successful ejections. One fatality.

That night, Rodolfo Castro Fox had the unenviable duty of informing Alberto Philippi's wife that her husband was missing in action. Just as Philippi had informed Castro Fox's own wife of his near-fatal failed ejection seat incident a few months before. He could only hope and pray that his friend had survived the ordeal.

* * *

After thirty-one hours manning his Bofors gun as the battle raged, Neil Wilkinson was finally stood down. It was dark again and the air attacks had stopped. Emotionally and physically exhausted, the adrenaline that had kept him on a high and able to concentrate now ebbed away. 'Time had gone in the blink of an eye. It had all been so fast-moving and intense. I began to understand that I might not survive the next few days.' He crawled into a small storage locker next to his gun and wrote a letter to his family.

> *We've been firing at the aircraft all bloody day. I'm knackered. No sleep for nearly two days. I can't see us lasting. That's why I'm writing now; after what I saw today it may be the last time. I know it's wrong to think like this, but I've got to tell you all that I love you and always will. It sounds easy to say it when life is at stake but you all know that I mean it. I'm going to have to stop for a moment because I'm upsetting myself ...*
>
> *Well, how's everyone? How's the weather? So that's it I'm afraid and if anything does happen to me, you all know that I loved you. Bye for now.*
>
> *All my love, Neil.*

'After writing that letter, I reflected on the day I had just witnessed,' he says. 'It was all too much to take in. In the last two years I had been sailing around the Med and the West Indies, now I was in a theatre of war. I was thousands of miles from home, fighting to save my friends and protect my ship and myself, trying to retake a distant land. I had seen friends die, tried to kill those who attacked them and watched aircraft go down. The emotion was far greater than I could ever describe. It was all too much. I sealed the letter up, said a prayer and began to cry myself to sleep.'

Five Royal Navy ships had been severely damaged. Thirty-two British sailors were now dead and more than twenty injured. Two British Gazelle helicopters had been downed, with three fatalities. An RAF Harrier had been taken out on a reconnaissance mission over West Falkland by a surface-to-air missile; its pilot ejecting successfully to become a prisoner of war. The Argentinian pilots who had reached the Falklands on 21 May had pressed home their attacks with great courage. But they had paid a higher price that day. Two helicopters and twelve aircraft lost; almost all shot down by Sea Harriers. Eight of their pilots had ejected safely, but another four had been killed.

The weather worsened the following day, and there were no air attacks. Shortly before sunrise, the offload of men and equipment from HMS *Intrepid* and her sister ships was complete. The liberation of the Falklands was underway.

Río Grande Naval Air Base, Argentina
23 May, 12.30

Rodolfo Castro Fox briefed his three fellow pilots on their next mission. After plotting a route across the hills and inlets north of West Falkland to avoid British radar, they walked out to the apron, where each jet had been loaded with four 500lb Snake Eyes, the bombs that had done for HMS *Ardent*. Rain and thick cloud smothered the airfield and a 34mph wind buffeted the runway as Castro Fox led the four pale-grey jets into the air. Still not fully fit, he should not have been flying at all, but once airborne, he reasoned, no doctor could stop him. And if he harboured any lingering doubts about strapping into what might be another faulty ejection seat, he wasn't showing them.

A flight of six Daggers and two flights of three Skyhawks had already completed their raids. On board HMS *Intrepid*, Neil Wilkinson clearly remembers 'the most nerve-racking thing was knowing they were coming and then suddenly from nowhere there would be aircraft all over the place'. He spotted the Skyhawks sweeping in towards him and immediately opened fire. 'I could see my tracer going off in front of the aircraft then just behind. It was like being in a movie.' There was a huge explosion just off *Intrepid*'s starboard bow and water cascaded high into the sky. 'We were all wearing ear protectors, but the noise of jets, bombs, anti-aircraft and machine guns, plus the incessant thump of my Bofors, was still head-splitting. I wondered how long we could endure it.'

Suddenly, he saw one of his shells hit the back end of a Skyhawk about a mile away.[20] 'I saw bits flying off but there was no euphoria as it disappeared trailing smoke. Had I shot it down? At the time, I didn't give it another thought. Had the pilot ejected or been killed? It didn't matter; we were all just living second by second, and fighting for survival. It was pure luck. Them or us.'

During infrequent lulls in the fighting, trays of food and chocolate bars were delivered to Neil's gun crew. They often ended up being jettisoned at the start of another attack. He wrote home:

When the Argies come, it's frightening. There's been a couple of times when I've nearly shit myself. I mean it. The speed they come in at, then start firing and dropping bombs! Tell my mates that I am trying to stay alive. Look after yourselves. I can't think of you all the time because I have to keep shooting at bloody aircraft.

Air deaths mounted on both sides that day. Three Argentinian helicopters were brought down, and another two jets. Neither pilot ejected and both were killed. A Sea Harrier crashed into the sea shortly after taking off from HMS *Hermes*. There was no ejection and the British pilot was also killed.

Rodolfo Castro Fox's own mission was not going well. An hour after taking off, one of his wingmen had trouble hooking up with a fuel tanker and had to head home to Río Grande. The remaining three flew on. Surviving missiles and heavy gunfire, Castro Fox and one of his wingmen managed to target HMS *Antelope*, *Ardent*'s sister ship, which had already suffered relentless attacks. Two bombs eventually drove into the frigate, killing one sailor, but they failed to explode.

And now Carlos Zubizarreta, Castro Fox's friend and second in command, was having bigger problems. His weapons failed to release from their under-wing pylons and he could not jettison them. He had no choice but to make for the mainland, low on fuel and hauling home 2 tons of unexploded ordnance.

Landing with bombs aboard was a whole lot riskier than taking off with them. Zubizarreta would have to bring down his over-weight jet on the icy runway in a strong crosswind. He touched down as gently as he could but suddenly found himself heading side-ways under the force of a 50mph gust. A tyre burst on his main port wheel and its rim sparked like a Catherine wheel. Still slewing side-ways, Zubizarreta was riding 2 tons of high explosive, with no way of reining in his mount. The Skyhawk's nose wheel ploughed into the soft ground bordering the runway and collapsed. Zubizarreta reached down and pulled his ejection handle.

As the nose pitched down, his canopy was blown clear and the seat began its sequence, blasting him upwards and out of the cockpit in a cloud of smoke. Tragically, like Rodolfo Castro Fox a few months before, he was the unwitting victim of yet another out-of-date ejection gun cartridge. But he did not share his friend and wingman's luck. The misfiring seat did not rise far enough and, when the drogue parachute deployed, the deceleration forced his seat downwards, smashing him face first onto the runway. The heavy metal seat then hammered onto his back and head as he lay prone on the wet tarmac. His main parachute canopy, now partially deployed, flapped uselessly in the wind. After an epic struggle to get safely home, no pilot deserved to end his life this way – killed by an ageing ejection seat with an out-of-date explosive charge.

Arriving at Río Grande a short while later, Rodolfo Castro Fox lowered his landing gear and switched to the control tower's fre-quency. 'They ordered me to land on the right side of the runway. When I asked why, they did not answer me. They just repeated that

I should use that half of the track.' Coaxing his Skyhawk onto the slick wet strip, he saw Zubizarreta's A-4 carcass to his left, without its cockpit canopy and ejection seat. 'The faulty ejector rocket was still burning and there was a lot of black smoke.'

His own near-death ejection off *Veinticinco de Mayo* the previous year made the loss particularly haunting. That evening he joined the funeral party as his friend's coffin was loaded onto a military transport aircraft with full honours and a rifle salute, to be flown to a nearby airbase where Carlos's widow was waiting to receive his body with their children.[21]

* * *

While his squadron was bidding Carlos Zubizarreta farewell, Alberto Philippi prepared himself for another cold night in the open after his successful ejection, unaware that he had lost two good friends and some other colleagues in a matter of hours.

Ravenously hungry, he spotted some nearby lambs, pulled out his service revolver and took aim. A dull click echoed from the chamber; it hadn't survived the seawater. He improvised a snare and caught six of them. Grabbing the best one, he slit its throat and skinned, gutted and butchered it with his hunting knife. He gathered a pile of peat, only to discover that his matches were damp too. Ever resourceful, he lit his fire with an emergency flare from his survival kit. He enjoyed the warmth from the flames and planned his next move as he watched a leg and shoulder of lamb start to sizzle. As he feasted, he recounted his favourite prayer. *'The Lord is my shepherd, I shall not want ...'*

Living on his wits and hunting skills, sheltering in abandoned buildings, Alberto Philippi avoided capture for three days. On the fourth he saw a group of men working near the south coast of East Falkland and signalled them with a mirror. Their leader turned out to be a local farmer. 'He was very polite,' Philippi says. 'He gave me sandwiches, cake and chocolate. He took me to meet his family and I had a bath. He was a one hundred per cent nice person.' Instead of handing him over to the British, the Falklander allowed his guest to be picked up by an Argentinian helicopter. Five days after ejecting, Philippi was flown back home, ready to rejoin the fight.

* * *

On 24 May HMS *Antelope* was ordered to move into more sheltered water in San Carlos Bay. As two bomb-disposal experts worked on Rodolfo Castro Fox's formation's unexploded bombs, still deep inside her, one of the bombs detonated, killing one expert and severely injuring the other.

Flames spread fast. The frigate's superstructure began melting into a tangle of twisted metal. Out went yet another command to abandon ship. As the last of the crew left, the magazine exploded. Searing white light silhouetted the stricken vessel against the Bible-black sky. Close by on HMS *Intrepid*, Neil Wilkinson was stunned. 'It was one of the most heartbreaking sights, sitting at my gun watching *Antelope* die. Seeing one of our own ships explode right in front of our eyes was very emotional. I looked around at my friends. They all had tears in their eyes.' *Antelope* broke in half and sank the following day. TV viewers across the world watched the iconic footage in horrified disbelief.

The news didn't get any better. The Cunard-owned container ship *Atlantic Conveyor* had also been hit by Exocet missiles. A dozen crew were dead. A number of the landing force's helicopters had been lost, including three giant Royal Air Force Chinooks.

And another RAF catastrophe was about to unfold 8,000 miles away.

CHAPTER TWELVE

FRIENDLY FIRE

RAF BRÜGGEN, WEST GERMANY
25 MAY 1982

Flight Lieutenant Steve Griggs was roused from a deep sleep in the single officers' accommodation with the help of a cup of strong tea. Showered and shaved, his fair hair still damp, Griggs headed down to the dining room for his usual breakfast of poached eggs and bacon on toast. Accompanied by more tea. He glanced through the papers for news from the ongoing battles in the South Atlantic, but by the time they had been sent via the UK to RAF Brüggen, 26 miles west of Düsseldorf on the Dutch–German border, they were several days out of date. Mobile phones, social media and 24-hour rolling news did not yet exist. 'We got most of our news via the BBC World Service and British Forces Radio. I was certainly following events down south, but it was all a bit distant to us in Germany on the front line of the Cold War.'[1]

Steve was not a complete stranger to the Falklands conflict unfolding 8,000 miles away. He was one of the Jaguar pilots who had buzzed the Task Force en route to the warzone so gunners like Neil Wilkinson could get their eyes in. 'Our job was to act as enemy fighters and anti-ship missiles. We flew tight under the wings of a Canberra bomber, then dropped away, heading straight for the navy ships at 600mph, 80 feet over the water, simulating Exocet missiles,' Steve remembers. 'So later images of ships like *Sheffield* going down, sunk by the missiles we had been simulating, certainly stuck in my mind; it was a harsh reminder of what our colleagues were facing. The previous few weeks had been a real wake-up call for everyone.'

But those at RAF Brüggen fighting the Cold War considered they were on the front line of a very different conflict that could erupt at any moment. The bare landscape of the Falklands was a world away from the heavily forested North German Plain that was home to huge numbers of NATO air force jets and troops on constant alert for attacks from the Warsaw Pact nations. This was the front line in a divided Europe where a nuclear Third World War was a clear and present danger. The tension between the two military blocs was electric. 'A conflict between East and West was always seen as a dreadful possibility,' Steve says. 'We trained hard in case the Cold War ever turned hot. Three or four times a month we were on Quick Reaction Alert duty, loaded with live nuclear weapons, ready to take off at a moment's notice to stem a sudden Warsaw Pact attack. For us, it wasn't a question of *if* the war was going to happen; it was more a case of *when*.'

Despite the ever-present threat from the East, it was the dream of every fast-jet pilot to go to Germany. Brüggen was a 'top of the tree posting'. Realistic flying, preparing for war. Duty-free cars, cigarettes and alcohol. And a full-on social life. 'There were official functions, great facilities and parties that were inevitably wild,' Steve says. 'We worked hard and we played equally hard. The daily training for a war no one wanted, but were all very ready to fight, was intense, challenging and very serious.'

It was always flying that came first for Steve. At school in Hampshire he was in the RAF section of the Cadet Force and joined the Royal Air Force in 1970 on a university scholarship. 'My mother had said I could sign up, but I wasn't allowed to be a pilot because it was too dangerous . . .' For somebody who had been obsessed with aircraft since childhood and had a bedroom stuffed with Airfix models, her words were bound to fall on deaf ears.

Steve checked his watch and headed for the car park. He drove from the 'Domestic Site' past endless barbed wire and sandbagged defences, towards the 'Operational Site' on the other side of the base. He passed the shadow of the tall, dense trees that hugged both sides of the narrow road. From here there were no visual clues that he was on an airbase capable of massive nuclear retaliatory strikes. Then, tucked discreetly into a clearing, he saw the semi-circular reinforced concrete roof of a hardened aircraft shelter (HAS) built to withstand Soviet attack. There were eight more of the muscular grey structures secreted between the trees, all just yards from the main runway. He could see the noses of Jaguar bombers protruding from each, all readied for a full day of training.

At 9am Steve pulled up outside a squat building on the south-east corner of the airfield, 14 Squadron's heavily fortified Pilot Briefing Facility (PBF), and joined Flight Lieutenant Paddy Mullen for a coffee. Paddy had driven over from his married quarters to lead the day's two-ship sortie. With several tours already under his belt, he had arrived at Brüggen in January 1981.

Like Steve, events in the Falklands had seemed a long way from Paddy's daily Cold War routine – until 21 May, when he heard a report on British Forces Radio. 'A Harrier had gone down off West Falkland. Then through the military grapevine I discovered it was a friend of mine.' Details were few. 'I had no idea if he had ejected and survived or tragically gone down with his jet. It was a very strange time. I felt almost detached from what I was hearing about a war thousands of miles away. He was operating in his part of the military, and I was in mine.' Paddy was no stranger to death. His best man had been killed in a flying accident. 'One moment we were sharing a drink, next he was gone and there was nothing I could do about it. Accidents – and deaths – were the norm; you accepted the job for what it was and just hoped it wouldn't happen to you.'[2]

Steve was a late addition to the sortie after Paddy's original partner had called in sick. As always, he jumped at the chance to get airborne. In the briefing room he learned their mission would be 230 miles to the north-east around Hamburg, close to what was then called the Iron Curtain – the border separating Communist

East Germany from the West.[3] Paddy was leading the trip with
Steve as his wingman. 'We were routing up past Hanover, towards
Hamburg,' he recalls. Then they would drop down and simulate
attacking advancing Soviet columns. Even though the large formations of invading armour and troops were imaginary for the purpose
of the training sortie, they practised everything for real. Minus
dropping real bombs, of course. Live weapons were only carried
under strict safety conditions: rules never to be broken.

They reached their jets at 11am and began their ritual safety inspection, crouching beneath the aluminium airframe, feeling tyres,
ensuring panels were secure, patting the landing gear and peering
into the twin engine intakes and exhausts, checking for anything
out of the ordinary.

The single-seat Anglo-French Jaguar had been the RAF's principal ground-attack aircraft since the late 1970s, and 14 Squadron
was the first outside Britain to be equipped with it. Air defence
crews flying the RAF's well-worn Phantom fighter, old hands on the
base and ever competitive, had welcomed the new Jaguar pilots to
Brüggen with saucers of milk, but the joke was soon on them as the
newcomer proved itself a Big Cat, winning flying accolades against
NATO's best.[4] Steve and Paddy were huge fans. 'The Jaguar was a
wonderful aircraft, technologically advanced, very dependable and
great to fly,' Paddy says. Steve agrees. 'It was a solid jet; I loved flying
it and totally trusted it.'

External checks over, Steve climbed the yellow steel ladder to the
cockpit. 'I was always meticulous in checking the seat.' It hadn't
always been that way. During his training he had flown 'once or
twice' without having removed the safety pin from the ejection
handle. 'A chill ran down my spine when I landed and realised that
I had flown a whole sortie on an unarmed seat. If I had needed to
pull the handle, nothing would have happened. I would have been
killed. That really concentrated my mind.' Like many aircrew, he
wasn't too preoccupied with the complex engineering of the ejection
seat. 'It was just *there*, an integral part of the jet.'

Steve's view of the ejection seat just 'being there' is one echoed by most aircrew who sit on them. The seats can remain in a jet for years, exposed to rain, sun, snow and ice. They are trodden on with muddy boots as the occupant scrambles aboard and thrown around the skies at high speed and high-g. While they are regularly checked in situ, depending on the version they might only be removed from the aircraft for fuller servicing every two years. And only be returned to the manufacturer every ten years to have the rocket motors and ejection guns, plus the parachutes, replaced.[5] As Steve says, 'You just strapped yourself onto it, and presumed it would work if you needed it.' He knew that, unlike many of the Argentinian versions, it was almost totally reliable. 'A number of my friends had ejected successfully, but you just never really imagined it would ever happen to you. Why would it?'

He made sure the yellow and black firing handle was fully down with the safety pin in position, then stood on the Martin-Baker Mk9, a fully automatic, rocket-powered seat, and began his safety checks. Whether he was in trouble while stationary on the ground, or doing up to 690mph at low level, or as high as 50,000 feet, it would blast him safely clear.[6]

*　*　*

The Mk9 'zero-zero' rocket seat had been developed by James Martin and pioneered in that first live testing in April 1961. Doddy Hay had been almost killed demonstrating it at the Paris International Air Show, but, glitches solved, it had been fitted to Harrier and Jaguar jets in the 1970s. The general design features of the ejection gun and guide rail were retained but a completely new gas-operated firing system had superseded most of the older cable and rod models. And the second ejection handle above the pilot's head, linked to a face-blind mechanism, had been removed. With modern flying helmets and face visors, it was considered redundant. It had also been decided that when thousandths of a second separated life from death, reaching for a lower, more attainable handle was crucial. The zero-zero Mk9 ejection sequence was still initiated by explosive charges firing in sequence in the ejection gun, but crucially, as the seat rose up the rail, a new multi-tubed rocket pack, fixed under the seat, was ignited by a lanyard attached to the cockpit floor, boosting the seat upwards to around 300 feet – depending on the weight of the occupant. This gave the parachute much more time to fully deploy, even if the occupant had initiated ejection while still on the ground – at 'zero feet' and 'zero speed'.[7]

Steve checked around the seat, ensuring everything was in order. From scissor shackles to drogue gun trip rod, parachute safety ties and the rocket initiator, the eighteen checks on his Flight Reference Cards were second nature. He made sure his leg-restraint lines were attached to the floor, his harness was secured to its anchor points and his cushioned personal survival pack was in place in the base of the seat. He double-checked that the manual separation handle on the side of the seat was securely in position, in case he and the seat did not part company automatically after ejecting. Fully satisfied, he sat down and called a member of the ground crew to climb the cockpit ladder and help him adjust and clip his shoulder, lap and crotch straps into the 'quick-release' buckle. Finally, he was handed his helmet.

Steve glanced around the multitude of gauges and dials wrapped around him like a high-tech cocoon and, after starting both engines, the two Jaguars were ready to go. The pilots closed their canopies, removed the safety pins from their ejection seats and held them up so the ground crew could see that they were now live before inserting them in their stowage point. It was yet another way to ensure all checks had been fully completed.

Paddy Mullen took off first. 'It was just a normal day in RAF Germany; a routine sortie. Nothing unusual.'

RAF WILDENRATH
LATER THAT MORNING

A few miles down the road from Brüggen, at RAF Wildenrath, now home of the RAF's air defence squadrons, Flight Lieutenant

Roy Lawrence of 92 Squadron and his navigator, Flight Lieutenant Alistair Inverarity, were all set for a low-level combat air patrol (CAP) in their F-4 Phantom. Like Griggs and Mullen, they were rehearsing procedures to counter a Soviet attack, but the pressure was increased as they were part of that day's Tactical Evaluation exercise. They were being examined by outside assessors to check their readiness for war and would have to respond to multiple changing conflict scenarios to test the base's capabilities. Air and ground crews had to be on their mettle, servicing, arming and getting aircraft airborne against the clock, replicating the stresses of a war against the numerical supremacy of the Warsaw Pact. Welshman Lawrence and Inverarity, a Scot, were sitting at battle stations ready to scramble when, without warning, they were told to switch aircraft. They climbed out of their cockpit and hurried to another Phantom.

A long-range supersonic interceptor and fighter-bomber, the F-4 had been around since 1961 and was one of the most feared aircraft in the Vietnam War. Sixty-two feet of destructive power with a top speed of 1,400mph, it was now commonplace patrolling the skies over West Germany, ready to engage and, if necessary, shoot down any intruders. A senior RAF instructor would later testify that 'flying a Phantom is like watching television, driving a car in a motor rally and working out the 13-times table all at the same time'.[8]

There was no room for error.

As an air defence Phantom, XV422 had been painted in low visibility camouflage – a dull matt-grey paint job with toned-down national markings and insignia. As part of the exercise to test the station's readiness for conflict, it was one of a number of jets that had been loaded with live missiles. While certainly not a normal occurrence, flying with live weapons did take place – with the proviso that a series of pre-flight and airborne safety procedures were in place to ensure they could never actually fire.

One safety measure was to ensure the cockpit 'Master Armament' switch – which activated the weapons system – had a cross of red sticky tape over it as a reminder to the crew not to enable it. The sticky tape had not been applied on Lawrence's jet so they asked for a roll to be sent out. Unfortunately, none could be found in the stores. In the urgency of the changing exercise scenarios, the crew decided they would continue the mission as they had another failsafe check to complete. In the rear cockpit, Inverarity pulled the cylindrical missile circuit breaker, on the panel by his right leg, as far out of its slot as it would go. The electrical circuit to the trigger on Lawrence's control column was broken and the Phantom would now be unable to fire its weapons.

Pilot and navigator were a highly experienced air defence team, and the rigmarole of flying an armed jet was second nature to them. The only shooting they would be doing that day was with a camera. Using the cockpit filming system which recorded simulated attacks, 'witness marks' would register on the film, pinpointing how a live hit would have impacted a target, for the benefit of the later sortie debrief.

Now fully prepared, the crew called, 'Cockpit ready.'

XV422 was 'on state' – fully serviceable and ready to launch.

Their radio crackled into life. *'Scramble! Scramble! Scramble!'* They blasted down the runway, afterburners blazing. Roy Lawrence, who had notched up 1,800 flying hours, hauled his Phantom into a hard turn after take-off from Wildenrath to avoid nearby Dutch airspace. Airborne ten minutes, and now 35 miles from base, 'we were flying east and I saw two Jaguars coming towards me, at about 1,000 feet'. They were ideal 'trade'; practice targets for a mock interception.[9]

NATO fighters and bombers filled the West German skies daily and combat was common. Aircraft were allowed to engage each other, with no radio communications, in order to practise their skills both as air defenders, and bombers trying to reach targets. It was a completely normal part of the flying routine. Most jets were fitted with 'radar warner' devices, which indicated if they were being targeted by other aircraft. Many were equipped with dummy missiles that gave all the indications necessary to prosecute an attack, but would never leave the aircraft. Any engagements were simply recorded on the in-cockpit film equipment. The system was tried and tested and allowed realistic practice combat, where both fighters and bombers could rehearse the skills required for war.

While communicating via radio with his ground controller, Lawrence completed his pre-attack checks, ready to intercept the Jaguars. Inverarity, in the rear seat, was not fully monitoring his pilot's activity; he was checking some unfamiliar routing and monitoring the aircraft's radar, which had a habit of playing up.[10]

Steve Griggs and Paddy Mullen had been in the air around an hour and a half. Mission completed, they were heading home. 'It had been the most enjoyable sortie,' Steve says. 'All routine. Nothing out of the ordinary.'

Paddy was leading. 'The familiar German landscape north of Düsseldorf around the Rhine was laid out in front of us as we began pulling from low level up to 1,350 feet,' he recalls. 'We weaved our way around villages to ensure our engine noise didn't disrupt people on the ground. With hundreds of jets in the air at any one time, it could be quite a problem. We were also keeping an eye out for other aircraft, exactly as we had done so many times before.'

At 12.47, Steve Griggs's radar warning receiver lit up, showing multiple fighters in the area. 'It wasn't unusual,' he says. 'We were completely used to being tracked or locked up as air defenders practised their intercept drills on us. And in that particularly busy air corridor, aircraft regularly used their radars to ensure they could keep track of everyone's position.'

In the Wildenrath Phantom, Alistair Inverarity was indeed tracking Steve's Jaguar on his radar. He wasn't concerned about avoiding him; he was guiding his pilot into a position behind Griggs to carry out a practice attack.

A ground radar controller, monitoring their return to base, came on the Jaguars' radios. *'Fast-moving traffic in your twelve o'clock, low, 5 miles.'*

Both pilots began visually searching in front of them, then rolled their jets to see if they could identify any aircraft beneath them. Steve spotted a German F104 Starfighter passing harmlessly below him. It was offering no threat and they were now close to Brüggen, almost ready to complete their pre-landing procedures. No worries. 'I was looking forward to the sortie debrief and a spot of lunch. I had some paperwork to complete in the afternoon, then probably a couple of beers in the bar after dinner.'

As he came back upright, Steve's radar warner now showed a Phantom coming directly towards him. Again, it was not unusual and, as Steve and Paddy were short of fuel and close to ending their sortie, they decided not to react as it sped by. Unlike the Phantom crew, they were no longer thinking about battle tactics. Then Steve spotted something unusual. 'I was flying at 1,500 feet and around 450mph when the indicator beam on my radar warning screen moved from the front, round to the side, then settled to the rear of my jet.'

Phantom XV422 had turned fully astern of Steve Griggs's Jaguar and was closing for a simulated kill.

'Target within three quarters of a mile,' Inverarity told his pilot.

'Tally-ho!' Lawrence called, confirming he was visual and had Griggs's jet teed up perfectly for the mock attack. A special camera fitted immediately in front of him in the cockpit to help record practice intercepts 'was reassuring me that I was on a normal exercise sortie'. Satisfied the practice attack would be registered on film, he pulled the missile firing trigger.[11]

Instead of the camera shooting a picture of the Jaguar, Lawrence recoiled in shock as a live, 9-foot Sidewinder streaked away from under the wing of his Phantom with a blast of yellow flame. Almost immediately, the heat-seeking air-to-air missile was travelling at close to 2,000mph, homing in on Steve Griggs's aircraft.[12] In another fraction of a second it had found its prey. There was no time to warn the Jaguar. The Phantom crew could only watch the horror unfold.

Still looking forward to landing and lunch, Steve had been oblivious of the events unfolding to his rear. 'The next thing I knew was a huge, violent bang,' he remembers. 'The aircraft was shuddering and everything was happening very quickly.'

His head buzzed. *What the fuck was that?*

The Sidewinder had a 'continuous rod' warhead; a casing of short rods that were welded together lengthwise and wrapped around an explosive charge.[13] Locked on to the heat of Steve's Jaguar's engines, the missile had detonated and the connected rods expanded outwards to form a huge destructive ring, chopping into the Jaguar's aluminium skin, cutting like a circular saw through the fuselage behind his cockpit, which began to fall away. A matter of feet further forward and it would have sliced through his ejection seat rendering it inoperative, or sliced through Steve, killing him instantly. His life was now being measured in thousandths of a second. 'I had no idea what was going on; the aircraft started yawing violently to the right and wasn't responding to my controls.'

Paddy Mullen glanced back in time to see it explode. 'I watched the rear of his jet separate from near the wing. I couldn't believe my eyes. Steve was still in the cockpit, now falling away from the rest of his jet which was being consumed by flames and smoke. What the hell was I seeing? A disintegrating, tailless Jag? It was unbelievable. There was no time for any emotions. Just action.'

Steve felt the same way. 'I had no time for fear; it was all pretty instantaneous and confusing. I was just reacting instinctively.'

'*Eject!*' Paddy Mullen yelled over the radio. 'Steve's detatched cockpit was still upright but falling.'

Through his helmet headset Steve registered Mullen's frantic call.

'*EJECT!*' he yelled a second time. 'The cockpit was now rolling rapidly through inverted, 30 degrees nose-down.'

'In my confused state I thought, *Paddy must know what's going on because I certainly don't*,' Steve recalls. 'I followed his wise advice, reached down with both hands and gave the seat firing handle a really good pull.'

'*EJECT!*'

'In the nanosecond between my second and third call to eject,

I actually saw Steve's canopy blast clear,' Paddy says. 'Then, as what was left of the cockpit rolled through upright, I could see the ejection seat – and Steve – travelling upwards.'

As Steve's ejection gun fired, gases from the first cartridge in the seat had been routed to the power retraction system attached to his harness, hauling him backwards and holding him firmly into the seat. The gases had also been sent to the canopy jettison unit, which immediately fired, blasting off the canopy. A quarter of a second after pulling the handle, the primary cartridges ignited and Steve's seat began to rise out of the cockpit. His leg restraints tightened and the aircraft's transponder was triggered into emergency mode, sending out a distress signal to monitoring air-traffic control agencies.

Like many ejectees, Steve was blissfully unaware of what was happening around him. 'All I remember was a violent explosion and my head being driven down hard onto the breast plate of my life jacket,' Steve says. 'Then I saw the floor of my cockpit dropping away.'

As the seat rose up the rails, the rocket pack ignited, blasting Steve clear. Around half a second later, the drogue parachutes deployed to stabilise it, and Steve's harness was released. His parachute began to deploy, allowing the seat to fall away. Just over two seconds after pulling the ejection handle, Steve felt the reassuring snatch of his main parachute opening. Then a searing pain in his right hand. 'I was still holding onto the firing handle – and hence the seat. When we parted company it was wrenched out of my grip.'

Paddy, still confused about what he'd just witnessed, had lost sight of Steve as he performed a hard turn through 360 degrees. 'As I came back around, desperate to see what was happening, I saw his parachute fully deployed. What a relief.'

Like Paddy, Roy Lawrence, his Phantom now missing a live missile, couldn't believe what he was seeing. Dumbstruck, he called out in horror to the radar controller monitoring the sortie from the ground, 'I just shot down a Jaguar!'

'Check your [missile] state, now,' came the reply. *How many missiles do you still have?*

'I don't care about my [missile] state! I just shot a Jaguar down.'

'Is that no-duff?' the controller asked incredulously. *This is not a drill?*

'No-duff,' came the grim reply.[14]

Steve experienced a few moments of blissful peace. 'I was under the parachute for just a few seconds so had very little time to think. As I looked down at the rapidly approaching green fields and buildings, I still didn't understand what had gone on.' He tried to remember what the physical training instructors had told him about rolling when touching down, to lessen the impact. Too late. 'I landed completely out of control in a field with a huge thump. To make matters worse, I hadn't had time to undo the connectors to the heavy personal survival pack and it was still attached to my backside when the ground came up to hit me. My legs crumpled as I fell hard onto my back.'

But he was alive, having cheated the Grim Reaper by a matter of feet. What was left of his Jaguar ploughed into the ground beside a huge fuel storage site on the River Rhine. If it had hit the tanks the result would have been catastrophic.

Paddy had watched in relief as Steve landed. 'Then he stood up and waved at me. Which was rather surreal.' He immediately put out a Mayday call telling the radar controllers where Steve was. Now, very low on fuel, he had to leave his friend to get back to Brüggen. 'As I pulled up, there above me was a grey F-4 Phantom. I thought, *What a jolly good chap, coming to the aid of a fellow airman . . .*'

Steve had also spotted it. It was now flying very low over him. 'I thought it was odd for an F-4 to be alone – they normally flew in pairs. Could I have had a mid-air collision with his mate? I simply had no idea what had happened and didn't realise just how close I had come to death.'

Nor was he aware that he was ejectee number 5,017 whose life had been saved by a Martin-Baker ejection seat.[15] Badly winded and gasping for breath, Steve's biggest concern was that his legs were hurting from his botched landing. Then something caught his eye. 'I thought I might be in heaven. A blonde vision of loveliness dressed in tight-fitting jodhpurs was running over to me.' Sadly for Steve, the girl's father, a former German cavalry officer turned farmer in his mid-sixties, was not far behind her.

'Did you see what happened?' Steve asked them. They had no idea. They helped him gather up his parachute and, taking his arm, walked him slowly towards their farmhouse.

Gazing down on the scene from the Phantom were the two most relieved men in West Germany. When Lawrence and Inverarity returned to Wildenrath, their jet was impounded immediately until engineers could establish what had caused the disaster. A British tabloid would later run a story: 'RAF Missile Bags One of Our Bombers!' An RAF spokesman saying, 'I don't think it was a case of the pilot making a mistake. It may have been an electrical fault, or something like that.'[16]

The German farmhouse was homely and comforting. Steve phoned the Duty Officer in the operations centre at Brüggen. 'I simply told him that there had been an explosion, that I had ejected and I was okay. That's all I knew.' Life looked even better when his host produced a bottle of brandy. 'We attacked it,' Steve recalls. 'This was in the days when there were no compulsory medicals with blood tests after accidents. One minute I had been airborne and now here I was drinking cognac with the farmer and his beautiful daughter.' In a combination of faltering German, English and hand gestures, Steve tried to explain his ejection. The farmer shrugged apologetically, put his unexpected guest's confusion down to shock and refilled his glass.

After half an hour and several more generous shots, Steve heard whirling rotor blades. He went to the window. A rescue helicopter was touching down in the paddock.

*

At RAF Brüggen, the station's senior executives had just assembled in the Mess for a formal lunch with visiting VIPs. Pre-prandial gin and tonics were interrupted when 14 Squadron's Commander was summoned to the telephone by a steward. To his horror he learned that one of his Jaguars had been destroyed. There was still no explanation as to why. Shortly afterwards, an air-traffic control broadcast resounded around the base. 'Emergency state 1! Emergency state 1! Aircraft reported lost . . .'[17]

With Paddy Mullen now just minutes from touching down, the Squadron Commander apologised to the assembled VIPs and rushed away, driving out to meet him as the Jaguar taxied in. 'On climbing down from his cockpit a shocked young Mullen told me that he had witnessed his number two suddenly explode in a fireball with front and rear sections of the aircraft falling to the ground separately.'[18]

'Nobody had any sense of what had gone wrong,' Paddy says. 'Back in the crew room they told me that Steve was inbound on board a helicopter. We all knew he had been extremely lucky. The sense of shock and utter confusion was palpable.'

When Steve Griggs arrived back at base he was rushed to the medical centre. 'Paddy and other members of 14 Squadron were there to meet me with bottles of beer stuffed in their pockets. We still had no idea what had caused the explosion. After a few drinks I was waved off on a rather wobbly car journey to the hospital on another RAF base for X-rays. Arriving in casualty covered in mud and with a cut chin, there was not a soul in sight, even though they had been told to expect me.' He eventually got the X-ray and waited an age to see a doctor when, hours after the incident, a good friend and fellow pilot came to see how he was. 'I'd had enough waiting around, so, giving up on the medics, we decided it was time for the bar and he offered me a lift home.'

After the Phantom had landed back at its own base, the crew had been able to explain what had happened, and details of the appalling accident were spreading rapidly around the RAF. On the way back to Brüggen, Steve's friend was able to tell him he'd been shot down by one of his own RAF colleagues, causing his jet to disintegrate and necessitating his ejection. 'I was simply astonished. How the hell could an RAF Phantom have shot me down during peacetime training? It didn't make any sense.' Paddy Mullen had also learned the truth. 'I was furious. How the hell did something like this happen?'

It was a question many were asking.

Steve hooked up with Paddy and other 14 Squadron pilots in the Officers' Mess. 'With only a slight cut and not even a sore back I decided to celebrate my lucky escape in the traditional way, by putting on a barrel of beer in the bar.' He also found out the names of the Phantom crew responsible for his situation. 'I telephoned their Squadron at RAF Wildenrath and invited them over for a drink to show there were no hard feelings. Even though they had nearly killed me and written off a £7 million jet. I met them in the Mess foyer. We shook hands and they offered their profuse apologies. Then, to their credit, they entered the lions' den. There were a few astonished faces when the Phantom crew who had just shot down one of their colleagues walked into the Jaguar bar.'

Later that evening, the Station Commander got wind that Roy Lawrence and Alistair Inverarity had been seen happily clinking pint mugs of German beer with Steve Griggs. 'He made his way quickly to the Mess where all of us involved were suitably admonished. He reminded us this would not look good for the inevitable Board of Inquiry, as we could be accused of collaborating and prejudicing the outcome.'

Further drinking continued at more discreet venues elsewhere on the base, into the small hours. 'I played golf the next day with a sore head and only as well as my hangover would allow,' Steve says. 'I was off flying for a week and the small piece of metal lodged in my chin from hitting the breastplate of my life jacket was cut out.' He couldn't wait to get airborne again. 'Back in the cockpit I checked the ejection seat even more carefully, but I had no lingering reservations. My ejection had shown me what a fantastic piece of kit it was. I presumed it was a one-off and would not be repeated.'

*

The same day Steve Griggs was blasted out of the sky by friendly fire during a Cold War exercise in West Germany, over the freezing South Atlantic seas Argentinian pilots were streaking back into the Falkland Islands' shooting gallery. Their own ejection seats were about to get yet another series of workouts amid a very real war.

CHAPTER THIRTEEN

'LONG LIVE THE HOMELAND'

SAN JULIÁN AIR BASE, PATAGONIA
25 MAY, MORNING

Around the time Steve Griggs's Jaguar was being downed by an RAF colleague, First Lieutenant Ricardo 'Tom' Lucero was one of four Skyhawk pilots taking off from a southern Argentinian airbase. Aged just twenty-six, he had already seen active service during the border conflict with Chile in 1978 and knew what it was like to face death in the air. At first, he had thought that there would be no conflict. 'At the time, the Malvinas were not a huge issue in the lives of many Argentinians,' Lucero later told a friend. 'But once the decision to take them back had been made, the military was totally behind the operation. We served our country with honour; it was our duty to fight for that cause.' He thought there was no way the British would try to retake the Malvinas. 'There would be some sabre-rattling and we'd fly over the Task Force in a show of strength,' he said. 'Then there would be a negotiated settlement. But when the *Belgrano* was sunk, and then the *Sheffield*, we knew there was going to be a bloody war. We were facing a baptism of fire; a severe battle where many lives would be lost.'[1]

It was now that full-blown 'bloody' war and the fifth day of the Battle of San Carlos. The previous day, the Argentinians had launched six attacks on ships and ground targets, losing four more aircraft. Two pilots had ejected to safety but two more had been killed. Earlier that morning, HMS *Coventry* had downed another Skyhawk whose pilot had also been unable to eject.

Lucero and his fellow crews had not been honestly briefed about what it might be like to encounter the heavy defensive fire from Royal Navy ships and British troops now dug in with surface-to-air missiles. It was May Revolution Day and Southern Air Force Command was keen to play down crippling pilot and aircraft losses. Like many of his fellow aviators, Lucero was concerned by the lack of quality intelligence. 'Every single mission was a learning experience; we had no idea it would be as deadly as it was. Around half of our squadron had not returned from previous missions, but we had just been told the missing aircraft had been redeployed.'

After refuelling, his formation approached their targets at ultra-low level. Lucero saw flocks of sheep beneath the unusually cloudless sky, running for shelter at the sound of the jets. Hugging the terrain, the Argentinian jets sped towards their foe. 'We had forty seconds to the combat zone,' Lucero said. 'It was impossible for them to intercept us with a hill between us and the target. We climbed over the top to find, about 550 feet below, six or seven ships in San Carlos Bay.' Another pilot, Captain Jorge García, called over the radio, 'There they are! In front of us. Come on!' It was 12.25.

The Royal Navy vessels opened up with a storm of shells and missiles. Lucero was surprised by their ferocity. 'We had been told to expect the minimum in the way of anti-aircraft fire.' Two Skyhawks attacked the ships with cannons and bombs. Like so many others, they did not explode. The wings of both aircraft were perforated by anti-aircraft fire, and, leaking fuel, they were forced to retreat.

Ricardo Lucero was travelling just above the waves at 550mph, ready to attack HMS *Fearless*. 'It was to my right with its bow pointed at me. I banked and turned towards it, fearing my wing would touch the water. I pressed the trigger but my guns didn't fire. Then the British opened up on me. I climbed and dropped my bombs. I knew I then had to get out of there.' The moment he flew over *Fearless* his aircraft started shaking violently. 'Outside my canopy I saw big parts of the nose of my plane flying by.' His port wing had been hit by a missile, with a fusillade of small arms fire thrown in for good measure. 'I tried to call García, then realised my radio was not working.'

Even if it had been, García would not have heard him. The ferocious barrage had severely compromised his hydraulics.[2] Then, as he struggled to control his aircraft and get to safety, HMS *Coventry* scored a direct hit with a Sea Dart and he crashed into the sea 40 miles north-west of San Carlos. Nobody saw him eject so it was assumed he had been killed in his cockpit. He hadn't.

Even though his Skyhawk had received a direct hit, he had been able to eject to safety and parachute into the sea where he boarded his emergency dinghy. Tragically, he was never rescued and a year later Jorge García's decomposed body was found in his dinghy, washed up on the beach of a remote island many miles from where he had initially gone down.[3]

Over San Carlos Water, Lucero's cockpit was filling with smoke. 'I felt strong vibrations in the engine,' he remembered. 'My aircraft wasn't flying any more. It could explode at any moment.' His instrument panel warning lights were blinking furiously. 'I saw another jet going down, trailing black smoke, then I closed my eyes and pulled the ejection handle. I felt a blast of air as the canopy was jettisoned.' And then a huge bang. 'The rocket in my seat began to shoot me upwards,' he said. It felt like having a steamroller on his chest. He couldn't breathe. His legs were shoved back hard against the seat. Then he felt overwhelming pain as his left knee dislocated. 'I had the sensation I was rolling backwards, out of control. I had managed to eject just seconds before my aircraft plunged into the sea. I was incredibly lucky to escape.' Three of the four Skyhawks in Lucero's raiding party had been shot down, with two pilots dying.

Surgeon Commander Rick Jolly, a senior naval Medical Officer, had watched the battle unfold from his slit trench in nearby Ajax Bay. He saw a missile hit Lucero's aircraft, then his parachute canopy crack open. 'He ejected just in time,' Jolly remembered. 'It only went through one or two swings before he hit the water.'[4]

Río Gallegos Air Base, Argentina
25 May, afternoon

HMS *Coventry* and HMS *Broadsword* had been a thorn in the Argentinian side all day, beating off attacks and protecting British ships off Pebble Island, north-west of West Falkland. Four Skyhawks left their Río Gallegos base determined to do something about them.

Leading one section was one of the air force's crack combat pilots, Mariano Velasco. Stocky and rugged, like many of his cadre the 33-year-old proudly sported a luxuriant black moustache. They spotted their targets at 3.20. 'The two ships were about 30 miles off the coast and the visibility was very good,' Velasco remembers. 'I could clearly see their contours on the horizon.' The warships were normally protected by two Sea Harriers, but they had been ordered to break off and keep away, allowing the ships' missiles free rein. The two pairs of Skyhawks would come in almost simultaneously, from different directions.

As he turned to attack, Velasco saw white smoke billow from HMS *Coventry*'s bow.

'Look out!' his wingman shouted. 'Missile.'

The destroyer had managed to get one of its Sea Darts away, its motor plume burning like a blowtorch. Velasco turned sharply away, even though he was sure the missile was flying too low. He was right. It had been fired at too close range. As it passed 500 yards away from him, Velasco turned back towards *Coventry*.

Both ships opened up, their guns creating a dense curtain of exploding fire. As the Skyhawks neared, geysers of water erupted in front of them. *Coventry* and *Broadsword* were shrouded in smoke with every shot they fired. Yet somehow they missed the advancing Skyhawks. The ships began a turn to the east, to minimise their profile.

Two jets zeroed in on *Broadsword*. The leader dropped his

bombs. The smoke was so dense it was almost impossible to see what he was doing. But he had damaged her. He said later that the experience was both terrible and fascinating.

Then he heard Velasco on the radio. 'I'm going in to attack.'

'*Viva la Patria*,' he replied. Long Live the Homeland.

Ignoring the defensive barrage, Velasco sped in at 500mph and dropped his bombs with unwavering accuracy. Two hit *Coventry* below the bridge, detonating a fraction above her waterline. A third hammered into the superstructure above the main deck. His aircraft now much lighter, Velasco screamed over the ship. His wingman shared his delight. 'They exploded very well, sir!'

HMS *Coventry* began to roll almost immediately. Captain David Hart Dyke, the father of comedian Miranda Hart, had no choice but to order his crew to abandon ship and her sailors scrambled to escape as she tilted to 45 degrees. Seventeen men had died in the explosion; one of them only eighteen years old. Two more drowned. HMS *Broadsword* picked up the survivors, many with severe burns. Half an hour after Mariano Velasco had bombed her, *Coventry* joined her sister destroyer HMS *Sheffield* as a war grave.[5]

But Velasco wasn't celebrating. He knew there was much more fighting – and dying – to be done before the war was over. And too many of his friends had already made the ultimate sacrifice. 'I always remembered my fallen comrades,' he says. 'But the Malvinas was a just cause.'[6]

Ejecting out of his aircraft earlier, Ricardo Lucero was now bobbing around in freezing, choppy water with jets screaming overhead, and missiles and guns loosing off all around him. Snared by his parachute lines, his dislocated knee was agonisingly painful. The wind was ferocious. He was half-drowned and helpless.

His target, HMS *Fearless*, anchored at the head of San Carlos Water, had spotted him a few hundred yards away and sent a landing craft to pick him up. As it approached, it didn't cut its engine in time and ran right over him. He felt its hull bounce on his helmet. Propellers spun past, dangerously close. Lucero was eventually hauled from the sea and floundered in pain on the deck. Suddenly he found some strength. 'I needed to continue to fight,' he said later. 'There had been many rumours back in Argentina that if any downed pilots were captured we would be killed out of hand.' Determined to defend himself, he pulled his .38 revolver from his survival vest, pointed it at his captors and yelled, 'Fucking gringos!'

The Marines in the landing craft quickly disarmed him and took him back to the ship he had been trying to bomb. He was dumped on the lower tank deck where he lay in a bedraggled heap, barely moving. Two more Royal Marines stood guard, training their rifles inches from his body. 'I was terrified of what they were going to do to me,' he said. 'I truly thought they were going to throw me overboard. When you have just been captured, you tend to be a bit paranoid.'

Frozen and terrified, Lucero gazed up at his captors. The ship's medical orderlies arrived, still wearing their anti-flash gear in case of another attack, and began to assess the prisoner. From that moment, First Lieutenant Ricardo Lucero ceased being the enemy they had been trying to kill, and became a man in pain, needing urgent treatment. 'I felt this great sense of relief; after all, I was still alive.' They gave him a shot of morphine and scrawled a large black 'M' on his forehead with a magic marker to prevent overdosing. It took five men to lift him onto a stretcher. A medic stabilised his damaged knee with a temporary sleeve and, as the drugs took effect, Lucero lay back in relief, big black circles around his dark eyes. But he would still not let them remove his prized orange silk squadron scarf. He was put on a helicopter that raced low over the water towards Ajax Bay, on the east side of San Carlos Water. His destination: a building that was once a slaughterhouse.

Rusting machinery, an overhead rail system for carrying carcasses, and tie-on meat labels strewn on the floor were testimony to its past, but the T-shaped structure had not been used for killing, butchering and refrigerating sheep for thirty years. Ajax Bay slaughterhouse had initially been used by the landing forces for ammunition storage, but was now an emergency field hospital, devoted to mending bodies and saving lives. The exterior walls had been bolstered with peat blocks topped with stones, in front of which reinforced slit trenches had been dug. The corrugated iron roof had been strengthened with old doors, sandbags and more peat clods. Now it was an A&E department, operating theatre and intensive care unit, all in one ramshackle space which was more *Mad Max* than *M*A*S*H*. Ingenuity and improvisation were the order of the day as surgeons worked through bombardment and air raids in dim light supplied by a small portable generator. Over the entrance in crude capitals were scrawled the words: 'WELCOME TO THE RED AND GREEN LIFE MACHINE' – red for the maroon berets of the Parachute Regiment and green for those of the Royal Marine Commandos, whose members were being treated by a dedicated round-the-clock team.

Surgeon Commander Rick 'Doc' Jolly was in charge. Although he had a team of 120 medics on land and aboard the hospital ships under his command, Jolly was very much frontline and hands-on. Born in Hong Kong, the son of a Polish prisoner of war held by the Japanese for five years, he was a combat doctor to his core. With a ready grin breaking through his battlefield beard, Jolly was strapping, brave, personable and eloquent. He looked forward to nothing more than a spicy curry and a gin and tonic once all this was over. But it was far from over. After he saw Ricardo Lucero ejecting, he heard HMS *Intrepid*'s whistle echo mournfully around San Carlos Water, warning of further air attacks. Jolly knew to expect more casualties.

Lucero's helicopter landed, and Jolly waited as their first prisoner of war was carried into the 'Life Machine'. 'He was a small dark man with large fleshy earlobes and was in obvious pain.' His flying suit had been cut open by the *Fearless* medics. 'I saw that his dislocated kneecap was about 4 inches from its customary position. I couldn't help but feel sorry for him.' The team crowded round the new patient, eager for their first look at an Argentinian adversary. Jolly shooed them away. They had work to do. With the help of a Spanish-speaking corporal, he told the ejectee he was among friends. 'Although he was a prisoner of war, he was also our guest, and just another wounded man that we were here to look after. I could see that, initially, he didn't really believe me and thought I was setting him up for torture.'

Lucero needed an immediate operation. 'We told him that he would wake up with his leg in a plaster cylinder,' Jolly explained. The worry lines clung to Lucero's face until the general anaesthetic began to work its magic. Many years later, Lucero would speak with affection about Doc Jolly. 'After I arrived at the hospital, the doctor told me, "Here, you are just another patient." And the truth is, that's exactly how I felt.'[7]

Doc Jolly watched the Argentinian pilot come round. 'He reached down to his left leg, found that it was still there and began crossing himself in fervent thanks to the Almighty.' Jolly then heard a radio report that more Skyhawks were passing to the north. He thought no more of it until the news of HMS *Coventry* – sunk earlier by Mariano Velasco's formation – filtered through. Soon after, the sound of clattering rotor blades announced the arrival of nine severely burned survivors. The Life Machine medics swung into action, covering their flash burns in a thick white antiseptic and pain-killing cream.

A young *Coventry* engineer, his flayed hands in plastic bags, eyed the sleeping Lucero and advanced on his enemy aggressively. As Jolly put it, 'The fire that burned his hands now burned in his soul. He simply could not understand our respect and friendliness for this colleague of the two pilots who had killed his ship.'

Awoken by the commotion, Lucero was unable to understand what the man was saying, but, recognising the anger and threat in his voice, he resigned himself to his fate. Medics held the wounded sailor back to protect the prisoner, and Jolly ordered a shot of morphine for him to defuse the situation. 'Then, like the rest of the burns victims from HMS *Sheffield*'s once proud sister ship, the young man fell asleep.' As Jolly got to know him, Lucero told him the Argentinian regime had lied to its pilots. The fact he was rescued and taken to *Fearless* was almost as big a surprise as surviving his ejection. 'We had been told it had already been sunk,' he said.[8] 'And they had been fed vile propaganda about how we treated prisoners,' Jolly says. 'It was even said we ate them. That's why they were very relieved to wake after surgery to find all their body parts intact.'

They had also been misled about the increasing loss of life, and the intensity of firepower they would face. Then there was the terrible state of the aircraft they were ordered to fly. While their fearlessness was obviously admired, many of their missions were near-suicidal. And the problem of expired ejection cartridges had never been solved. Only the technical staff knew which aircraft had fully functioning seats and which did not. But they weren't letting on. The allocation of pilots to the jets with a full set of working cartridges was simply left to chance. Those who managed to eject safely – and then be rescued like Lucero – were the lucky ones.[9]

San Julián Air Base
27 May

Mariano Velasco strapped himself into his ejection seat once more and closed the cockpit canopy with his squadron motto ringing in his ears. *Regresad con Honor* – Return with Honour. After sinking HMS *Coventry*, he had every intention of doing just that. His was one of four Skyhawks that took off at 3.30 in the clear afternoon. Their target: the vicinity of the slaughterhouse and refrigeration plant turned field hospital at Ajax Bay where – although they did not know it – their comrade Ricardo Lucero was being treated for his ejection injuries. Although the building itself was no longer being used for ammunition storage, the immediate area was crowded with troops, pallets of supplies and ammunition – all a legitimate target.

An hour later he found what he was looking for. 'We arrived from the south and saw a lot of containers, one or two helicopters, and many pallets,' Velasco says. 'I saw everything very fast; the aircraft was on full power, carrying four bombs.' A hail of shells, small arms fire and missiles rose up from the nearby ships and almost every person onshore. The first two Skyhawks dropped their bombs and raced away. As Velasco climbed to drop his, he felt impacts on his jet. 'It was not very loud – as if a hammer was knocking the fuselage. Then my wingman yelled, "Look out, missile!"' It passed harmlessly between them.

Velasco let his bombs go. Now for the getaway.[10]

Neil Wilkinson was relaxing in the wintery sun by his Bofors with a puzzle magazine. There had been no air attacks that day, but he was still in his flash gear, ready for action, his headphones and helmet hanging on the gun. As the ground forces started their yomp across boggy moorland, all was quiet out on San Carlos Water, but *Intrepid*'s bridge doors were open so Neil could hear any orders. His gun crew were talking to another battery on the port side when a shout rang out.

'STAND TO! STAND TO!'

Velasco and his wingman were streaking away from their attack on Ajax Bay and about to pass *Intrepid*.

Neil raced up onto his platform. Because his fellow gunners would not get there in time, he had to load it himself. He started the motor. 'By then I saw two Skyhawks coming into my firing arc.' He moved the Bofors around until he was sure he had them lined up in his sight. He only had six shells in the magazine. By the time his colleagues returned to reload, it would be too late. He had seconds to get the jets in range. By now it was second nature. He aimed fractionally in front of the first jet, so it, and his shells, would arrive in the same piece of sky at the same time. He fired the first shot, then five more in quick succession.

It had been a good week for Velasco. First *Coventry* and now this bombing run. As he headed swiftly down San Carlos Water, he saw tracers almost brush his wingtip. 'Then I felt two very big impacts coming up from underneath the aircraft,' he says. '*Bang! . . . Bang!*' He didn't have time to count the rest. Neil Wilkinson on *Intrepid* was finding his target.

Experienced fighter pilot that he was, Velasco calmly radioed his wingman: 'I have hydraulic system failure lights on.' Leaking hydraulic fluid was highly flammable.

A few seconds later: 'My fuel emergency lights are on.'

'Sir! Your left wing's on fire,' came the reply.

His wingman saw flame gushing out from under the liquid oxygen tanks. The aircraft could explode at any moment.

'Mariano,' the wingman called. 'Eject now!'

Velasco tested the throttle. It was working well. The aircraft was still flying. In his mirror, he saw smoke spewing from his fuselage. Determined to reach dry land, he started to climb.

His wingman was yelling at him again: 'Mariano! I order you to eject now.'

'Wait a little,' Velasco told him. 'I'll go after I cross the water.'

Neil Wilkinson saw the Skyhawk disappear over the hill, thick black smoke trailing behind it. 'I had the thumbs-up from the lookouts on the bridge that I had hit it.' This time there could be no doubt that the kill was his. He heard a huge explosion. The jet had crashed. 'There was no way that anybody could have got out of that. There was certainly no thought of him ejecting – that never crossed my mind. I just presumed he had crashed and I had killed him.' There was no celebration. No time to ponder the harsh realities of war. Neil's crew was back, and the gun was reloaded in readiness for the next attack.

* * *

As Velasco's formation had sped over Ajax Bay's Red and Green Life Machine seconds earlier, Rick Jolly had been taking a quick break from surgery to eat and rest. He heard repeated, urgent blasts on the soccer referee's whistle which hung on the front door. Air Raid Warning Red – the enemy was inbound. Everyone jumped into the trenches or huddled in boltholes. Most had started blasting upwards with their personal weapons as Velasco and his team had approached, throwing up a curtain of lead. The jets screamed overhead and a bomb hit an ammunition pallet, which had been moved to another part of the complex to make way for the operating theatre. Rifle and mortar rounds detonated ceaselessly.

A bomb-disposal expert approached Jolly after the 'all-clear'. 'Excuse me, sir, will you come and look at this.' A few metres from where the surgical teams were in action, he shone a torch behind the refrigeration pipework. The beam picked out a greenish metal cylinder. One of Mariano Velasco's weapons? 'When he told me it was a high-explosive bomb, my first instinct was to turn and run,' Jolly admitted. 'He grinned at my evident discomfort, and then told me of a second device lying on the other side of the bulge in the ceiling above our heads.'

The noise of the blasts from the exploding ammunition pallet continued. Jolly worried about any time-delay fuses on the unexploded ordnance but kept his fears to himself. He had a decision to make. Shut up shop and evacuate, or carry on. He had been informed there was to be an attack on a heavily armed Argentinian garrison at nearby Goose Green the following day. 'Their assault was bound to be bloody, and there would be a butcher's bill to pay.' He decided they would carry on prising out bullets, staunching wounds and fixing bones under the threat of the two bombs where Ricardo Lucero was also still a patient. There was such demand on helicopters to ferry the wounded to hospital ships that he had become marooned. He also had a problem with blood circulation in his foot. Rick Jolly – or 'Capitán Holly' as Lucero called him, unable to get his tongue around the 'J' – was worried they would have to crack open the cast to ease the swelling around his knee.

During Mariano Velasco's bombing raid, Jolly's crew had given Lucero a steel helmet and laid his stretcher on the floor for protection. Now fully lucid, the Argentinian knew exactly what drama was unfolding, and when he saw the bodies being brought in, ravaged by his compatriots' bombs, he burst into tears. Lucero called over one of the medical staff, took off his blankets and said, 'They must be used for the wounded.'

Rick Jolly was in no doubt. 'This was a good man.'[11]

* * *

Mariano Velasco climbed to 1,000 feet. Once over West Falkland he reduced his speed below 300mph and pulled his ejection handle. Safely under his parachute, he could no longer see his aircraft but heard it explode. 'A pilot can feel a strong bond with his aircraft,' he says. 'But when the time comes, you have to think about trying to save yourself so that you can continue to fly missions. But without a doubt you feel sad when you lose your machine.'

He landed on rough ground. When he stood, he realised he had hurt his ankle. He limped away and found some cover. He heard two aircraft circling overhead but couldn't see them. Were they British jets looking for him?

He weighed his options. Port Howard garrison was closest, only a few miles away. 'But I had heard that maybe they had surrendered.' He couldn't risk it. He waited until it was dark and then started south for Fox Bay, the second largest settlement on West Falkland, and home to 900 men of an Argentinian motorised infantry unit. It would be an agonising walk of around 30 miles. Velasco dragged his injured leg over brutal terrain in temperatures well below freezing, stopping every forty minutes to draw breath and rest. He struggled on, every mile feeling like 10, limping through the night and the following day.

On the evening of 28 May he heard large explosions and saw flares lighting up the sky over East Falkland, close to Ajax Bay. Just a week after the British landings at San Carlos Water, their ground forces were attacking the garrisons at Goose Green, the first land battle of the Falklands War. More casualties would be flooding into the field hospital where his colleague was still being treated.

Velasco pressed on, determined not to be captured after his successful ejection.[12]

* * *

On HMS *Intrepid*, Neil Wilkinson wrote home. The adrenaline was still coursing through him from shooting down Velasco's Skyhawk. '*They dropped their bombs right on top of the hospital. We pumped that much lead into the sky they didn't get away.*' Still believing he had killed the pilot he had shot down, he continued, '*they are learning to play a musical instrument now: the harp.*' He was also in the grip of exhaustion.

> *I'm bloody knackered. I pray for night to come and my six hours of sleep. My beard is coming along fine because of fright. Every time I jump, it grows another millionth of an inch (it's nearly down to my ankles now). I just wish this experience would end and I could rip it out like the page from a book. But you can't. This has been a week I'll never forget.*

On the morning of 29 May, Mariano Velasco stumbled up to a small corrugated iron shepherd's hut on a bleak moorland ridge. With peeling white paintwork and a red-tiled roof which had seen better days, it was unlocked and unoccupied. He slipped inside to find it stocked with food and fuel. After walking some 10 painful miles, this felt like paradise. Unaware that British soldiers were relentlessly fighting their way hill by hill, bog by bog, in frostbite temperatures towards Port Stanley, Velasco was keeping his head down in the warmth. On 31 May he chanced a look out of the shepherd's hut window and saw three men coming towards him on horseback.

Even if he had been physically capable of it, he had nowhere to run. As they came closer, he emerged. They were islanders. He presumed his time was up. 'Instead, they gave me tobacco, rolling paper and matches.' They refused to assist him further but told him they would alert the Argentinians at Port Howard, which had not actually fallen to British forces. Next day, a Land Rover arrived carrying an Argentinian doctor. On the way back to Port Howard, his rescuers detoured to the crash site. His Skyhawk had made a sizeable crater in the moor. Jagged bits of fuselage and landing gear surrounded it. As they poked through the wreckage, Velasco knew how lucky he had been to eject in time. Then he spotted something wedged in a tussock. A small metal box. He knelt down and retrieved his cockpit's compass and clock unit. The hands had frozen at ten minutes to five, the time he had ejected and the aircraft had crashed. One of the soldiers asked if they could take the cockpit canopy as a souvenir. Velasco rested his feet on it for the journey to Port Howard.[13]

Velasco stayed at the garrison, under nightly British bombing, until 6 June, when he was put onboard an Argentinian hospital ship where he was joined by Ricardo Lucero. Both were reunited with their families in Argentina and would return to flying after the war.[14] Like many pilots, Velasco would later admit, 'I never liked talking about the war.' Only those who had been in it could possibly understand.

'At the start, I was apprehensive about what we faced,' recalled Lucero. 'But I was keen to be part of it, to join the fight for what we all believed was right. It was my job to fly, to fight. And, if necessary, to kill or to die. That is the job of a fighter pilot.'

* * *

British victory in the Falklands was not long in coming. White flags greeted the entry of British troops into Port Stanley. The large Argentinian garrison – mostly terrified, bedraggled and hungry conscripts – laid down their weapons and surrendered at 9pm local time on 14 June 1982.

Just over 10,000 Argentinian prisoners of war were repatriated, many of them aboard Neil Wilkinson's HMS *Intrepid*. A total of 255 British servicemen lost their lives liberating the Falklands, and 649 Argentinians were killed. Three Falkland Islanders were killed in a tragic 'friendly fire' incident. The Argentinian forces lost forty-five aircraft in the skies, and countless more on the ground. Twenty-nine of their aircrew were killed; seventeen ejected successfully. Four Harrier pilots died. Five others ejected and survived.[15]

The Task Force was on its way home. Neil wrote to his grandmother:

> We are sailing at slow speed to soak up all the nice warm weather. And why not? We've earned a good breather. I should be home in a few weeks. I've been offered a lift home if British Rail are still on strike.

Grateful to have made it, to have been at the heart of the action and even a small part of the liberation, he soaked up the hero's welcome in Portsmouth. The memory of blasting hostile aircraft from the skies and the images of burning ships were far from his mind as the small boats came out to greet them. Sailors' caps were raised. Flags were waved. Bands played 'Rule, Britannia!' and the national anthem. The crowds cheered in relief as much as celebration.

The war in the Falklands was over and life, seemingly, went back to normal. But, as is so often the case, once the bunting, marching bands, cheering crowds and triumphant politicians melted away, those who had witnessed the trauma of war first-hand had to adjust to normality. For many, life would never be the same. And something soon began to gnaw at Bofors gunner Neil Wilkinson. The image of a burning Argentinian jet plummeting over a hillside, no sign of an ejection, remained imprinted on his mind.

He had taken another man's life.

Dornoch Firth, Tain, Scotland
Monday, 13 September 1982

Despite being accidentally shot down by an RAF Phantom crew a few months earlier in a still-unexplained incident during a training flight in Germany, normality for Jaguar pilot Steve Griggs meant being back in the place he loved best. The cockpit of his jet. And today, miles of uninhabited Scottish peat bog stretched out in front of him. He had just completed a practice mission while deployed from RAF Brüggen to RAF Lossiemouth near Elgin, where he had enjoyed the weekend catching up with old friends over a curry and a few beers. Now it was back to serious work. He was returning to the Scottish base in a formation of Jaguars after rehearsing bombing techniques. 'Suddenly, my number two engine fire alarm illuminated on my warning panel. I went through my checks and pressed the fire extinguisher button. The light went out and I began a gentle climb to assess the situation.'

Then his number one engine fire alarm lit up. 'I radioed the formation and went through my drills again. Before the warning light had a chance to dim, my wingman came on and told me that the jet had fire spurting out of both engines. There was not much assessment to be done; this was going to be a major problem.'

A fracture in engine two had caused a fuel leak to ignite at 1,600°C. Steve checked his mirror. 'I saw flames enveloping the rear fuselage. I had a bit more time to think than on my first ejection, but I realised I had no option. The jet could explode at any moment. I would have to eject for the second time in less than four months. I simply couldn't believe it.' He had time to brace against the back of the seat. 'As I pulled the handle and the seat fired I blacked out, but I came round quickly under my parachute. This time, I was ready to make a proper parachute roll on landing. It still wasn't to be. As I hit the ground I disappeared straight into a soft peat bog up to my calves, my legs and feet firmly stuck.'

Eventually freeing himself, Steve Griggs was dragging his equipment to higher, drier ground when the rescue helicopter arrived. The Jaguar he had ejected from was enveloped in flames. Like the first one, it had been borrowed from RAF Brüggen's 14 Squadron. In the aftermath, a new Station Flying Order jokingly informed the base that, *with immediate effect, Flt Lt Griggs is no longer allowed to fly ANY 14 Squadron aircraft!*

In the sixteen weeks between his first and second ejection,

another fifty aircrew had used Martin-Baker seats. Steve was now ejectee number 5,017 *and* number 5,068.[16]

RAF Wildenrath, West Germany
10 January 1983

The court martial of Roy Lawrence and Alistair Inverarity, the Phantom pilot and navigator responsible for Steve's first ejection, was held in a building recently vacated by the RAF Regiment's Rapier anti-aircraft missile squadron. It was now on duty in the Falklands, protecting the airfield at Port Stanley, deterring any repeat Argentinian invasion.

Looking strained, the two officers attended the hearing in their best uniforms. Their wives and families sat nervously behind them. Facing four charges of negligence, they could serve up to two years in jail. Both pleaded not guilty. The British and international press were there in force, to feast on every detail of how the 'RAF Missile Men' shot down Steve Griggs. The two words Roy Lawrence had shouted when he spotted Griggs's Jaguar jumped eagerly from the headlines.

Tally-ho!

It was going to be a bad week for the RAF. Leading the prosecution's case, the Deputy Director of Legal Services for the RAF in Germany opened proceedings. 'If proper precautions were taken by the aircrew,' he charged, 'the inadvertent firing of a Sidewinder missile was impossible.'

It wasn't quite as simple as that. The findings of the RAF's own official accident report, now shared with the court, revealed a long chain of personal and organisational errors which led to the near-fatal incident.

RAF regulations at the time stipulated that if aircraft were to be flown with live missiles, the missiles themselves must be fitted with 'safety locks' to prevent accidental firing. But no such devices were ever available or cleared for *airborne* use. They did not exist, so the order simply could not be complied with.[17]

Then, when boarding the armed Phantom, the crew had discovered that the Master Armament switch in Lawrence's cockpit – which rendered the weapon system 'live' – had not been secured in the safe position with the regulation cross of red sticky tape to provide a visual and tactile reminder to a pilot that he was carrying live missiles. Inverarity had sent a member of the ground crew to get some tape to secure the arming switch, but was told there was none available in the store.

'Some of our red tape is missing,' said one gleeful tabloid headline. A commentator asked, 'Could the day be coming when a British Prime Minister has to phone Soviet President Andropov and say, "So sorry about Moscow, Andy. We forgot to put red tape on the firing button."'

Lawrence was asked why he had not aborted the mission when he saw the Master Armament switch had not been taped. 'We were the last aircraft away,' he explained. 'We were conscious of an operational requirement to get to battle stations.' RAF Wildenrath had been in the middle of a crucial NATO tactical exercise and the pressure on crews and commanders to prove themselves ready for war was intense.

In the hushed courtroom the revelations, most of them almost unbelievable, but tragically true, kept coming. Another regulation stipulated that the 'trigger isolate' electrical circuit breaker in the rear cockpit should be pulled out – this would mean the firing trigger on Lawrence's control column was inactive, providing yet another means of preventing a missile firing. But Alistair Inverarity *knew* he had pulled the circuit breaker before taking off. How had *that* safety feature failed?

When investigators examined the Phantom, they discovered the circuit breaker by Inverarity's seat, though still disengaged, was partially depressed. He had been carrying his copy of the station's bulky *Aircrew Guide*, and the pistol he had been issued because of the tactical exercise, in the leg pocket of his anti-g suit. A reconstruction showed that the movement of such bulk in the right lower leg could fractionally depress the circuit breaker and make electrical contact. The missile trigger had been – unknowingly – still live.

The litany of embarrassing revelations continued.

Once airborne, their ground controller in charge of the practice interception should have made a safety call of 'Check switches safe' before continuing the mock attack. But the information that they were now controlling a live-armed Phantom had never been passed to the radar centre, so the safety call was never made.

There were many other minor technical and personal links in the chain of events, but the scene was already set for a terrible accident. In simple terms, even though Roy Lawrence had sixteen years in the RAF, and had flown thousands of training sorties, under the stress of the constantly changing and unusual exercise requirements, once airborne, he had simply forgotten he was armed with live missiles. Of course, it may seem incredible to a non-aviator that this could have happened, but 'standard' aircrew drills, endlessly repeated, day after day, year after year, were occasionally messed up. As Steve Griggs himself had admitted, when he flew without removing the safety pin from his ejection seat.

Griggs's near-demise was a simple – if expensive – human error. A veteran Squadron Leader who had logged some 2,600 hours on Phantoms, one of a phalanx of expert witnesses, was asked: 'Is there a risk that a pilot who is used to flying sorties without weapons may forget that he has armed weapons?'

'Yes.'

After five days of evidence, a newspaper report declared, 'Missile-Kill Crew Are Guilty.' The court martial agreed with the RAF's own Board of Inquiry report that the main cause of the accident was *Aircrew Error*. The ultimate responsibility for the loss of Jaguar XX963 lay with the crew of Phantom XV422. The pilot had selected his weapons switches live and pulled the trigger. The navigator had not monitored his pilot's actions or warned him not to proceed, so was complicit in the accident. The Board of Inquiry also acknowledged that a contributory cause of the disaster was *Organisational Fault*.

Roy Lawrence and Alistair Inverarity were sentenced to a 'severe reprimand' which would remain on their service records. Their families left the court fighting back tears. Lawrence, chastened but relieved, told journalists, 'We both deeply and sincerely regret and repeat our sorrow for the serious consequences of this incident in respect of the loss of the Jaguar and risk to Flight Lieutenant Griggs.'

The *Daily Telegraph*'s editorial of 15 January pronounced:

> It is a sad irony that at the same time Sidewinder missiles were being used with such devastating effect by the pilots of the Royal Navy and the Royal Air Force in the South Atlantic, two of their colleagues in West Germany should have accidentally used the same missile to bring down an RAF Jaguar.

As a parting shot it added:

> The suggestion that the only thing which reminds a pilot as to whether his plane is armed is a piece of tape over a small switch, is not a comforting one.

Steve Griggs and Paddy Mullen had both asked after the shooting down how it could have possibly happened. Now they knew, and, to the embarrassment of the RAF and the Ministry of Defence, so did the world. Steve simply wanted to put the whole thing behind him. 'I thought the decision to court martial Roy and Al was an overreaction and unfair. They were merely links in the chain of an inevitable accident that could have happened at any point and involved any other crew. Yes, it was unbelievable, but it showed how pressure on individuals combined with poor organisation and inadequate orders leads to mistakes being made.'

The dust settled. Steve held no grudges. Shortly after, he had dinner with Roy, Al and their families at RAF Wildenrath. 'We didn't really talk about the incident and ejection. There was no point. We had moved on. It was just one of life's experiences that I had no control over.'

He was, though, certain of one thing. 'The ejection seat is a fantastic invention that saved my life twice. If it wasn't for James

Martin's work, I would not have met my wife two years later or become a father.' Steve still has the yellow and black seat handle from his first ejection, mounted on a wooden base. It sits in the downstairs toilet at his home next to a framed newspaper cutting describing the incident. 'It's a treasured souvenir and a vivid reminder.'

Steve's son, himself now an RAF officer, echoed the gratitude many relatives express. 'I have two sisters and, in brutal terms, if it wasn't for those Martin-Baker ejection seats and their brilliant engineering – working precisely as they were designed to do – we wouldn't be here today. And it's even more strange to think how many more thousands, probably tens of thousands, of others like us there must be.'[18]

Some years later at a training exercise in Italy, Steve met the pilot of the German Starfighter that had flown close to him on the day of the accident. 'He had seen the missile leave the Phantom and hit my aircraft. He flew back to his base where he told his colleagues, "The Brits have gone mad! They are shooting each other down!" It sounded so incredible that at first they did not believe him . . .'

It was not long before Paddy Mullen was also able to say, 'I am lucky to be alive because of that invention.' In December 1985 he was firing high-explosive rounds on a weapons range when a ricochet hit his cockpit, which then filled with smoke. The radio went dead. The explosion had cut the electrics and flying controls. 'I was thinking, *Shit, this is not good*. The right engine was failing and I was about to shut it down when, suddenly, the control stick slammed forward against the instrument panel. The aircraft was now going down almost vertically.'

Twelve seconds after the explosion, he pulled the ejection handle and then blacked out. His eyes shot open when his parachute jerked him upwards. 'I was just above the fireball but luckily I drifted to one side and hit the ground quite close to where the jet had crashed.' Paddy had had a very lucky escape. 'The ejection seat was amazing,' he says. 'An amazing piece of engineering. A life-saver. I had never given them much thought before I had to use one for real. But I'm now truly thankful to all those people involved in the design, manufacture and production of that magnificent invention.' Paddy joined his friend Steve Griggs in Martin-Baker's roll of lives saved. Number 5,524.

Reflecting on his ejection nearly forty years later with the author, himself ejectee number 6,089, Mullen is in no doubt that 'the only reason you and I are alive and chatting today is that our aircraft were fitted with ejection seats; an invention early flyers were very suspicious of'. He pauses. 'That's quite a curious concept, isn't it?'

PART V

1988 ONWARDS

THE MODERN SEAT
IN PEACE AND WAR

CHAPTER FOURTEEN

COMMAND EJECT

The sun was warm over the Irish Sea, the sky clear. A perfect scene for flying. The only things missing were the Tornado's cockpit canopy and the navigator's rear ejection seat. Test pilot Keith Hartley was about to take to the air in the two-man bomber on an unusual and potentially hazardous mission. As the sortie progressed, Keith increased his speed to 690mph. He did not look like a man dicing with death in vicious headwinds in the open cockpit. An accompanying aircraft photographed him flying close to 5,000 feet; he can be seen, elbow resting nonchalantly on his cockpit sill, as though he was going for a Sunday afternoon drive in a cabriolet.[1]

But the trip was deadly serious, months in the planning. Keith's Tornado, the RAF's newest addition to its combat fleet, was equipped with the latest Martin-Baker Mk10 ejection seat, the next radical step change in ejection systems. The fully automatic Mk10 used a cartridge that released gases to operate the harness retraction system, ignite the canopy jettison motors and trigger the initial ejection gun cartridges. Considerably lighter than its predecessors, in addition to leg restraints, it also now incorporated arm restraints to prevent flailing limbs. It could put an ejectee under a fully deployed parachute just over two seconds after firing.

But what made the complete system truly revolutionary was possibly one of the most important developments in ejection seat technology. A 'Command Eject' system had been incorporated into the two-man jet.[2] Traditionally, if a two-crew aircraft got into trouble, each crew member physically pulled their own handle. The back-seater (or student in a side-by-side cockpit) would leave first, and the pilot should notionally wait until he was clear before pulling his own handle. At low altitude, where almost all Tornado flying took place, this small time delay could prove fatal.

Now, in the Tornado, the right side of the rear cockpit boasted a stubby black and yellow 'Command Eject' selection lever. When it was selected backwards to the 'REAR' setting, if the pilot initiated ejection, both crew would go, but if the navigator pulled their handle, only they would eject, leaving the pilot in the front of the aircraft.[3] An unlikely but not unknown scenario. When the lever was lifted and pushed forward to the 'BOTH' position, either crewman pulling their *own* handle would initiate *both* seats ejecting in a predetermined sequence as soon as the detonators had blown off the cockpit canopy, the navigator exiting first. In short, if the pilot was incapacitated, the navigator could eject them both. The ingenious system would go on to save countless lives over the coming years of military operations, both in war and peace.

However, in the early days, some experienced pilots joining the Tornado force with its shiny new Command Eject-equipped jets were initially suspicious of giving a navigator the ability to eject them without so much as a by-your-leave. This was especially true of those who had retrained from the single-seat jets like Jaguars and Harriers and had never flown operationally with a crew member. As one ex-pilot acknowledged, 'By nature, pilots were a conservative bunch and enabling "Command Eject" at that time had not yet become standard practice.'[4] While RAF recommendations for its use were not yet fully established, some pilots displayed the same kind of mistrust of the system as that with which their forebears had greeted the very first ejection seats in the 1940s.

But, in the early 1980s, two pilots had been killed while flying Tornados with, for various reasons, the Command Eject system set in the 'REAR' position. In both incidents, their navigators had ejected alone after a serious aircraft malfunction, the front-seater had not and had been killed in the crashing jet. It could never be fully established why their pilots had not ejected. Had they been struck by the canopy as it blasted clear? Were they already incapacitated? Was the jet even flyable without the cockpit canopy after a navigator's ejection? Many aircrew were concerned by the uncertainty and Keith Hartley's test flight was designed to, hopefully, provide some answers.

Monitored by scientists back at the airfield, he was determined to discover what would happen if the Command Eject system was left off, and only the rear-seater had ejected, leaving the pilot in the aircraft without the protection of the hefty Perspex canopy. 'The only question to be answered was whether there was something in the front cockpit environment, after canopy jettison, that could result in injury and incapacitation of the pilot.'

In July 1988, Keith got the green light to put his cabriolet Tornado to the test. The canopy was removed and the holes where it was normally hinged were faired over with wood. Not an absolutely accurate representation of a post-canopy-jettison situation, but hopefully it would make little difference. Mesh was fitted over some of the rear cockpit equipment to ensure it was not blown away in the expected windblast.

Once airborne, and without a canopy to protect him, although airflow noise increased considerably, Hartley was still able to hear air-traffic control and talk to his engineers. 'There was a steady draught which tended to blow under the visor and into my eyes: not a big problem, more of a mild distraction. The loose ends of the shoulder straps tended to waft around in the breeze as well.' Another striking effect was the clarity of vision he had without the thick Perspex canopy. 'It's surprising how much contrast and clarity is lost through even a well-maintained and polished canopy transparency.' However, 'there was one characteristic that the boffins hadn't anticipated', he says. The heating effect of the circulating air blast at increasing speeds. At just over 500mph his cockpit was heating as fast as a fan oven. Approaching 700mph he was sitting in a 'decidedly unpleasant' 60°C (140°F). The fact that he was wearing full winter clothing designed to protect him from the predicted freezing temperatures at high speed didn't help. 'Once back to 400mph the cockpit felt very comfortable. Time to review the test with the engineer and then home for beer and medals.'

This highly unusual test flight had proved that it was possible to fly a Tornado without a canopy. Nothing came loose in the cockpit that could have injured its occupant. 'The front seat environment was not only habitable but surprisingly benign,' Keith Hartley reported. 'After the results spread through the squadrons, the concerns quickly melted away.'

Although almost all Tornado crews did begin flying with the Command Eject system set to 'BOTH' there were still occasional exceptions. When the author began his Tornado training in early 1989, some older pilot instructors – perhaps understandably – still didn't fully trust inexperienced navigators, and preferred to fly with the system unset. Though once on his operational squadron later that year, every pilot he crewed with wanted the system active, one saying, 'If the shit ever hits the fan, I want *every* possible chance of getting back home to my wife and family!'

However, few other RAF jets at that time were equipped with the Command Eject technology. And lives were still being lost.

RAF Leuchars, Fife, Scotland
Monday, 9 January 1989

After spending a restful Sunday with his family, Flight Lieutenant Gordon Moulds joined his pilot Dave and another Phantom crew in the briefing room.[5] Now the extended Christmas holidays were over and everyone was back to work, they couldn't wait to lead the two-jet training sortie over the North Sea. With more than 4,000 hours, of which 1,500 were on the Phantom, Dave was assessed as 'an exceptional pilot and instructor'. Gordon Moulds knew they didn't come any better.[6]

The two crews changed into full flying kit and walked to their aircraft. Moulds was full of praise for the now ageing jet. 'The Phantom was phenomenal and brilliant to operate,' he said. 'It was still an incredibly capable machine. Functional, but very powerful and very noisy.' He climbed into the navigator's cockpit and began strapping into his Martin-Baker Mk7A, an older, rocket-assisted model with two firing handles. Either the pull-down face screen or the one between his knees could be used to exit.

When they were both strapped in and ready, they lowered their canopies and removed the ejection seat safety pins. The two Phantoms took off at 3.30. The weather was good, with visibility a little over 6 miles. There was no flight control column in Gordon's rear cockpit and, unlike the modern Tornado, the older Phantom did not have a Command Eject system; regardless of the circumstances, each crewman needed to pull their own handle to escape.

On their way out to the exercise area over the North Sea, the aircraft split apart to perform their first exercise. RAF jets regularly intercepted Soviet aircraft straying too close to the UK, shadowing them and monitoring their actions. And if the Cold War had ever turned hot, they could use a variety of weapons to shoot them down. The second Phantom was playing the Soviet 'intruder' and Gordon's pilot climbed to 10,000 feet and then swept down to 1,000, heading at 575mph to intercept the intruder who would then evade, simulating an attack profile and making the fighter's job much harder.

The Phantoms prepared for a second intercept. Gordon's pilot took his jet back up to 10,000 feet. Up until that point, Dave had responded coherently to all his calls. Everything was going to plan.

It was 3.57.

Then the timbre of Dave's voice on the intercom began to tell a different story. 'This is no use,' he said. 'We'll have to go back. I'm feeling ill.'

They were 90 miles north-east of Leuchars.

'It was the first indication things were going wrong but I didn't really understand what he meant,' Gordon recalls. 'I told him to turn west and I immediately put out a "Pan" call [a step down from a Mayday emergency call] to alert our controller, telling them we had a "pilot problem".'

The other Phantom now came alongside on the left. Its pilot told Gordon that Dave was sitting normally in his ejection seat and looking straight ahead. 'At this point, I had no idea how quickly events would unfold.'

Dave told the second Phantom pilot, 'There's something wrong with my eyesight. I can just about maintain straight and level flight.'

Gordon pulled his mask to one side to make sure there were no toxic fumes in the cockpit that might have caused Dave to suddenly feel so ill. 'Everything seemed normal.' But it was far from normal. 'Dave then told me he was feeling numb. I was now very alert to this becoming a serious problem. I began talking him through the manoeuvres – telling him where to turn, when to level out, to keep the nose up. Dave responded "yes" to each instruction. His voice was normal but slow, and it quickly became obvious that he wasn't following them properly.'

Gordon's internal alarm bells were now ringing loudly.

'Dave rolled the aircraft out but was not heading in the right direction. And we were still going down.' Gordon tightened his seat straps in case they had to eject. The Phantom plunged through 4,000 feet. 'Pull up!' he screamed.

'Dave finally reacted and pulled the nose up.' But in his confused state, he had yanked the stick too far back. 'We were now going up near vertically and the speed was reducing rapidly,' Gordon says. 'I had no doubt the jet was about to stall.' The Phantom was still climbing, but increasingly slowly. 'I knew that things were not going to end well,' he recalls. 'It was now totally clear Dave could never get the aircraft back under control.'

At 4.04pm Gordon upgraded the Pan call to a Mayday – the highest classification of emergency.

At 9,000 feet, with the speed down to 172mph, the Phantom 'departed' from controlled flight and began to tumble. As the jet sliced down through the sky like a falling leaf, Gordon had only one thought – 'how to ensure Dave ejected'.

'Stand by to eject!' Gordon shouted to his friend.

There was no response.

'All I needed was him to pull his own handle and get out of the aircraft safely. I really felt I shouldn't go before he was clear. But I was helpless.'

Over the radio, the second Phantom pilot implored Dave to eject too. Four times.

'I was really struggling with the reality that I knew I'd have to leave him,' Gordon remembers. 'But I didn't want to. How was I meant to do that? I had so much flashing through my mind, desperately hoping for a solution.'

There wasn't one. And Gordon's own chances of survival were rapidly diminishing. The second Phantom's pilot now began shouting at Gordon to eject.

Eject in time.

'He had to yell at me three times. I have no doubt those calls brought me back to reality and saved my life.' Over their recorded radio transmissions, a deep audible sigh could clearly be heard on the tape. 'That was me realising time had run out.' Gordon still relives the emotions he went through that day, but he had no choice if he was to save his own life.

'I reached down for my own handle, wondering if I'd left it too late.'

A matter of seconds before impact, as the jet fell below 4,000 feet, he pulled the handle between his legs. As he came out of the cockpit, his chin was forced down hard towards his knees. When his parachute jerked him upwards he felt an agonising pain in his back. 'I saw the jet falling beneath me. My rear cockpit was empty, but the front cockpit was still intact. The jet pancaked into the sea and exploded. Then disappeared beneath the bubbling surface. I knew Dave was dead.'

It was 4.05. Just eight minutes after the incident had begun.

*

The sea was coming up to meet Gordon fast. He fumbled behind his waist to release the catches on his personal survival pack connectors but the pain in his back was so severe he couldn't reach them. He knew that landing with the heavy yellow box still attached was never the preferred option but he had no choice. Instead of dropping on a lanyard, the PSP, containing his life raft, emergency signalling flares and a few basic rations, was still fixed firmly under his backside. He crashed into cold, heaving, 20-foot waves, and somehow was able to release his parachute harness, inflate his dinghy and grasp the handles. Despite the agony in his back, he managed to haul himself in and disentangle himself from his parachute lines.

The sea temperature was less than five degrees. Even though he was wearing a sea immersion suit and gloves, fluffy internal suit, long johns and his helmet, he started to lose the feeling in his hands. Cold water robs the body's heat thirty-two times faster than cold air and, without his waterproof survival suit, Gordon would have succumbed in perhaps ten to fifteen minutes. First, he would begin shivering. Then experience extreme tiredness and rapid breathing. As the body temperature drops, the shivering becomes more violent; however, it stops just before you become delirious and struggle to breathe. Finally, you lose consciousness. Hypothermia is fatal unless the victim is rapidly removed from the water and properly treated.[7]

Even in full kit, the freezing temperatures, brought even lower by the wind effect, soon began to take their toll and Gordon's dinghy was now taking in an alarming amount of water. 'I'd been in the water about fifteen minutes and it was now dark and the cold was taking over.' Gordon could feel himself fading; losing awareness of his surroundings and feeling detached from reality. 'I concentrated on my children to stay awake, telling myself, *If you ever want to see them again, you have to stay conscious.*'

'I began to get quite frightened, wondering if I'd get rescued before I succumbed to the cold.' He spotted a search helicopter in the distance. 'My hands were like blocks of ice, but I just managed to activate a flare.' The helicopter did not see him. He had now lost all feeling in his fingers and with what strength he had left launched a second red mini-flare.

This time they saw him. Half an hour after ejecting, he was winched into a yellow Search and Rescue helicopter.

'The first thing they asked when I got inside was, "Do you know where your pilot is?" I simply had to tell them he had gone down with the jet. It was a truly terrible moment.' There was no sign of Dave or the Phantom and the air-sea search was eventually called off.

Dave's body was recovered with the wreckage three weeks later. The post-mortem showed he had been killed by multiple injuries on impact. Losing control of the aircraft and his failure to eject was put down to incapacitation. But what caused it? Many medical reasons were explored by the RAF's subsequent Board of Inquiry, including heart attack, severe migraine and irregular heartbeat, but none of them were conclusive. The only thing the board could say with certainty was that Dave had been 'totally incapacitated' and unable to fly the aircraft.

Gordon had suffered three crushed vertebrae in his upper back during the ejection. The pain gradually ebbed, but not his anxiety over having to leave Dave. He did rediscover his love of flying and went on to have a long career in the RAF, but what still haunts him to this day was his inability to save a friend and colleague. As he says, if only the Phantom had been fitted with Command Eject, they both might have survived when he pulled his handle.

* * *

The Tornado's Mk10 seat and Command Eject system were revolutionary. Most of the mechanical rods, cables and firing bolts of earlier versions had been dispensed with, and there was only one ejection handle between the aircrew's legs. With the system set to 'BOTH', when either pilot or navigator pulls their individual handle, the system is activated. The cartridge in the firing unit under that seat is triggered, releasing gases which are routed via a labyrinth of valves and pipes to begin the whole ejection sequence.

1. Headbox containing parachute
2. Combined seat and parachute harness
3. Seat cushion
4. Leg restraint lines
5. Arm restraint lines
6. Seat firing unit
7. Rocket pack
8. Thigh guard
9. Manual separation handle
10. Fibreglass personal survival pack box
11. Seat safety pin
12. Ejection handle
13. Harness quick-release buckle

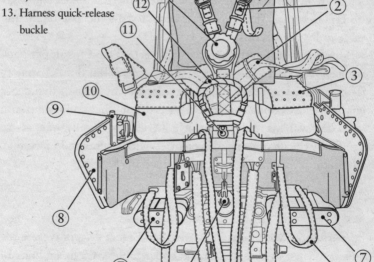

MARTIN-BAKER MK 10 EJECTION SEAT

Instantaneously, the gases route via pipes from one cockpit to the other, setting off the same sequence in the other seat, even if that handle has not been pulled. The gases also route to the aircraft's canopy jettison system, unlocking its catches and firing its rocket motor. The canopy, freed from its locks, is forced upwards until the hinges fracture and it flies away. If for any reason this system fails, a secondary system consisting of an explosive cord routed around the base and through the middle of the Perspex would be detonated like a small bomb, shattering the canopy open.

Gases also go to the harness 'power retraction unit' which fires and hauls the seat occupant back, tightening and locking the straps.

0.3 seconds after the handle is pulled, the gases reach the first ejection cartridge in the navigator's seat ejection gun which ignites so the seat begins to rise.

0.4 seconds later, they trigger the pilot's own cartridge and, as both seats rise, two further cartridges ignite to sustain the thrust.

Unlike older seats, which only had leg restraints attached to garters around the airman's legs, the Mk10's arm-restraint cords ensured even greater aircrew safety. This had involved a complete redesign of the old waistcoat-style life jacket to include sleeves. Two white cords were attached to the cockpit floor, then routed up through grips in the seat. While strapping in, the crew would clip that other end of the cord to special connectors on their life jacket sleeves at the shoulders. During normal flight the pilot and navigator would tuck the straps under their harnesses out of the way, allowing them to move their arms freely and attend to their cockpit duties. On ejection, as the seat rises up the guide rails, the loose restraints would instantly pull tight down a simple Velcro seam in the sleeve, bringing the arms within the frame of the seat, holding them tightly to avoid flailing and injury. Leg restraints are similarly tightened. An emergency oxygen supply is also activated and personal equipment tubing and communication leads automatically disconnected.

As the seat reaches the top of the guide rail, a lanyard fixed to the floor of the aircraft pulls the sear from the rocket pack under the seat which ignites, burning and delivering 4,500lb of thrust for

0.45 seconds. The rocket packs under each seat are slightly offset in opposing directions so that each rising seat diverges from the other to avoid collision. One second after ejection initiation, as the rocket burn subsides, the drogue gun primary cartridge fires to release the seat's stabilising drogue parachutes.

The barostatic unit senses what height the seat is at, and that it has decelerated sufficiently, and, if below 10,000 feet, fires its own integral gas cartridge immediately. The gas-operated shackle holding the drogue parachute to the seat opens, releasing it to extend and begin withdrawing the occupant's personal parachute. At the same time, the gases route to the harness release mechanism and leg-restraint lines, and the guillotine system that cuts the arm restraints, freeing the occupant. As the ejectee's personal parachute is withdrawn and begins to inflate, they are pulled clear of the seat which falls safely away.

The whole process takes around 2.5 seconds.[8]

It was the Mk10 seat that gave the author his second chance at life.

* * *

Although lifesaving and revolutionary, Command Eject systems were not always foolproof. In 1984, a Tornado crew was flying a low-level training sortie in southern Germany when the pilot saw a US Air Force A-10 'Warthog', best known for the fearsome Gatling gun mounted in its nose, crossing in front of him at very short range. 'All I could see was the A-10 turning directly towards me,' he recalls. 'Straight towards my bloody face.' If he didn't take immediate, drastic action, the two jets would collide in seconds. 'I banked hard left, rammed the throttles forward and pulled on the stick harder than I had ever done in my life.' There was no time to warn his navigator who had his head down in the back seat, checking his navigation kit.[9]

'The next thing I felt was a huge bang and I presumed we had hit the A-10. My hands were still holding the stick and throttles. I blacked out and, when I came to, I was floating down in my parachute and maps from the cockpit were fluttering down alongside me.

It was all incredibly confusing. I glanced over to see my nav under his parachute. I just didn't understand what had happened; had our seats somehow been fired during the collision?' Unfortunately, the explanation was rather more simple, though perhaps more difficult for a non-aviator to comprehend.

As the pilot had hauled on his controls to avoid the collision, the violent high-g manoeuvre and sudden dramatic acceleration had forced the navigator, head-down in the rear seat, lower in the cockpit. When he looked out to his left, he saw the ground filling his view and no visible horizon. They were rolling rapidly and seemingly descending. Completely disorientated, he was convinced the Tornado was out of control and seconds from crashing. He pulled his yellow and black handle, ejecting them both from a perfectly serviceable aircraft. To his shock and surprise, his unknowing pilot in the front seat, still fully in control of a normally functioning, multimillion-pound jet, was also shot out of his cockpit. The Command Eject system had functioned exactly as designed.[10] The Tornado crashed into the German landscape below, exploded and was totally destroyed. The pilot suffered a major back injury during the ejection, which kept him away from flying for three months. His navigator only had superficial wounds.

Command Eject may not have been *entirely* perfect, but there are many aircrew alive today more than grateful for its existence.

RAF LAARBRUCH, WEST GERMANY
9 JANUARY 1990, MORNING

A year to the day after Gordon Moulds had ejected and his pilot was killed, Tornado navigator Neil Johnston was enjoying breakfast with his wife Ivy in their married quarters in the German town of Weeze, close to his RAF Laarbruch base on the Dutch border. They both hailed from Northumberland, near the market town of Hexham, and had been married just over a year. Neil was looking forward to his twenty-sixth birthday the next day, and a skiing trip the following week in southern Germany, part of a military winter survival exercise.

The young Flight Lieutenant kissed his wife goodbye and cycled off to work. Laarbruch was set amid sprawling forest and during the Cold War was home to many British frontline squadrons. As he pedalled through the base, past the whitewashed barrack buildings, the NAAFI supermarket and the Astra cinema, the weather was not at all promising. A typical German winter day – cold, misty and raining. Dense cloud hung like a heavy duvet over the airfield. 'I suspected there would be little flying that day.'[11]

Rubbing his hands to restore warmth to his fingers, Neil shared a brew in the Squadron's crew room with his pilot, Flight Lieutenant Ian McLean, a couple of years his senior. Previously a Jaguar pilot, back in 1983 he had been forced to eject as a trainee at RAF Lossiemouth, near Elgin in the Scottish Highlands, when he hit a flock of birds and lost all power. He had always remembered one of his instructors hammering home the three words so many aviators held dear: *Eject in time*. 'He told us that we were all young lads and, whatever the cost, a jet could be replaced. Don't try to be a hero. Ejecting and being carried away on a stretcher was far better than being carried away from a burning wreck in a coffin.' Ian had ejected seconds before his jet hit the ground and returned to flying within days.

Ian and Neil suddenly got some good news. They would be airborne after all, but would bypass the poor weather by transiting across the North Sea. They were to fly a first sortie over to the UK, landing to refuel at RAF Wattisham, north-west of Ipswich, and then a second low-level sortie around the UK after lunch, before heading back to Germany.

They landed at Wattisham later that morning.

RAF COLTISHALL, NORWICH
9 JANUARY 1990, LUNCHTIME

Sixty miles north of them, Wing Commander Dim Jones was engaged in what he described as an 'idyllic tour' as a Jaguar pilot and Squadron Commander at RAF Coltishall near Norwich. Highly experienced, he was getting ready for his own sortie that January day. He and two other Jaguars would be flying a low-level sortie at 250 feet, routing north through Yorkshire and past Newcastle, before returning home to Norfolk.

The three aircraft would be carrying out a 'two-versus-one' sortie: Jones and his wingman would be flying as a pair of attackers hitting simulated targets, while the third Jaguar would act as the aggressor – known as the 'bounce' – carrying out a series of intercepts against the pair as they tried to hit their targets.

Low-level tactical flying so close to the ground was hazardous. There was no 'air-traffic control' system to keep high-speed jets apart. Aircrew had to rely on their own 'lookout' – constantly checking the skies around their jets using their eyes alone. This also had to be done while keeping a check on cockpit instruments, fuel, engines, hydraulics, navigation, weapon systems and more. And, of course, still flying manually and avoiding the ground, radio masts, power lines, civilian airfield air-traffic control zones and countless other dangers.

As Tornado pilot Ian McLean says, 'It's a bit like driving – no one is *telling* you to avoid the car on the other side of the road, or the one you are overtaking on the motorway. You are simply looking out of the windscreen, ahead and around you, adjusting your speed and manoeuvring as required to drive safely. That's what we do when flying at low level. Except we are doing it in three dimensions. And all the traffic can be travelling in different directions. At nearly 600mph and at 250 feet. And there are no crash barriers, no warning signs and no police patrols monitoring the route. In reality, the chances of two aircraft being in that same, tiny, three-dimensional piece of sky are infinitesimally small. But it can happen . . .'

Crews kept their eyes 'out of the cockpit' as much as possible, looking for other aircraft or hazards in the area. This could be anything from other fast jets, to helicopters, transport aircraft, light civilian aircraft, even hang-gliders and, occasionally, parachutists. The single rule governing this kind of combat flying was 'see and avoid'.

As Dim Jones knew to his cost.

In 1984 he was in a formation of three pairs of Jaguars heading

for the Norfolk coast, all crews were constantly visually scanning
the area around their formation, to ensure they kept position and
that they stayed well away from other jets. 'After checking the skies
in front of my own wingman, I resumed my normal forward lookout
just in time to see my entire windscreen full of a Tornado. I can still
see it there clearly today,' he says, holding out his hand in front of
his face. 'There was nothing I could do, no time to react.' The two
jets came together at a combined speed of 1,000mph, and the tail
was sliced off his Jaguar. 'It became an uncontrollable metal tube
with engines.'

A thousand feet up, both he and the Tornado crew ejected safely.
The RAF's Board of Inquiry deemed both pilots 'culpably negligent
to a serious degree', deciding that – at high speed and with just
seconds to spare – they still should have 'seen and avoided' one
another. The board also emphasised the warning that: 'all aircrews
are reminded of the paramount importance of lookout in relation
to all other tasks. Moreover, that specific emphasis is given to short
timescales available for collision avoidance when small, high-speed
aircraft are head-on.' It was a harsh judgement involving two aircraft closing on each other at almost 3 miles every ten seconds.
The admonishment had not harmed Jones's career and he was now
vastly more experienced, had flown Phantoms with the US Air Force
and was the Commanding Officer of his Squadron. But as he walked
to his Jaguar six years later, he still felt that the inquiry had been
unfair to blame him for not seeing the Tornado. He believed that
there were other factors that should have been taken into consider-
ation. Visually identifying and avoiding a collision with fractions
of a second to spare was a very tough ask.

As he was about to find out again.

RAF WATTISHAM
LUNCHTIME

After a packed lunch at RAF Wattisham, Ian McLean and Neil
Johnston climbed back into their Tornado and went through their

ejection seat checks. Like all aircrew, they knew the twenty-step strapping-in process by heart. They made sure their seat firing handles were free of any obstruction, that leg garters had their quick-release connectors on the inside of the leg, any loose objects in the cockpit were safely stowed and their seats were not set too high in order to avoid spinal and head injuries if they had to eject in a hurry. They sat down, adjusted and tightened their lap and shoulder straps and checked their leg-restraint lines. Oxygen tubes, life preserver and radio connections were all good to go. Both were now sitting on cushioned personal survival packs (PSPs) located in the bottom of the seat. The bright yellow, solid fibre-glass boxes contained all the safety equipment they would need after ejection over land or sea. Ian, as he always did, checked that the Command Eject lever in Neil's cockpit was set to 'BOTH'. 'I never strapped in thinking, *I might be using this seat today.* You took it for granted; it was just *there*, ready to be used – again – if I ever needed it.'

They were airborne at 1.55. The sortie was planned to head up to the Lake District, across to the east coast near Berwick-upon-Tweed, before turning south to route past Newcastle. Once they had completed their training mission they would climb back to high altitude and return home to Germany.

Twenty minutes later, Dim Jones took off from RAF Coltishall and descended to 250 feet in the Harrogate area alongside his wing-man. The 'bounce' Jaguar had already pre-positioned to the north, ready to attack the pair of 'invaders'. Ten minutes in, 'all was going swimmingly well', Dim recalls. 'We had both hit our first target and beaten off the bounce.'

They were now approaching the 'Hexham Gap', a narrow cor-ridor of airspace across the Tyne Valley. To the east lay the dense sprawl of Tyneside and Newcastle Airport's air-traffic control zone. To the west was the market town of Hexham. Overflying any town at low level was strictly prohibited. And flying into an airport's control zone without specific permission was a rule never to be knowingly broken. Therefore, aircraft transiting through the area were all funnelled into a narrow gap.

On their second sortie of the day, Ian McLean and Neil Johnston were nearing the end of their mission and heading south towards the gap with Hexham on their right. Neil was not that far from his parents' house. 'The sun was low in front of us,' he remembers. 'The ground was glistening as light reflected off the frost.' Visibility was 5 miles – perfectly safe for low-level flying – and they were travelling at 500mph. Just over 8 miles a minute.

Dim Jones was busy in his own cockpit. Not only did he have to keep looking out for other aircraft and the bounce, and keep in formation with his leader, he also had to find, identify and overfly a fixed point to update his navigation equipment. All while monitoring his aircraft systems. This intense level of workload was routine on most training sorties, especially in a single-seat jet. But what was not in the script was the fact that his flight path was taking him inexorably closer to Neil and Ian's Tornado, heading in the opposite direction, and at the same altitude. Oblivious of the other's presence, the two crews were speeding towards one hell of a rendezvous, approaching each other at a combined velocity of 1,035mph. Or 17 miles every minute.

Sixty seconds away.

Come right, 10 degrees, Neil told Ian. 'I wanted to ensure we were completely clear of Newcastle's control zone.'

Thirty-five seconds from impact.

'I remember Neil telling me that we had a light civilian aircraft above us and out to our left,' says Ian. 'I glanced over to see where it was. It didn't pose any threat.' He turned his head forwards.

Twenty seconds.

Dim Jones was entering the narrow gap with his wingman. 'We were clear of Hexham on our left and I was looking out as normal, concentrating on the area ahead of us.'

Ten seconds.

Experts later calculated that five seconds, or 1.7 miles, was the minimum needed for the human eye to register an oncoming aircraft, the brain to then process the fact that it was a threat and then send a signal to manoeuvre the aircraft to avoid a collision. It was too late.

Four seconds.

'Suddenly, through the front of my cockpit, all I could see was another windscreen filling my own,' Ian McLean remembers. 'There was no fuselage, wings or fin, just another windscreen. I simply ducked.'

Dim Jones's experience was similar. 'I became aware of something grey, just a blur. It flashed past my port wing. At the same time I felt a thump and the aircraft started to roll left, but not uncontrollably.' He called to the other jets in his own formation: *I've hit something, probably a large bird.*

The Jaguar bounce pilot, who was flying above them, had just witnessed the collision and knew differently, laconically replying, *It's just crashed behind you.*

Dim looked out to his left. The last 3 feet of his wing were no longer there.

'*Some seagull*, I thought.'

It was 2.49.

'I still had no idea that I had hit another aircraft. A Tornado. Again.'

Neil Johnston had glanced in to check his radar/map display when the cockpit was plunged into darkness. 'Although I obviously didn't know it then, but this was presumably Dim's Jaguar passing inches over my head.' Then an almighty bang. 'I had a momentary sense of the jet flying sideways, slewing and rolling. Instead of looking up at sky, I was now looking *up* at the ground. Never a comforting sight, so I started moving my hand down to the yellow and black handle.'

Ian McLean's warning panel was 'lit up like a Christmas tree, the jet rolling violently left. I pushed the stick fully right. Nothing happened. We were out of control.'

The Jaguar's wing had sliced off the Tornado's fin. Without a fin, the jet had no vertical stability. It was rolling and rotating uncontrollably. Seconds from crashing, it was time to get out. 'I just let go of the stick,' Ian says. 'I reached down and pulled the ejection handle. I couldn't believe it was happening to me again.'

As the sequence started, their gas-operated shoulder harnesses hauled them back into their seats and ensured they were sitting correctly. Because their aircraft was equipped with Command Eject, they both exited automatically, Neil first. As their seats began to rise, their limb restraints pulled tight. With everything now happening in hundredths of a second, it is difficult to establish the exact sequence, but Neil probably pulled his own handle as he was already ejecting. He could feel his feet moving upwards but his body was compressing under the force so his head was still stationary, his radar/map display still in front of his eyes. For a split second he wondered why he was not ejecting. 'Then, there was a *whoosh* as the rocket motor fired and that's the last thing I remember.'

Ian recalls, 'As the Perspex canopy was blasted off by its own rocket, bloody great flames were coming down either side of my body. I wondered if they would burn my legs.' He experienced the blast of the 500mph windspeed as he cleared the cockpit. It was the last thing he remembers.

The jet was falling while also rolling violently. It was going to affect their exit as they went up the rails. Just 1.75 seconds after the collision they shot out of the diving Tornado, around 300 feet off the ground, but because the jet was rolling, they ejected almost sideways, parallel to the horizon. Their seats were at the limits of their capacity to save them. They were in their parachutes for fractions of a second and were still swinging under them like pendulums when they hit the ground.

Eight seconds after the collision, the stricken Tornado plunged to earth and exploded. Twisted shards of flaming jet cartwheeled 400 yards, ploughing up fields off the busy A68. Part of its fuselage, including its radars, crashed over the road and into the Tyne, spewing toxic fuel into one of England's best salmon rivers.

*

A teenager in the village of Riding Mill, near Hexham, a mile down-stream from the collision, had seen the jets 'almost strimming the tops off the trees' at the end of her garden. She said the 1,000mph collision and explosion was so violent 'it banged the front door shut and blew the soot out of our chimney'.

AA Patrolman Bill Fletcher was returning from a routine job. 'I saw one jet shoot over a car. Next, I saw another climbing full throttle and it was obvious something was wrong,' he later told a newspaper reporter. Then he saw thick black smoke rising from Styford Bridge, which carries the A68 over the River Tyne.

One woman had been driving to pick up her children from school. 'I had heard the sound of jets just above me,' she said. 'I know low flying goes on quite a lot, and I just thought, *Here we go again*. Suddenly, the roar was unbearable. One of the planes was right above me.' She gazed through her windscreen in horror. Burning wreckage crashed down only feet in front of her. She stamped hard on her brakes. 'It was so close I couldn't see out of the car. Now huge chunks of flaming metal were dropping from the sky. Another second, and the plane would have landed on top of me.'[12] It was sheer luck that no civilians on the ground were killed.

Bill Fletcher abandoned his van and ran to the bridge on foot. 'There was a wall of flames in the middle of the road and burning wreckage was strewn everywhere. If anyone had been anywhere near, they would never have survived.'[13] He was trying to put out the flames on the bridge without success when the fire service and police arrived. He went in search of the missing airmen to see if he could help.

Still airborne, Dim Jones was struggling with his badly damaged aircraft. 'I recall having great difficulty concentrating on one thing at a time.' His warning lights were telling him that hydraulic fluid and fuel were leaking out of the remnants of his Jaguar's part-missing wing. 'It didn't take the brains of an archbishop to work out that this was a major fire risk. I knew I might have to eject in the next few minutes, so I quickly hauled on my straps to ensure everything was tight and sat back in the seat. Then I began to think, I might have a sporting chance of getting this thing back on the ground. Unlike the previous time I'd hit a Tornado, could I get away with it?'

He decided to try for RAF Leeming in North Yorkshire, which had extensive emergency facilities. If he couldn't make it down at the correct landing speed he would fly the jet into the wilds of the Yorkshire moors, where he could eject and dump the jet safely. Then his left engine stuttered and shut down. He did not dare lower his 'flaps', which helped control the aircraft at low speeds, in case those on the left wing were damaged and no longer worked symmetrically with the right. It could render his jet uncontrollable. 'Whether I liked it or not, I was now committed to a single-engine, part-winged, flapless landing. The cognoscenti will tell you that any of these constitutes an emergency in a Jaguar, and a combination of the three is decidedly undesirable.'

He flew on and, despite the parlous state of his jet, managed to land safely at RAF Leeming.

The next thing Ian McLean remembered after the windblast as he ejected from the Tornado was coming to in a deep ditch after he had catapulted through a barbed wire fence. He glanced down at his right leg and foot. 'They were twisted and sticking out at right angles to my body which looked really strange. But I couldn't feel any pain. I took my helmet off and tried to move, but my brain wouldn't allow me to shift position.' Lying deep in the ditch, he couldn't see where he was. He started shouting for Neil. There was no response.

AA Patrolman Fletcher suddenly peered over the lip of the ditch. 'You okay, mate?' he asked.

'I'm fine,' Ian replied, ignoring his own injuries. 'Please see if you can find my navigator, Neil.'

More people started to appear. He heard a helicopter in the distance.

Neil was not far away. He had also been knocked unconscious by the ejection and, although he had come round, he was in a bad way, in intense pain. His face was badly cut and he could see his left leg and foot were at right angles, halfway down his shin. 'I knew I was in trouble. The pain was indescribable. What I didn't know was that my right leg was also badly broken, but the agony from the left was masking it.' Both Neil's legs were snapped in half, bones poking through the skin.

He looked down the hill towards the bridge over the Tyne. He could see that the traffic was stationary and burning wreckage was strewn everywhere, with police and fire officers swarming over it. 'I still had no real idea of what happened,' Neil says. 'Although it was dawning on me that all that wreckage and fire was something to do with me.' Then he had an appalling thought. 'Had we hit a passenger aircraft coming out of Newcastle Airport? I was horrified.' But commercial airliners did not fly low level, he reasoned. 'We must have hit another fast jet.'

Bill Fletcher rushed up to help him. 'I'm quite badly injured,' Neil gasped. But he was more concerned with establishing the facts. 'Did you see what happened?' he asked. 'What did we hit?'

'I think you hit another fast jet going the other way, but it managed to climb away,' Fletcher said, then confirmed that Ian was alive and injured, and that help was on the way.

Neil heard emergency sirens wailing ever closer. 'I had time to reflect that I was lying only a few miles from where my mother lives. Soon she would be hearing all about the accident. I was now in shock and really concerned about my broken leg sticking out. I was in the most overwhelming pain.'

Before putting him on a stretcher, two firemen were told to pin him down, holding him across his chest. 'The rescue crew decided they needed to sort out my legs before they could put me on the stretcher. It got really bad from this point.'

He remembers somebody telling him, 'We are going to straighten your legs. We'll do it as quickly and painlessly as we can.' It was neither. As he was held down, a medic pulled hard on his shattered legs and, bones grinding together, returned them to their rightful position.

'Oh my God, the pain was unbearable, simply indescribable. I'd never wish that on my worst enemy.' Nobody had given him any pain-relief drugs and, reliving the experience thirty-two years later with the author, an old friend, both still wince at the description. Neil's legs were wrapped in an inflatable splint and, assisted by AA man Bill Fletcher, he was lifted into the helicopter which then went to pick up Ian McLean.

RAF Laarbruch
Early evening

Back in Germany, it had been a miserable day. Constant rain, not a hint of winter sun. Ivy Johnston was gift-wrapping a book for her husband Neil's birthday when her doorbell rang. 'I presumed it was Neil and he'd forgotten his keys. I was a bit annoyed that he was home early and messing up my routine.' She shoved the partly wrapped present under a cushion on the sofa and went into the hall.

Neil's Squadron Boss's wife, Sally, stood on the step. With senior officers still trying to establish exactly what had happened, and who had survived, it fell to her to deliver the dreaded 'knock on the door'. 'Her eyes were full of tears,' Ivy recalls. '"It's Neil," she said. "He's had an accident, but he's alive."'

Ivy was furious. 'I presumed that flying had been cancelled because of the poor weather and he'd been to the bar for a few pre-birthday beers and fallen off his bike on the way home.' With almost no mobile phones in those days, he'd had no chance to tell her that he was spending the day flying in England.

'He's all right,' Sally reassured her.

Ivy was certainly not reassured. She was still livid at her beloved husband's presumed stupidity after drinking. Not an uncommon occurrence in his group of young friends in those days.

'I was just thinking, *He won't be all right when I get hold of him.*'

Sally could see that Ivy was not taking the news in as she had expected. 'He's been involved in a mid-air collision,' she explained.

'So now I thought this was more serious, but I still wasn't really processing what I was being told,' Ivy remembers. 'He must have been doing something daft on his bike, come off it and been tossed through the air?'

Seeing her ongoing confusion, Sally spelled it out. 'He was flying over in the UK; he had a mid-air collision with another aircraft and ejected.'

'Then it hit me ...'

As it grew dark, more details dripped through. She heard how Neil had ejected a few miles from his mother's village and fifteen minutes' drive from her own parents. 'At first they told me that he'd broken one leg. Then that he had broken them both. Things were getting worse.'

A neighbour whose husband was a fellow navigator came over to sit with her. At some point she was handed the phone, Neil was on the other end – although he does not remember the conversation. 'He said he was in pain but okay, and was heading into surgery.

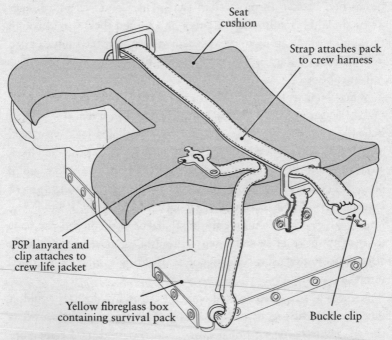

Seat cushion

Strap attaches pack to crew harness

PSP lanyard and clip attaches to crew life jacket

Yellow fibreglass box containing survival pack

Buckle clip

EJECTION SEAT PERSONAL SURVIVAL PACK (PSP)

Speaking to him made me think he wasn't too badly hurt so I felt much more relaxed about everything. Little did I know.'[14]

Next morning, Ivy and Ian McLean's wife were flown to the UK.

The surgeon at the Royal Victoria Infirmary in Newcastle confirmed that Neil's and Ian's breaks were serious. As the aircraft was rapidly rotating, falling and slewing when they had ejected, they had been at the limit of their seats' ability to fully operate. Unconscious due to the 500mph ejection blast, they had also been unable to lower their heavy PSPs on their lanyards and out of harm's way. And in conjunction with the pendulum effect of ejecting almost parallel with the ground at low level, the PSPs – 10-kilo rigid containers[15] – had acted like hammers, smashing into the backs of their legs when they landed. The medic told them that they had suffered 'comminuted compound fractures': bones had broken in more than two places and were no longer aligned. Additionally, the fractures had forced their way through the skin. He was going to repair them with a new technique called a 'Hoffmann Fixator'. The two airmen had no idea what he was talking about.

When Neil woke up he felt totally battered. At the end of his bed, his sheet was draped over a large frame, like a tent. He peeked beneath it, expecting to see plaster, but both his legs were held together by what looked like scaffolding. A series of posts had been fixed through the skin and screwed into the bones to hold them in place. They were attached by brackets and wing nuts to lengths of external metal rods running the length of his shins, keeping the limbs stable. This technique meant that the surgeon did not have to open up the fracture site to reposition bone ends; he could simply adjust positioning using the array of 'scaffolding' connections bolted into the bones.

Ian was as surprised as Neil when he lifted his sheet. 'I couldn't quite believe the metalwork protruding from my leg.' It looked as though civil engineers, not doctors, had repaired it. 'I remember the consultant sticking up three X-rays on the light box, one for me, two for Neil. It showed three legs, all neatly snapped in half around the shin.'

When Ivy arrived at the hospital she couldn't disguise her shock. 'Neil looked truly terrible. He was drawn, deathly pale, his face badly scratched. It really hit me just how badly injured he was, and how lucky he had been to survive. I also knew how lucky I was to still have a husband.'

Ten days later, Neil was flown home to Germany. Ivy drove him straight to RAF Laarbruch Officers' Mess to collect their post. 'It was early evening and the bar was packed,' Neil remembers. 'So it seemed churlish not to pop in for a quick pint. It was amazing to be back in the fold with mates.'

The author well remembers his friend's return. 'I watched him hobble into the bar wearing a pair of shorts. We were all totally shocked to see his legs held together by metal posts and rods like a kid's Meccano set. We had never seen anything like it. Anyone with broken limbs had always been encased in plaster.'

The severity of the airmen's leg breaks was a stark reminder of the urgent need to lower the solid personal survival pack when ejecting. Their accident was to pave the way for an automatic lowering system to be fitted to aircrew PSPs.

But how had the collision happened? The RAF's Board of Inquiry into the accident ruled: 'The sole cause of the accident was that the crew of the Tornado and the pilot of the Jaguar were unable to see each other in time to take avoiding action.' The low winter afternoon sun had significantly reduced visibility. To make things more challenging, both combat jets were camouflaged for the low-level environment and came out of dark, wooded backgrounds. In simple terms, in those conditions, at the speed and angle they were approaching, they were never going to see each other in time to react. The accident was seen as an acceptable hazard of military flying, though if the Jaguar had been flying just a few feet lower, its wing would have sliced through the Tornado's cockpit rather than its fin, killing both Ian and Neil.

As far as Dim Jones was concerned, the board's conclusions also set the record straight on his first collision. Scientific and medical specialists had proved that RAF investigators could no longer put collisions solely down to pilot error for not seeing a conflict early enough.

Ian McLean's and Neil Johnston's rehabilitation continued through the summer of 1990. Despite catastrophic leg injuries, their seats had saved their lives and they became ejectees number 5,991 and 5,992 in the Martin-Baker 'Hall of Fame'. While they worked hard in the gym to recover from their injuries before returning to flying later that year, it had not escaped their attention that 3,500 miles away the Iraqi dictator Saddam Hussein had invaded neighbouring Kuwait, setting in motion a chain of events that would have momentous repercussions for their friends and fellow aviators.

As in Korea, Vietnam and the Falklands, ejection systems would once again come to the fore in combat.

CHAPTER FIFTEEN

STORM IN THE DESERT

USS *SARATOGA*, RED SEA
21 JANUARY 1991

When Lieutenant Larry Slade joined the US Navy in 1985, his flight training class had been given a lecture by a Vietnam veteran. 'He told us in no uncertain terms that we needed to be fully prepared to fight, and that the military wasn't a game,' he remembers. 'And if we were not prepared to put our lives on the line, ready to die for our country, we should leave the navy. Immediately. I never forgot that.' Growing up in Lebanon, where his parents were professors at the American University in Beirut, he knew that the next explosion of hostilities was likely to be in the Middle East. 'I was all too aware that it could come at a moment's notice.'[1]

Larry and his colleagues had already been in that region for nearly six months on board the giant aircraft carrier USS *Saratoga*. In August 1990, Saddam Hussein's Iraq had invaded and occupied its much smaller neighbour, Kuwait. Amid ongoing violence, destruction and brutality, the UN had given the despotic Iraqi regime until 15 January 1991 to order its 300,000 troops home. Sanctions were applied and a US-led task force of a million service personnel drawn from thirty-two nations, including just over 53,000 from the UK, assembled in the region, ready to drive Saddam's forces out if he did not comply. It was the largest military alliance since the Second

World War. On 16 January a devastating air assault began against military, economic and communications targets in Iraq. Day one of Operation Desert Storm saw some 2,700 sorties launched.[2]

Larry, a Radar Intercept Officer (RIO) and award-winning graduate of the US Navy's 'Top Gun' training school, had already flown six missions from the carrier with his pilot, Lieutenant Devon 'Boots' Jones, in their supersonic, dual-engine, twin-tail F-14 Tomcat, call sign Slate 46. Four days into a relentless air war, 'we'd lost a number of aircraft, including two from *Saratoga*', Larry remembers. 'We knew the dangers were very real.'

Despite their often superior aircraft technology and training, by day four the allied coalition had already lost fourteen aircraft in combat. Thirteen aircrew had ejected, been captured and were now suffering as prisoners of war; four had ejected over friendly territory and been recovered, and eight had been killed. In the same period, coalition aircraft had shot down fifteen Iraqi jets. It is not known if any of the Iraqi crews ejected or survived.[3] Iraq had an astonishing defence network of 154 fixed surface-to-air missile (SAM) sites, alongside almost 10,000 anti-aircraft guns and countless mobile and handheld SAM systems,[4] but Larry, who had often demonstrated ejection procedures to trainees in Florida, had complete faith that his ejection seat would look after him if he found himself *in extremis*.

Leaving the carrier's deck under cover of darkness, Jones, with Slade in the back seat, were to escort a strike on Al Asad Airfield 100 miles west of Baghdad. Target time was 6.10am. Timing was tight. Larry and Devon had been briefed that sunrise was at 7.30. If they had to eject, cover of darkness would be in short supply. The mission went smoothly and, as the attackers departed the area, it was time for the Tomcat crew 'to get the hell out of there' and return to the safety of their carrier. 'We were in a left-hand turn at around 30,000 feet,' Larry says. 'I was looking up and out to the right when Devon suddenly called out, "Missile inbound!"'

It was an old Soviet surface-to-air, high-altitude SA-2. Dating back to the late 1950s, it had been a menace to American aircraft in Vietnam and was one of the most deployed air defence systems in history.[5]

Devon Jones piled on the power. 'I rolled, as briefed, to give the missile tracking problems.' The manoeuvre was so violent and sudden that Larry banged his head on the cockpit canopy. 'But the SAM wasn't so easily fooled,' Devon says. 'As we rolled down almost inverted, it tracked us, came up toward our tail and detonated with a bright, white flash. The F-14 shuddered and kept rolling right.'[6]

'It was a massive explosion,' Larry says. 'The jet was going into a flat spin.' The impact had been so powerful it had ripped off Devon's mask so there was no communication between pilot and his back-seater. 'I couldn't tell if he was trying to get the aircraft under control,' Larry says. At 24,000 feet he reached towards his ejection seat handle.

In the front seat, Devon was fighting hard to save the jet, but he was being so badly buffeted as they fell that he couldn't see his aircraft attitude, altitude and airspeed instruments. 'The only thing I could sense was dark or light. It was light in the clouds, dark when we came outside.' With no idea what was going on, Larry made a decision. 'I didn't know if Devon was even conscious. Or alive. So told myself that I would give him until 15,000 feet before ejecting both of us with Command Eject.' For a few seconds, he closed his eyes. 'I imagined I was back on *Saratoga*. Telling myself it was all a bad dream and, when I woke up, I'd be in my bunk.'

When he opened his eyes again, the nightmare continued. 'I was still falling. There was no time for fear, but I was bent forward out of position with my face against the instrument panel. All I could see was the altimeter spinning down.

'The last thing I saw was 15,200 feet.'

With their jet spiralling earthwards, Devon Jones also knew that time had run out and was already pulling his own yellow and black handle. 'I wasn't in a very good sitting position and expected to get hurt since I was slumped over trying to get the handle. It took me a while to find it because I was getting tossed around, but the seat worked like a charm. I was conscious through the entire ejection; I remember the blast and the parachute opening shock.' Because of the Command Eject system, Larry was already on his way once Devon initiated the ejection. 'I recall everything clearly, especially deciding to pull the handle. But as I tightened my grip on it, the canopy blew off which was somewhat surprising, and scary,' he says. 'I felt the seat go up the rails. I closed my eyes tightly and didn't open them again for some time. There was no tumbling; just a blast of cold air and an almost instantaneous upwards jerk as my chute deployed.'

Larry looked up to his 28-foot, multicoloured parachute and, as he had automatically done so often during training, shouted out loudly, 'It's a good canopy! I looked around and saw Devon's parachute had also fully deployed, but then lost sight of him as we went into clouds.'

During his descent, Larry took out his radio and called the overwatching AWACS (Airborne Warning and Control System) aircraft. *Mayday, Mayday! Slate 46 is down.'* He thought he heard a reply but it was difficult to tell. In fact, one of the AWACS controllers in the high-flying, specially modified Boeing 707, stuffed with radar that could detect, track and identify aircraft hundreds of miles away, had indeed acknowledged him. Larry's radio signal had allowed them a brief opportunity to assess his position and they now knew the area where he and Devon had ejected. Combat Search and Rescue (CSAR) resources were immediately alerted.

Larry saw the Tomcat hit the ground and explode. The billowing flames helped him estimate how far he was from landing. Steering away from the crash site, he hit the ground fast and hard, five minutes after ejecting. Not the elegant shoulder roll he had practised in training, but straight onto his backside. He gathered up his parachute quickly and folded the orange, white and green panels under the brown one to avoid detection when dawn broke.

It was around 6.30am. He had an hour of darkness to distance himself from the remains of their aircraft and wait to be rescued. 'Despite the incredible change in my circumstances, I was clear-headed. I had ejected to safety from a doomed jet but the next part of my ordeal was just beginning. I was no longer F-14 aircrew, I was an evader, alone in hostile territory.' Then he remembered that during the run-up to the war, he had heard stories that any captured aircrew would be torn limb from limb by an enraged population.

Forty-five minutes after Larry's Mayday radio call, US Intelligence assets reported that Iraqi helicopters were already out looking for him and Devon. A flight of American F-15 'Eagle' fighters was scrambled to provide cover for the downed aviators until the rapidly assembling rescue package arrived.

Finding two men in 64,000 square miles of desert was never going to be an easy task, especially in the days before detailed satellite mapping. Few aircraft had military-grade satellite navigation systems,[7] and they were still in their infancy. Saving Larry Slade and Devon Jones would depend on using co-ordinates often supplied by search pilots dodging missiles, which were tentative at best. And the maps, according to the Commander of 20th Special Operations Squadron, 'pretty as they were, were not worth a ding-dong. There are things on the map that are not there. There are things that are obviously there, but not on the map. We thought we should be able to fly to a co-ordinate, hover above the guy, drop the hoist down and pick him up. We didn't know.'

Their survival, essentially, depended on the skill and bravery of fixed-wing pilots and helicopter CSAR crews.

Devon Jones had come down hard. He could see the flames devouring their Tomcat lighting up the desert darkness, and reckoned it was 5–8 miles south of him. He was in shock, haunted by what the future held. 'Reality hit me in a big way. I thought, Geez, I'm going to be a POW! They'll probably rip my fingernails out and shoot me. My family's going to go nuts.' Then his 'SERE' training began kicking in: Survival, Evasion, Resistance and Escape. He had to get rid of his brightly coloured parachute before it gave him away. 'I started looking for a bush or tree to hide this stuff under, but there was nothing.' The desert was as flat and featureless as a parking lot. He balled up his parachute and shoved it under his survival pack. 'I figured that if the Iraqis were flying around looking for me in a helo, they might not see the green seat pack. I made a radio transmission. I said I was on the ground, okay and heading west.'

Was anybody listening? He then tried contacting Larry. No response. It was time to get moving, as far away from his burning jet as possible. 'I wasn't thinking rescue, only evasion.'

Larry Slade heard metallic clacking at every stride. 'What the hell was that? I looked down and saw I still had my leg restraints on. The metal loops were rapping against each other. I quickly ditched them.'

He had been walking for more than two hours, using his radio every hour to try to call both CSAR and Devon. With no reply. He felt dangerously exposed in the barren moonscape. The horizon was starting to glow. The sun was not yet up but it wouldn't be long. 'I couldn't believe that any rescue would come in daylight. It was too risky.' As dawn broke he came upon a small, round knoll. 'I needed to find a place to hide.' He pulled out his survival knife and tried to dig a foxhole, but the ground was rock hard. 'My hands were bleeding at the effort as I managed to scrape a few inches down.' He lay down on his back and dusted dirt over himself to try to blend in. He weighed up the odds of evasion. He had still received no response on his radio. Was it really working? He had water in his bottle, but it was not going to last long in the desert heat. 'There was still a chance of rescue if I could survive until dark,' he says. 'But I realised my prospects were not looking too good.'

As the sun came up, Devon Jones's spirits were also low. 'I had expected helos to be in the air looking for me.' In spite of his repeated radio calls, he had heard and seen nothing.

He did not know until later that aircraft and helicopters had already been searching for him and Larry, but inaccurate co-ordinates, a communications glitch and unclear mapping had directed them 30 miles too far south. Devon trudged on. Around

9am, he spotted something blue and cylindrical 2 miles away. From that distance it looked like a car. Getting closer, he realised it was some kind of storage tank, 20 feet long. If it contained fuel, it meant traffic and increased his risk of being captured. Three hundred yards east he found a wadi where the ground was softer.[8] He fell to his knees and started digging with his knife and hands. An hour later he had made a hole 4 feet long and 3 deep, with a mound of sand and stones between him and the road. Then he realised, 'I had dug up a tremendous amount of fresh dark dirt, which would probably draw attention from the air.' So he put it all back in the hole, trying to slide under it like a blanket. 'Of course, that didn't work, and I had to take it all out.' Next, he sprinkled the surface of the mound with lighter, sandy soil and hunkered down in his hole, placing his radio on the ledge. He tried to get comfortable, presuming he would have to wait many hours behind enemy lines.

At 10.30 he heard two jets high overhead. He suspected they were Iraqi. 'I hope they shoot them down,' he mumbled.

At that very moment Larry Slade, still lying flat on his back, heard a vehicle about a hundred yards away on his right. It was now four hours since he had come down. 'The engine note changed as it drove past me, parallel to my position, and speeded up.' A white Datsun pickup was approaching. 'I could still see the smoke from our wrecked jet and presumed the truck was heading towards it.' Larry breathed out slowly. They hadn't seen him. He had got away with it.

'Then, it came to a screeching halt, tyres spitting dirt. I realised, *Fuck. I haven't made it.*'

Two men jumped out. 'One had an AK-47 and wore a shabby uniform, like he had gotten it from an army surplus store. It had no insignia. The other man was a Bedouin in black robes who was carrying a 12-bore shotgun.' Their body language suggested they were not a regular military search party. It had been Larry's bad luck that they happened to be driving by and spotted him. He was carrying a .38 snub-nosed revolver. For a split second he considered taking them on. 'Even if I was a crack shot, which I wasn't, trying to take down two armed men before either could get some shots off was unrealistic. I simply didn't have the firepower.'

He threw down his weapon and surrendered, with no time to tell rescue forces that any attempt to find him would be in vain. 'Now, the *next* part of my story was beginning,' Larry says. 'I had no idea how long it would go on for. I just hoped it wouldn't be my final chapter.'

Devon Jones's heart was pounding. A farmer's truck had rumbled up to the tank, and he watched two men filling containers with what he was now sure must be water. He reached for his pistol. The truck seemed to offer him a chance of escape. Was he willing to try to kill the two men and steal their vehicle? Shooting a man at close range was a very different prospect to killing someone in aerial combat. He didn't find out if he could go through with the plan; the truck drove quickly away.

Devon found a few bushes and twigs to camouflage his hiding place. 'I kept making radio calls. I didn't expect a day recovery, so I wasn't thinking of rescue.' All he could do was hope for the best.

A sudden movement by his left shoulder made him flinch. He was sharing his hiding place with a black scorpion. 'Not a big one, but still a scorpion.' It was trying to climb out, but the sides were steep, and even with eight legs it kept falling back. As it struggled to leave, Devon willed it on. 'C'mon, buddy.'

Again it hauled itself up to the lip, and then fell onto Devon's arm. 'My survival instinct took over. I flicked him off and flew out of my hole, my cover blown, but I went "one-versus-one" with him and killed him.' It was like the poisonous predator had been sent by the enemy to flush him out. Devon's hide never felt the same again.

At 12.05 he switched on his radio to have another attempt at contacting Larry. Suddenly, he heard American voices on the emergency frequency. And, amazingly, his mission call sign.

'*Slate 46, how do you read?*'

'*Slate 46, this is Sandy 57.*'

'Sandy, this is Slate,' Devon answered. 'Go ahead.'

Larry Slade's captors made him strip off all his gear. He was shivering and shaking. 'They bound my hands, but they were very polite. They sat me between them in the pickup and drove me to their tent, where they had a surprise.'

A leisurely lunch. They didn't seem in any hurry to go anywhere.

'Through sign language and pidgin English, they asked me if I wanted to go to Baghdad or Saudi Arabia. Of course, I told them Saudi and named a town.' The truck moved off. Larry knew the drive time to Saudi was around eight hours. Baghdad three or four. Which would it be? Three and a half hours later they pulled into an army camp outside the Iraqi capital.

'I now had little expectation of living through the experience,' Larry remembers.

Devon Jones thought Sandy 57 might be a rescue helicopter. He had in fact been talking to a pilot of a pair of A-10 Thunderbolt ground-attack aircraft, desperately trying to find the downed crew. The early versions of the A-10 had earned it the nickname 'Warthog' – a mean and ugly aircraft. Despite its looks, 'Hog' pilots loved it for its dependability over sleeker fast jets, and especially for the rotary cannon mounted in its snub nose which could fire 3,900 armour-piercing shells a minute.[9]

Talking to Devon was one thing. Pinpointing him under mounting cloud in a largely featureless landscape was very much another. The signal Sandy 57 was getting from Devon's radio was very weak. 'I began flying west while talking to him,' the pilot recalled. 'When the signal faded, I realised that wasn't right and returned to my start point.' Next, he flew east until the signal weakened once more. As soon as he headed north, it was stronger. 'Although I knew I was headed in the right direction, I had no idea how far north I had to fly. Ten miles, maybe 15? Fortunately my map was folded so I couldn't see just how close I was to Baghdad. Ignorance really is bliss.'

Sandy 57 was getting closer but flying in and out of cloud. Devon couldn't see the jet and its pilot couldn't see the ground properly. 'When I was close to where I thought he was, I dropped down and popped out some very bright flares.' Devon saw one of them burning. He looked up, heard a rumble, then saw the Hog. He began guiding it in. It was six hours since he had ejected. 'You're right above me,' Devon told him. The pilot marked the spot in his navigation system, then climbed higher for a gap in the clouds so he could memorise the terrain. 'A couple of dirt tracks crossing each other, a blue tank off to one side . . .'

Devon knew a rescue was on. But not quite yet. The thirsty Hog, now dangerously low on fuel, had to find an airborne tanker and refuel. It would be back in thirty minutes. For a lone aviator trying to avoid capture in hostile territory this must have seemed like an age. Devon couldn't know that Moccasin 5 and 6, two Pave Low Sikorsky Search and Rescue helicopters of the 20th Special Operations Squadron, had already taken off from a covert base in the far north of Saudi Arabia, 260 miles south of Baghdad. They were being routed by AWACS controllers, who, like them, had been briefed on Devon's location by Sandy 57 and were monitoring their radio conversation.

However, due to the urgency of the situation and lack of secure communications at that time, the calls were all being broadcast on standard, open, emergency frequencies, and Iraqi forces in the area were also eavesdropping. They too were on the move. The two A-10s were back on station. Spotting the rescue helicopters, they dropped to 1,500 feet and flew daisy-chain patterns to protect them from Iraqi guns and SAMs. F-15s took station above them, to deter any Iraqi pilots from interrupting the rescue mission.

Moccasin 5's pilot had taken his helicopter down to 20 feet and was speeding across the desert towards Devon Jones. The radio frequency was now full of chatter – much of it mundane and irrelevant – especially from the overflying F-15s. Sandy 57 had had enough. He told them to shut it. Only the right players should be talking – himself, Moccasin 5 and Devon.

Moccasin 5 was finally in contact with the downed airman. 'For the first time I looked to the east and saw the Pave Low about 5 feet off the ground,' Devon recalled. 'I have never seen such a beautiful sight as that big brown American helicopter.' He guided it in. The rescue crew made ready. They should have him safely off the ground in minutes. Or so they thought. Then Moccasin 5's left gunner spotted two Iraqi army trucks kicking up dust clouds as they sped across the desert.

'Slate, confirm they are headed in your direction,' said the helicopter pilot.

'Headed straight for me.' Panic gripped the downed airman. He was seconds from being captured.

Moccasin 5's pilot called the A-10s. 'We've got movers approaching from our eleven o'clock.'

'What do you want me to do?' Sandy 57 replied.

'Smoke the trucks! Smoke the trucks!'

Sandy 57 rolled in and fired at the lead vehicle. He missed. His wingman followed with a volley from his 30mm cannon. Sandy 57 rolled in again and hit the truck's fuel tank. It exploded and the occupants of the second Iraqi vehicle took one look at the fierce orange flames and thick black smoke consuming their comrades and fled the scene. No one in the burning truck could have survived.

Moccasin 5, circling at a safe distance, immediately refocused on picking up their fugitive.

'Where is he?' the pilot asked.

'Just land next to the burning truck,' Sandy 57 told him.

The two A-10s orbited the area, providing cover. Beneath him, Sandy 57 saw 'this guy jump out of his hole just 100 yards from the burning truck'.

'I grabbed my gear and just ran,' Devon Jones said later. The helicopter barely kissed the ground 40 yards from him, engines roaring. Another hovered close by to repel attacks. Special Forces soldiers jumped out to cover Devon as he sprinted forwards. Every stride seemed like five as he forced his aching body over the last few yards. 'The Special Forces guys waved me on,' he says. 'They sure had a lot of guns hanging off them.' Finally, gasping for breath, driven by adrenaline and the survival instinct that had saved him from capture, he reached for the hands that would haul him in. Moccasin 5 sped him the 170 miles south to safety, at 160mph, 20 feet off the ground.

Around 100 people had been involved in planning and executing the daring daylight CSAR mission, including the Special Operations ground staff, four A-10s, two F-15 fighters, two seven-man rescue helicopters, a refuelling tanker and its crew, and an airborne AWACS control crew. All to save one ejectee. It would go down in Gulf War legend – the first combat rescue mission behind enemy lines since Vietnam and employing some of the same direction-finding radio equipment that had been used in that war. It was 2.30 in the afternoon when they scooped up Devon Jones. He had ejected nearly eight hours earlier. Arriving on friendly soil back in Saudi he said, 'I don't know who was happier [with the rescue], me or everyone else.'[10] A few days after his desert ordeal, Devon Jones – Martin-Baker ejectee 6,102 – was back on USS *Saratoga* and would fly many more combat missions before the end of the war.

* * *

Larry Slade's war on the ground was only just beginning. Iraqi soldiers had shuttled him, now blindfolded and handcuffed, through six different military camps. He was interrogated at each, but the questions were haphazard and disorganised. Some were soft-sell, with the interrogator taking the 'I'm-a-pilot-too' tack. Then there were the hardcore questions about his aircraft, his base and his mission. When he did not supply detail, punishment was swift. 'They didn't use any classic torture methods. Instead, they tied my hands behind my back and double-blindfolded me so I couldn't even blink, and then beat me up. The violence was shocking in its intensity.'

The threats were often as terrifying as the beatings. 'At one point they undid my flying suit and exposed my genitals, screaming at me. I had no idea what they were going to do to me, or my genitals, which I was rather fond of.'

They beat him when he didn't answer their questions and they beat him when he did. They strapped him to a chair. He heard a pistol being cocked, felt the cold steel of its barrel against his head. They slammed him into a wall and broke his seventh vertebra.

He sensed these first six interrogations were but stepping-stones to something far worse. After some of the inquisitions, and while being moved from base to base, hostile crowds gathered outside. 'They'd grab at me, kick and punch me. They shouted, "You kill our children."' It was an action replay of what happened to Guy Gruters and others shot down and captured in Vietnam, and Ron Guthrie's experiences as a prisoner in the Korean War.

Larry's captors tried to protect him from the mobs but ended up arguing among themselves about how to do it. 'At one point, the situation turned really ugly as a crowd was trying to get at me. My guard was yelling and screaming, pointing his weapon at them instead of covering me. I felt the whole thing was out of control and that I was in mortal danger. He tried to bundle me into the vehicle but I bounced off the doorframe and broke my nose and a couple of teeth. I ended up lying on the floor of the car, covered in blood.' The unpredictability of truly angry people made Larry fear that he would not make it through the next few hours, let alone what might be coming his way in the days that might follow. 'Day by day,' I told myself. 'Day by day.'

Talking to the author about their shared experiences, from his home in Virginia Beach on the east coast of America, Larry admits to still being affected by those days of brutality. 'Thirty-two years on, John,' he says, 'I am reluctant to revisit the experience.'

* * *

One of Larry's fellow POW ejectees, now being gathered in a military interrogation centre in Baghdad, was American F-16 pilot

Major Jeff 'Tico' Tice of 614th Tactical Fighter Squadron, the 'Lucky Devils'.[11]

Tico's luck had run out on 19 January when he had been shot down by a SAM on a major US daylight raid on Baghdad. After ejecting he landed to be greeted by a hail of bullets from a nearby encampment. 'As soon as my feet touched the ground I got up and started running hard,' he recalled. 'I sprinted four or five steps, but that was it. They shot up the mud in front of me, a line of automatic fire just missing my toes. Ten nomadic men huddled around me. Most of them didn't have teeth or shoes, but they all had brand-new AK-47s.' One kicked him in the groin. Another beat him with his rifle butt. They stripped him, took his radio, some of his clothes and even his wedding ring.

One of his captors rolled a bullet around Tico's face, kissed it, replaced it in his AK-47 and then shoved the barrel against the American's temple.

Welcome to Iraq.

Embarking on the same ordeal as all the captured airmen, he had also been interrogated by Saddam Hussein's henchmen. Deprived of sleep, denied food and water, and chained to a chair and viciously pummelled. He felt he had endured 'a couple of rounds with Mike Tyson'. The goons were very keen that he make a video that would be shown on TV news, admitting he was a war criminal. In the first few days of the war all the POWs were being beaten to force them in front of the cameras as part of Saddam's propaganda campaign.

'Make the tape,' they commanded.

'No, no,' he insisted. 'I'm not going to do that.'

They clubbed his head so hard he feared they had ruptured his eardrum.

After his repeated refusals to co-operate, one of his interrogators lost control. 'He hit me with the butt of his pistol and dislocated my jaw.' Jeff could hardly speak as they hurled him out into a hallway. 'I put my face up against the wall and managed to ram my jaw back into position.' Like Larry Slade and the other ejectees, Tico didn't know how much more he could take. 'These guys were really going to damage me.' His jaw was dislocated a second time, and he couldn't hold his head up properly.

Ramping up the brutality, an interrogator wrapped a piece of wire around one of his ears, under his chin and around the other, then connected it to a power supply. When Tico refused to answer his next question, he flicked the switch. 'There was a flash in my head and my jaws slammed together with incredible force.' Every muscle contracted in agony. Pieces of his teeth were broken off. He suffered around a dozen shocks during the next twenty minutes, interspersed with the same order: *Make the tape.*

Drenched in sweat and urine, the session with what he called the 'Talkman', after the iconic Sony Walkman cassette player, had almost wiped him out.

'Okay,' he gasped, 'I'll make the tape.'

Afterwards, decrying himself loudly for giving in, he heard a distinctly British voice, another captive airman. It was the author. 'Don't worry, mate,' he said. 'We all made the tape.' He had appeared onscreen with his pilot, both beaten and badly bruised, to the horror and outrage of the British public in what became one of the enduring images of the conflict.

Larry Slade faced the same dilemma. 'We had been taught that it was something we should avoid at all costs. But there was the slim hope that it would show the folks back home I was alive.' In the end, after being beaten senseless, he too went in front of the camera, hoping the videos would become 'documentation, where, at the end of the war, if we weren't in the same condition as at the start, someone would have to answer for it'.[12]

The images of the bruised and battered POWs were flashed around the world, sparking outrage and condemnation. British Prime Minister John Major declared the treatment of prisoners as 'inhuman, illegal and totally contrary to the third Geneva Convention'. American President George Bush echoed his words. 'If Saddam Hussein thought this brutal treatment of pilots is a way to muster world support he is dead wrong and everyone is upset about it.' Dick Cheney, the US Secretary of Defense, condemned it as 'in effect a war crime'.[13]

Even so, all the POW ejectees involved never forgave themselves for succumbing to the violence and pressure to appear on camera.

* * *

Dale Zelko, an F-117 Stealth pilot with US 416th Squadron, the 'Ghostriders', will never forget the sight of his fellow airmen paraded on Iraqi TV. There were twenty-five fast-jet POWs and the first eight were forced to make broadcasts, under obvious duress, to denounce the war around the 21 January.[14] 'It was a real shock to see them beaten and subdued. It brought the reality of war firmly into focus; a real wake-up call and a reminder of what we faced.' But, as Dale said, 'hearing about the survivors meant that at least they had ejected safely'.[15]

Dale's Squadron had been on the second wave of strikes on the first night of Desert Storm, in the early hours of 17 January. Based in the mountains of south-west Saudi Arabia, he and his colleagues had trained hard and were at peak readiness.

The F-117 Nighthawk, the world's first Stealth attack aircraft, looked like a black dart with a V-shaped tail. It could deliver over 2 tons of laser-guided bombs and its forward- and downward-looking infrared technology, used to home in on targets, would be one of the most effective systems deployed in the war.[16]

Dale had joined the F-117 programme in 1989. 'The aircraft was a fantastic accomplishment for the US Air Force,' he says. It had been developed under the strictest security and only the best were invited to join the Nighthawk community in the high desert of Nevada. 'Not only that, the USAF had two fully operational squadrons as far back as 1983, but only acknowledged its existence in 1988.'

In the depths of night, Dale strapped himself into his American ACES II lightweight ejection seat. The latest 'Advanced Concept Ejection Seat' was much admired by those who trusted their lives to it. 'I had great confidence in that seat,' Dale says. 'I hoped I'd never need it, but it was always there in the most extreme moment of danger. It gave an aviator another roll of the dice in life.'

As he taxied to the runway, how much danger was he likely to face? The Nighthawk's state-of-the-art coating of radar-absorbent material, and shallow-angled airframe, meant it was hard for an enemy to lock on. Many said it was invisible and invincible. 'That just wasn't the case,' Dale says. 'It employed technology that made us difficult to see and we were certainly well protected, but it could be defeated by a skilled and knowledgeable adversary.'

After the first wave of ten F-117s had dropped a salvo of bombs on Baghdad at 3am to launch Desert Storm, the Iraqis had woken up. Following in their wake, it was time for Dale and his Ghostriders to 'stealth up'. He switched off all lights, retracted his antennae and shut down the radios in his spacious, fully digital cockpit.

'I didn't think too much about the reality of what I was about to do until I approached the Iraqi border,' he says. 'Then a wave of terror washed over me. It was both psychological and physical. I felt it start at the top of my head, flow through my body and out through my feet. It wasn't a fear for my life, my survival or myself. It was for my family back home. What were they going through? If I didn't make it home, what anguish would they face, what suffering would they endure?'

People assumed the Nighthawks would be operating safely above the fray. Not so. Dale's first combat target run, and those that followed, were at 10,000 feet. Hitting the capital, the reception he received was ugly. 'We flew right smack into the heart of some of the worst triple-A [AAA; anti-aircraft artillery]. They were putting up a massive barrage.' The F-117s had come under fire from thousands of AAA guns firing from the Baghdad rooftops. 'I was manoeuvring around jagged lines of fire like a kid trying to run through a sprinkler without getting wet. I thought, *There's no way I'm not going to get hit and downed by this stuff*.' Unleashing his weapons and turning for home, 'it seemed that the entire Baghdad skyline was ablaze'. The first question Dale asked after landing was, 'did we all make it?' All the Ghostriders had survived. No kills and no ejections.

By the first week in February, the air war had seen few Iraqi Air Force sorties. Those that did get airborne caused little damage and coalition jets had shot down thirty-seven Iraqi Air Force fighters. More than 100 enemy aircraft had been destroyed on the ground. Around 140 Iraqi military aircraft had fled to the safety of bases in Iran, its long-time enemy. Twenty-four allied aircraft had been lost, with twenty-five personnel ejecting to be captured. So far.

* * *

Larry Slade and his fellow prisoners in Baghdad had learned to live with the stench of their own sweat and excrement, with the body lice and the interminable cold. 'But there was comfort in knowing there were others in the same situation,' says Larry. 'That I was not the only one who didn't complete a mission.' Even though he was banged up on his own in a cell 6 feet by 9, 'I could hear friendly voices from adjoining cells.' One of those belonged to the author, who had been shot down and captured four days before Larry, the first day of Desert Storm. 'I regularly chatted to him in the darkness as we compared experiences of ejecting, capture and interrogation,' Larry recalls. Suffering shared eased the emotional strain even if it did not heal the wounds.

Towards the end of February, after nearly five weeks of solitary confinement, Larry was moved into a slightly larger cell, in a new prison, with the author. Delighted to see a friendly face after weeks of violence and hostility, the two were chatting away when they made a big mistake. They looked out of the barred window. On the other side of the jail, three Arab prisoners waved at them.

Then a yell rang out in the courtyard. Larry and the author had been spotted by a dozen guards. 'They were gesticulating furiously in our direction,' Larry recalls. Inside a minute, he heard them hammering on the main door, to be let into the wing. 'Shit,' Larry said, 'they're coming to get us.' One burst into the cell. They started on Larry first. 'He advanced towards me, holding his pistol out. I thought my time was up.' Larry was pistol-whipped and beaten to the ground. 'What was so terrifying was the clear loss of control of our captors. All bets were off as the boots and fists flew. They were in an absolute rage.'

Now it was the author's turn. They punched and kicked him until he fell and then started lashing boots into his face. He felt the blood streaming down his cheeks.

A guard took his pistol out and snapped the slide sharply back to charge the weapon.

'You are now going to die,' he said and moved closer to Larry so he could see down the barrel of the gun. The Iraqi stood there quivering with rage, the knuckles of his fingers white, the slack taken right up on the trigger.

Larry realised that 'I was a heartbeat away from death'.

The guard pulled the trigger.

Click!

There was no bullet in the chamber. A mock execution. He walked out, laughing.

'After ejecting and being captured, the whole experience as a POW was hideous,' Larry recalls. 'Minute by minute, you simply never knew if you would live or die.' And then he adds, with considerable understatement, 'It wasn't a particularly pleasant way to exist.'

* * *

Saddam Hussein's war was not going well. His tanks, aircraft and weapons systems suffered a daily battering. On 25 February, 20,000 of his troops surrendered. The next day, thousands of Iraqi soldiers retreating from Kuwait were killed on Highway 80, when coalition aircraft bombed them in their tanks and commandeered civilian trucks and cars. It became known as the 'Highway of Death'. The body-strewn devastation and twisted, burning metal would become one of the most iconic images of a war that was nearing its end. On 28 February 1991, Iraqi forces surrendered and, after forty-three days of brutality, Larry Slade and the other POWs dared to dream.

By the final day of the war, coalition aircrew had flown over 100,000 sorties and dropped more than 88,000 tons of bombs. Forty-three allied fast jets went down on combat sorties and twenty-eight aircrew ejected into captivity. Including downed fixed-wing aircraft and helicopters, fifty-eight allied aviators were killed.

Iraq lost forty-one aircraft in combat. There are no figures for how many of their aircrew ejected or survived. It is estimated they lost another 255 aircraft on the ground. Overall coalition military deaths were put at 250, most of them American. Forty-seven were British.[17] In the air and on land between 20,000 and 30,000 Iraqi soldiers had been killed, plus an estimated 3,600 civilians.[18]

A week after the ceasefire, all the POWs were released. For many, however, the ordeal had long-term implications. 'One of the enduring emotions I, and most of the other POWs have,' says Larry, 'is the nagging question: *Did I do my job to the best of my ability? Could I have done better, as an aviator, an ejectee, an evader and a captive?* The only answer I have to that question is to stare into the mirror and look at myself. The mirror can be an unforgiving judge. I came close to death in that tumbling F-14 and my life was saved by the ejection seat. *But*, it propelled me into a completely different, though just as lethal, situation. Ejecting was just the start of my survival, but I was lucky to get through relatively unscathed. Thirty-two years on, I often reflect on that. I suppose I did okay.'

Of all the coalition aircraft involved, the F-117 Nighthawk had perhaps the most high-profile, stellar campaign. 'We didn't lose a single pilot or aircraft,' Dale Zelko says proudly. 'It was truly a miracle.' However, he would never forget the momentary terror he felt crossing the enemy border, and would certainly experience it again. 'I saw action again over Iraq a few years later, and we were always ready to deploy into any combat zone around the Middle East. But the concept of ever going to war in Europe was something I never, ever imagined,' Dale says.

He had been born and brought up in America but his father's

family had originated in Slovenia on the northern coast of the Adriatic. It bordered a region in Europe which would be riven by brutal conflict in the coming years. 'I had no idea how my own life was going to interact with that part of the world in the not too distant future.'

CHAPTER SIXTEEN

'OH MY GOSH, YOU'RE A GIRL!'

RAF WITTERING, PETERBOROUGH
25 SEPTEMBER 1991

While the dust was still settling on the Gulf War, RAF Harrier pilot Ashley Stevenson felt England's warm autumn sun on his face. He glanced up contentedly at the Cambridgeshire sky. Small puffs of cloud drifted over the airfield on a light wind. The ground crew had his jet ready for the sortie. Ash loved flying the 'vertical take-off and landing' aircraft or, as the media regularly called it, the 'jump jet'. 'It was a formidable machine,' he says. 'Able to operate from virtually anywhere on land or sea. It was also technically very demanding, which is why I wanted to fly it. I had this dream of being the best.'[1] Ash had joined the RAF as an engineer, but quickly changed direction. Servicing jets would never have the same allure as flying them. After being commissioned as an officer and completing pilot training, he had first taken the controls of a Harrier over West Germany during the closing years of the Cold War in 1982.

Like all fast-jet jockeys, Ash had his feet on the ground where danger was concerned. 'You didn't fly long in the RAF before losing a friend in a flying accident. I remember going to a lot of funerals in the late 1980s. It was just the reality of fast-jet flying.' And he had had his own brushes with death. In October 1990 he had been flying a Harrier at high level over Denmark when the aircraft was rocked by a massive explosion.[2] 'The engine suffered a catastrophic mechanical failure and simply blew apart. I had no power and was descending slowly with the dawning realisation that I wouldn't be taking the jet home. I knew I had to ditch it.' He pulled the handle on his Martin-Baker Mk9 ejection seat at 2,000 feet. 'It was a text-book ejection. I landed, stood up and gathered my kit, then walked away with nothing more to show for it than hurt pride and a split lip.' He felt he had got off lightly. 'I had total trust in the ejection seat and had no qualms about getting airborne again a couple of weeks later.'

On today's sortie from RAF Wittering, 8 miles north-west of Peterborough, Ash was piloting an older version of the Harrier, a two-seat trainer. Kate Saunders, a 21-year-old classics student at Cambridge University, would be in the rear seat for a 'passenger flight'; the opportunity for a civilian to experience frontline flying in a combat jet.

Like other large companies, the RAF regularly recruited its next generation of officers direct from educational establishments and had formed many 'University Air Squadrons' to aid the process of persuading students that the military offered an attractive career option. 'I knew nothing about the RAF when I went to uni,' remembers Kate. 'But I was doing the tour of the various clubs and societies and saw something about a "University Air Squadron" with an advert saying I could learn to fly for free. It was a very different time back then and I initially thought, *Girls don't fly*. But as a student, anything for "free" seemed like a good idea. And learning to fly sounded like a fabulous idea.' Kate had soon accumulated almost forty hours in light aircraft with RAF instructors. 'I eventually qualified to fly solo; flying over the university on my own and look-ing down on Cambridge as life went on beneath me.'[3]

As part of their time on a University Air Squadron, students would often spend a few weeks of their holidays at RAF bases expe-riencing military life. But even though they were training female students to fly, women pilots in the RAF were as rare as hen's teeth. Astonishingly, the restriction banning them from flying had only been lifted in 1989 and, even then, females would not be allowed to fly operational fast jets. Determined to have an RAF career, Kate was well and truly bitten by the bug, but frustrated in equal measure. 'If I qualified as a pilot, I would not have been allowed to fly combat jets on the front line. There was still much doubt about the female physiology; how it would be affected by fast-jet flying and, of course, if they ever had to eject.'

In America, female pilots were ahead of their British counterparts, but only just.

Linda Shaffer-Vanaria, the daughter of a Second World War aviator, had known she wanted to be a pilot since the age of five and had been in the first generation of American women to join a combat squadron, and the ninth in the navy. She voiced her ambition to become a test pilot from day one, to be told that her chances were one in a million. 'I ignored comments such as "You're pretty enough not to have to do that". It was my passion. When you are doing something you really enjoy, you feel very alive. No amount of negative comments would stop me doing it.'[4]

During a solo training sortie in 1981, her aircraft went into an extremely violent inverted spin. 'It's a bad day at the office when your head's getting bashed against the canopy. I couldn't reach the flight controls. I was upside down, blood rushing into my head, seeing stars, and at the same time I was thinking, *I can't allow this to be annoying*. The awareness of having to eject was very much with me.' She pulled the handle on her American-produced ejection seat. 'That's when the real challenge started. The seat didn't leave the aircraft.'

Because Linda had been shoved up against the canopy as the aircraft tumbled, she could not get a strong enough purchase on the handle and hadn't pulled it hard enough. 'By all reason, I should have died that day,' she says now. With seconds to live, she hauled on the handle again as hard as she could, dislocating her shoulder in the process. She became the first female aviator in history to eject. 'Now I occasionally find myself buying a lottery ticket,' she says, 'as a reminder of having only seconds to live, and what it truly means to beat the odds.'

Ten years later, in February 1991, US Navy aviator Linda Maloney was flying an ancient twin-seat EA-6A Intruder on a training sortie a hundred miles off the coast of Florida. 'We started our climb to 15,000 feet,' she recalled. 'The plane appeared sluggish. One of the hydraulic warning lights illuminated, indicating failure.' Within seconds the jet shuddered and started to roll. More warning lights glared. The pilot could no longer hold it. Linda remembers him shouting, 'I don't have control. EJECT!'

'Initially stunned to hear those words, I paused.'

'EJECT!' he repeated. Linda pulled the handle on her Martin-Baker Mk7 seat. 'I remember a flurry of yellow papers flying around and then my seat exploding through the canopy glass. I passed out briefly then the tug of the parachute brought me round and I realised I was heading out towards the ocean.' A Search and Rescue helicopter was soon on the scene, dropping one of its rescue swimmers to help the extraction. 'When he swam over to me his first reaction was, "Oh my gosh, you're a *girl*!"' Linda was the third 'girl' to use the ejection seat to save her life. Many others would follow as militaries across the globe began to reverse their out-of-date policies.

'Unbeknown to me,' says Linda, 'I was Martin-Baker's first woman ejectee – number 6,110. I am extremely grateful to them for that seat, which worked when I needed it. Not only did it save my life, it gave me the opportunity to finish my navy career, write a book and have a family, which is really the best thing I've done in my life!'[5]

Regardless of the evident dangers, Kate Saunders was determined to join those early female aviators in their battle for equality in the skies. Before walking out to the Harrier that day alongside Ashley Stevenson, she had been through a stringent medical, given an emergency briefing about the ejection seat and watched an explanatory video. 'Although I was a member of the University Air Squadron, I was just a civilian, flying in the Harrier as a "passenger", which often happened. Squadrons might fly other members of the station staff as a "thank you" for a job well done. But there were also passenger flights of civilian VIPs such as MPs, journalists or local dignitaries; demonstrating what the RAF did and how it operated. I had the same standard of training on the ejection seat as any other civilian passenger.'

She had been handed g-suit, helmet, life jacket, flying boots and gloves. 'Most of it didn't fit properly,' she remembers. 'All the kit had been designed for men. Even the smallest gloves were too big. The Safety Equipment specialist told me he didn't have a flying helmet small enough for my head – but that I'd be fine as long as I never had to eject . . .'

* * *

In those days, no aircrew equipment had been specifically designed for women.[6] Flying suits were designed for men with a minimum height of 5 feet 6 inches; flying boots were a minimum male size 6. Even with women of the requisite height, other dimensions did not match up to men's. Feet, hand and neck sizes all tended to be smaller in women than in men of the same height. In consequence, even the smallest aircrew boots, gloves and helmets were too large for some women. 'My hands were a lot smaller than most men's,' remembered the first female pilot to go through the system. 'In flying training, they tried to sew some gloves up for me, but it was never really satisfactory, and the flying suit was always too big.'[7]

Fast jets provided another challenging twist to the usual dilemma of toilet facilities for women working in male environments. Trying to urinate in a fighter was to be avoided at all costs but was occasionally required on very long sorties. A system had been designed for men whereby, after flying to a safe area at height, they could make their ejection seat safe by inserting the pin. They would then actually unstrap from the seat, undo various layers of clothing and extract the required appendage before peeing into a small plastic bag fitted with a non-return opening and containing absorbent sponges. The full bag – or bags – would then be stowed around the cockpit until landing. It was neither pleasant nor particularly safe – especially if the aircraft suddenly developed an emergency which required rapid re-strapping into the seat. Needless to say, due to the female anatomy, a different solution was needed and incontinence pads would be provided for women to wear while flying. They were never particularly popular.

The female aviator who would eventually become the RAF's first fast-jet pilot – once the rules were changed – was going through her training at that time. She told the author she spent the first six years of her RAF career wearing baggy male Y-*front* long johns under her flying suit.[8]

And it wasn't just underwear and toilet facilities that would cause problems. Although it was not widely understood around that time, ejection seat parameters were another critical factor. Some women were simply too light for the powerful seats designed to eject bulky men, so there was increased risk of serious injury if they had to pull the handle. Later, as more women entered the training system, the RAF merely seemed to accept that risk. The first female to qualify as an RAF fast-jet flying instructor thought that, even as the issues became better understood, none of the women of her era would have dared to withdraw from flying duties on the basis of increased risk of ejection injury through being too light. She recalled flying with bulky flight reference books in her g-suit pockets to increase her weight.

* * *

Neither Kate Saunders, nor indeed any of the frontline RAF squadrons flying passengers at that time, knew any of this and she was just excited to be getting airborne in a fast jet. Strapping in, she checked the ejection seat firing handle between her knees was properly seated with the safety pin firmly inserted. She remembered vividly the words of one of the briefing pilots. 'He said, "If you ever have to eject and nothing happens, it's probably because *you are a girl* and haven't pulled the handle hard enough, so make sure you really yank it." It was part in jest and part a warning, which I took to heart. I knew what to do in an emergency, but I just didn't think I'd ever have to do it for real. My mother had been a bit worried about me flying fast jets. I told her there was no need – the aircraft was completely safe and, if anything did go wrong, there was always the ejection seat to save me, but that I'd never need it.' She pulled on her helmet, piling her thick blonde hair inside it, hoping it would fit a little more snugly.

Ash Stevenson, who had flown many passengers in the two-seat Harrier trainer, completed his standard instruction brief over the intercom when the canopy closed: 'In an emergency, I will call, "Eject! Eject!" Pull the handle, shut your eyes and go. Don't ask questions, or you'll be sitting there on your own.' His experience over Denmark eleven months earlier was still fresh in his mind, but he had no concerns. 'Ejections may not have been a particularly *regular* occurrence, but they did happen and the simple fact was that the ejectee normally stood up, walked away, had some medical checks, then went back to flying.'

Kate eyed the yellow-striped handle between her legs and wondered how strong a yank it would need. There was no Command Eject system incorporated into this early version of the Harrier – although it was now deemed desirable, military authorities had decided that the cost of retrofitting the complex system into existing jets meant it was 'impractical'. If anything happened, she would be pulling the handle herself. The last thing she had to do before the Harrier could taxi out was remove her safety pin, hold it up in full view of the ground crew to show them the seat was now armed, then stow it in position for the flight. She had every confidence in Ash. 'It was the last trip of my stay with the Squadron, and I was being given the chance very few were ever offered: flying in a combat jet. I was going to make the most of it.'

'All set?' Ash asked.

'Roger,' she smiled. 'All set.'

At 9.30am Ash Stevenson released the Harrier's brakes and

rumbled towards the runway. Kate felt the power of the jet surge as it left the ground.

For forty-five minutes, Ash put the Harrier through its paces at 250 feet and 500mph. 'All was normal as we entered the home stretch, heading south past Scarborough and across the Yorkshire Wolds.' He asked Kate if she'd like to take the controls. She didn't hesitate. 'She knew how to fly a light aircraft but a combat jet was a very different prospect so I'd shown her the rudiments of how to fly the Harrier. We were on a long straight leg over unchallenging terrain and I could keep a close eye on everything.'

Kate was thrilled. 'I was flying a modern combat jet at high speed and low level. Life couldn't get any better. We were just chatting normally – about my possible career as a pilot, enjoying the view, and watching the world flash past.'

Ash picked up his map to confirm their position. Unseen beneath them, a farmer was ploughing his fields and hundreds of birds were following his tractor in the hope of finding a tasty worm in the freshly exposed earth. 'As I wedged the map back between the windscreen and instrument panel, there was an almighty bang and something slammed into my visor. I felt as though I'd been hit in the face with a sledgehammer.'

Impact + o seconds

Still flying the Harrier, from the back seat, Kate saw a bright flash and felt the huge thud, but the protective blast screen between the front and rear cockpits prevented her from hearing anything, or being injured. 'I yelled, "Ashley, what's wrong?" There was a deathly silence. Up to then he had been talking to me normally. I didn't have a clue what was going on.'

Although neither of them knew it, a black-headed gull weighing around 9 ounces had smashed through the Perspex canopy above the front cockpit and into Ash. It is difficult to comprehend how so much damage was done, but one can imagine a cricket ball (weighing around 6 ounces) hitting someone in the face at 500mph.

The Harrier's speed gave the gull the wrecking power of a cannon ball. Ash had been knocked unconscious, his oxygen mask had been ripped from his face, and his helmet dislodged. The left side of his visor had splintered and his chin had been sliced open. His bottom teeth had been smashed sideways and a flap of skin now hung down over his jaw.

He was incredibly lucky to have survived the initial impact of the bird. But both he and Kate now faced further mortal danger.

Impact + 1 second

'What the hell's happening?' Kate shouted again. Still no reply from the front cockpit.

In the pit of her stomach, she knew they were in serious trouble. It looked as though Ash had slumped forward in his seat, but she couldn't be sure. Maybe he was preparing to take back control? Although she didn't understand, Kate Saunders was in exactly the same jeopardy as Phantom navigator Gordon Moulds had been two years before, with an incapacitated pilot and no Command Eject system.

Still flying, she yelled to get Ash's attention, unaware that his microphone and headphones no longer functioned. She shouted louder, but the blast screen between them muffled her cries. She had no idea what to do. 'Everything was now unfolding in mere fractions of a second. I decided I shouldn't be fighting him if he was trying to take control of the jet,' she says. 'So I deliberately let go of the control column.'

Impact + 1.5 seconds

Ash began to come around. Hearing Kate's muffled yells, he sensed her fear, 'but I was in a semi-conscious haze, disorientated, blinded, doubled over, my head resting on my left leg. The noise and buffeting from the airstream blasting through the gaping hole in the shattered canopy added to my confusion.' The impact had been so violent he was convinced they had been hit by another aircraft. 'Unable to see, I had no idea if the jet was steady, going up, or descending rapidly towards the ground.'

As he regained his senses, he felt strangely calm, despite the pain savaging his face and blurring his vision. Training and gut instinct kicked in. His mind raced as he began working through a process of logical thought in order to head off disaster.

'*How long have I been unconscious? What happened, how did we get here?* We had been flying at 250 feet over the terrain, straight and level, before impact. *How much time has passed and what is our height and altitude now?* I was in an out-of-control cabriolet at nearly 500mph, how long did we have before we hit the ground? Seconds?'

Impact + 2.5 seconds

'*Can I fly the aircraft?* No. I can't see anything. *Can Kate fly it?* All she needs to do is get it away from the ground and hold a safe height and attitude until I can fully regain my senses. She had control before impact, but she hasn't the experience or skills to take command of the situation, or to land. Besides, I can hear her shouting, so she's probably let go of the controls and expecting me to sort us out. So no, she can't fly it. I realised that for the second time in less than a year, I was going to have to ditch my Harrier. But the aircraft is not fitted with a Command Eject system, so Kate has to pull her own handle. As she's in the rear seat, she should really go first, otherwise we could hit each other on the way out.'

Impact + 3 seconds

'*How can I tell her to eject when I have no mask or microphone, and my mouth is severely damaged?* The truth was brutal. The only way is for me to eject, and hope to God she takes the hint and follows. *But what if she doesn't? What if she crashes with the jet? What if I survive and she doesn't?* I could never live with that, the guilt would consume me – I'd rather be dead, too. But if I don't pull the handle she'll die anyway, as will I, so what's the point of not giving it a go? Catch-22.'

Impact + 3.5 seconds

'I had worked through my predicament and there was only one option left for me: I *have* to pull the handle. It's the *only* way that offers Kate a chance to survive. And I have to do it *now*. Every second's delay reduces our chances significantly.'

He reached down and pulled the handle.

'Strangely, I did so with little emotion and, as the seat fired, two inauspicious thoughts flashed through my mind. First: *I can't believe I'm doing this again just months after my previous ejection.* They say once is unlucky, twice is careless – could I have avoided this?

'Second: *You've left it too late. There is no way you're going to survive this.* I still couldn't see anything and presumed the aircraft must be inches away from the ground by now. I thought I'd hit the deck still strapped to the seat. *It's gonna hurt*, I thought, but it'll be brief, then it'll be over.

'*Bugger, I was really looking forward to going to the pub for a beer tonight.* I felt quite calm, logical, but really disappointed.'

Impact + 4 seconds

In the rear cockpit, Kate struggled to comprehend what she was seeing. 'I watched the back of Ash's ejection seat rising, the rockets flaming beneath it and then disappearing above me. I was alone in a jet that was going to crash at any moment. I kept thinking, *This really can't be happening to me . . .*'

Remembering the safety brief – and the words of advice about girls' supposed strength – she straightened her back, shut her eyes and yanked the ejection handle upwards for all she was worth. 'There was a huge *whoosh* and I was moving upwards. That's the last thing I remember.'

Tomcat pilot Devon Jones sprints towards a rescue helicopter after being
shot down and ejecting during the 1991 Gulf War.

Jones and his RIO Larry Slade (left), who
was captured, reunited after the war.

A Martin-Baker Mk10 seat.

An F-18 pilot ejects after losing engine power during a Canadian air show in 2010. The full ejection sequence is captured on camera.

The aircraft canopy has been jettisoned, the pilot's ejection seat rocket motors have fired and he exits the cockpit.

The white drogue parachute has deployed and begins to stablise the seat. The jettisoned canopy can be seen on the left of the photograph.

The main parachute begins to deploy and the pilot can be seen just released from the seat as the F-18 plummets towards the runway.

The pilot's parachute opens and the seat can be seen falling away as the jet hits the ground and explodes.

The pilot descends on a fully deployed parachute, very close to his burning aircraft.

Mark 'Skids' Richardson and Mike 'Elvis' Costello visit the crash site after their Tornado F3 collision. Wreckage can be seen strewn over a large area and in the deep crater.

Navigator Ian Weaver, who was severely injured in the Tornado collision in 1996, with his daughter Jenny in 2000.

Tornado F3s similar to the ones involved in the mid-air collision are seen here in very close formation. The crew can clearly be seen strapped into their ejection seats.

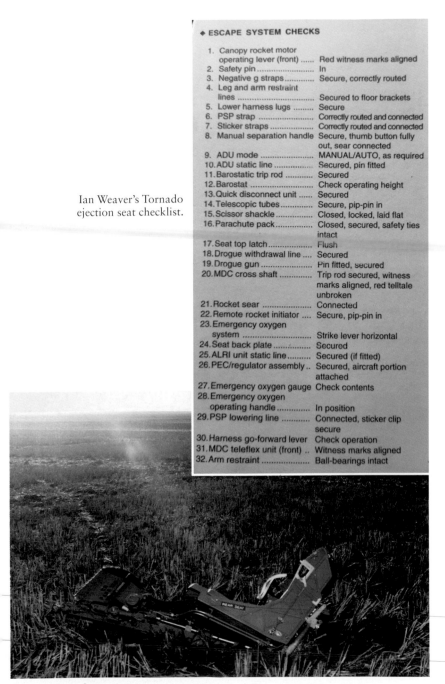

◆ ESCAPE SYSTEM CHECKS

1. Canopy rocket motor
   operating lever (front) ...... Red witness marks aligned
2. Safety pin .... In
3. Negative g straps .... Secure, correctly routed
4. Leg and arm restraint
   lines .... Secured to floor brackets
5. Lower harness lugs .... Secure
6. PSP strap .... Correctly routed and connected
7. Sticker straps .... Correctly routed and connected
8. Manual separation handle Secure, thumb button fully
   out, sear connected
9. ADU mode .... MANUAL/AUTO, as required
10. ADU static line .... Secured, pin fitted
11. Barostatic trip rod .... Secured
12. Barostat .... Check operating height
13. Quick disconnect unit ...... Secured
14. Telescopic tubes .... Secure, pip-pin in
15. Scissor shackle .... Closed, locked, laid flat
16. Parachute pack................. Closed, secured, safety ties
    intact
17. Seat top latch................... Flush
18. Drogue withdrawal line .... Secured
19. Drogue gun .... Pin fitted, secured
20. MDC cross shaft .... Trip rod secured, witness
    marks aligned, red telltale
    unbroken
21. Rocket sear .... Connected
22. Remote rocket initiator .... Secure, pip-pin in
23. Emergency oxygen
    system .... Strike lever horizontal
24. Seat back plate .... Secured
25. ALRI unit static line.......... Secured (if fitted)
26. PEC/regulator assembly .. Secured, aircraft portion
    attached
27. Emergency oxygen gauge Check contents
28. Emergency oxygen
    operating handle .... In position
29. PSP lowering line .... Connected, sticker clip
    secure
30. Harness go-forward lever Check operation
31. MDC teleflex unit (front) .. Witness marks aligned
32. Arm restraint .... Ball-bearings intact

Ian Weaver's Tornado ejection seat checklist.

Weaver's used ejection seat in the field where it fell only yards from farm buildings.

An F-117 Nighthawk similar to the one flown by Dale Zelko.

A SAM 3 missile test firing. This was the system used to shoot down Zelko in March 1999.

Zelko on landing after his rescue, about to present the American flag he had been carrying back to his mission planner.

A Serbian teenager sitting in Dale Zelko's used ejection seat.

Civilians dancing on the wreckage of Zelko's jet.

Zelko with his daughter, reunited with Zoltan Dani, the man who had shot him down. They are attending the 2012 premiere of the film that tells their story.

Harrier pilot Martin Pert ejects after crashing on the runway
at Kandahar Air Base, Afghanistan, in 2009.

A Martin-Baker test ejection.

Impact + 5 seconds

Ash had ejected at just 150 feet as the jet descended. He too, had been knocked unconscious, 'by a 15g kick up the backside courtesy of Martin-Baker'. Seconds later, the jolt of the parachute opening and the flash from its orange canopy were enough to bring him around. 'The first thing I saw was the ground, about 10 feet away, rushing up to meet me. No time to release and lower my heavy PSP survival box. I put my feet and knees together and hit the ground with a monumental thump.' He was knocked unconscious for the third time.

Impact + 6 seconds

The Harrier hit the ground, exploding and disintegrating. Its remaining fuel ignited, torching 40 acres of corn stubble. Ash had landed some 50 yards short of the aircraft's impact point and came to winded, trying to take stock. A wall of flame and smoke from the burning field stretched several hundred yards in front of him. He hauled himself groggily upright, flying suit dripping with blood, when a stab of pain in his foot almost floored him again. His heavy landing had crushed a bone in his heel. 'My mouth was a mess, with blood pouring from a huge gash on my chin; I could lift the skin up like a catflap. I could feel some of my lower front teeth were now at right angles to the gum, pointing inwards and sitting on my tongue.' He clenched his eyes against the blistering heat and peered into the dense black smoke.

No sign of Kate.

Maybe she didn't get out ...

No! He couldn't accept that. He had to find her. He limped towards the inferno, praying she was on the far side of it. 'I may well have ejected and survived, but it was quickly becoming clear that the ordeal was far from over.'

Kate had ejected from the Harrier just 90 feet from the ground, as it descended rapidly and was rolling over through 40 degrees from horizontal. Because of the downward vector of the jet, and the hazardous angle of ejection, there was not enough time or height for her parachute to fully open. Kate should not have survived, but incredibly the blast wave caused as the Harrier crashed and exploded just after she ejected had helped inflate her parachute and broken her fall. When she recovered consciousness, she saw that the fireball had also set her parachute alight, and molten droplets of its artificial material had scorched her forearms. That was the least of her problems.

'One of my legs looked as though it wasn't attached to me any more.' She had suffered open fractures to the tibia and fibula of her right leg, a compression fracture of the twelfth thoracic vertebra and fractured her pelvis. As would later be discovered, Kate was too light for the seat and the force of ejection had broken her pelvis. Her leg was probably broken when, still unconscious, she hit the ground – like Ian McLean and Neil Johnston the previous year – with her heavy PSP still attached behind her backside.

Somehow, she managed to sit up. 'I was in a hell-like inferno. I was surrounded by flames and grey smoke. I just shut my eyes because I didn't want to be there. It seems stupid now, but I wanted to do something to make that terrible situation go away. It was too much to take in.'

But it was all too real; Kate had landed in the blaze created by the crashing Harrier. The fireball had just saved her life, now it threatened to burn her to death. The ground beneath her was on fire and the back of her legs and buttocks were burning. The paper maps tucked into the front-leg pocket of her g-suit burst into flames. Her flying boots were melting. Her nylon life jacket was on fire and as the flames reached the nape of her neck she began screaming. But immobilised by so many fractures, evasive action was impossible.

She started to panic, closed her eyes, praying again that it was all a bad dream, but then took herself to task. *Come on, Kate. You have to deal with this. Now!* 'I tried to beat out the fire on my body with my hands.' She had lost one of her ill-fitting gloves during the ejection. 'I had to get my life jacket off which was now melting onto my back.' She struggled to unclip it, then tore off her remaining glove, helmet and oxygen mask, so she could see more clearly. Remarkably, her blonde ponytail was only slightly singed.

'As I was trying to slip my arm out of my life jacket, I held a hand in front of my face and watched as my skin melted, burning through to the bone. I now realised I couldn't save myself.' She started screaming louder. 'Help me! I'm dying . . .'

A strange calm descended. 'I just accepted my life would soon be over. It was quite peaceful really.'

Nearing the wall of flames, Ash heard her cries for help. He felt a surge of relief. *She'd made it!* 'Keep screaming, Kate. I can't see you, I'll follow the sound.'

Thinking that they were now only separated by a narrow ribbon of fire, Ash hobbled to the upwind end of it, hoping to see Kate on the other side. 'When I got there I was faced with another wall of flame, perpendicular to the first and just as long. I was standing at one corner of an edifice of fire and smoke, and Kate was somewhere inside.' He shuffled back to the point where Kate's agonised cries had sounded closest, and yelled at her again to keep calling out, so he could get a bearing on her.

'I still couldn't see her, but now I knew where she was – in the middle of the blaze. *Shit!* I remember standing outside the flames thinking, *I'm okay out here, if I go into that fireball I'm not going to be okay any more. And things could end really badly . . .*'

Reliving the experience with the author Ash pauses as his emotions get the better of him. He's facing the inferno once more. 'I consciously had to tell myself it was something I needed to do. Something I had to do. I pulled myself together and hobbled into the fire.' He found he could dodge some of the flames by picking his way across patches of blackened, smouldering stubble. But he was soon choking on the impenetrable smoke. And he still couldn't find her.

'Kate, keep screaming . . .'

He heard her, but more faintly now. Ignoring his own pain, he forced himself to move faster. Seventy yards further on, he finally saw her, only 15 feet away. 'She was sitting upright on the ground, lightly slapping the flames licking at the back of her neck, as if lethargically shooing flies on a hot summer's day. I could see her right leg was badly broken, sticking out sideways below her knee, the foot lying flat to the side. Her whole body was burning, her hands, her neck, her back and legs ... all on fire.' Ash falters again as the memories flood back. 'I can still see it clearly today. It wasn't an easy sight to take in. But I had to try and save her.' Raising his forearms as best he could to protect his smashed face from the searing heat, he rushed towards her.

Kate spotted a figure appearing through the conflagration. 'I was so relieved to see Ash coming towards me like a scene from a Hollywood movie. The first thing I said to him was, "You didn't say *Eject! Eject!*" which was a bit ungrateful, as he was putting his own life at risk to help me.'

There was no time to reply. Her nylon g-suit was melting to her legs and 8-inch flames were now consuming her life jacket. Ash threw himself on top of her in an attempt to smother them, then did his best to beat them out. 'In the end the jacket more or less disintegrated and I managed to tear it clear of her,' he says. 'Skin from her back came away too. Her hands looked like molten plastic.'

Kate's pain was now close to unbearable. 'I'm burning,' she screamed. 'I just want to lie down ...'

The smoke was catching at the back of their throats.

'I'm going to have to move you, Kate,' he gasped. 'If I don't, we both die.'

But did he have the strength? And would he cause her further injury if he tried to lift her off the ground? There was only one way of finding out. Standing behind her, he slipped his arms under hers and around her waist, and began dragging her backwards through the fire. Her broken leg and burned backside bounced and scraped across the rough ground, prompting further agonised screams. 'Leave me here,' she whimpered. 'It's hurting *so* much.'

Ash knew it wasn't an option. Battling against the smoke, the searing heat, the stabbing pain in his heel, and close to exhaustion, he had to stop three times to rest, each time picking her up again and dragging her on.

Leave me here, Kate implored. *Leave me here, I beg you.* 'It all became a bit of a blur. I just wanted it all to be over.'

Ash barely had the strength to shake his head. 'We were still in real danger. The wind had changed direction, and fire was speeding towards us.'

As they neared the edge of the cauldron a small van approached. A decorator had been painting the exterior of a house a few miles away and seen the Harrier crash. Scrambling down his ladder, he'd told the owners to dial 999, then sped towards the smoke.

'He helped me lift Kate onto the bonnet,' Ash says. 'And while I steadied her, he gently reversed away.'

Once they were clear, they eased her back onto the ground and made her as comfortable as they could. Ash felt the first tremors of shock. 'I began to think that we had got away with it, and would probably survive.' The police and fire service were soon on the scene. While some dealt with the blaze, others removed Kate's right boot. 'There were further waves of incredible pain,' she remembers. The leather had stuck to her skin. 'I was begging them to leave me alone.'

An RAF Search and Rescue helicopter touched down twenty minutes later. 'It seemed like a very long time,' she says. 'I was delirious by then. A doctor put me on a drip but didn't give me any painkillers. I was yelling, "Put me out! Put me out!" I just wanted to escape the agony.' Kate and Ash were airlifted to Hull City football ground, from where an ambulance rushed them to the Royal Infirmary. Once the medics were satisfied she didn't have a head injury, she was given morphine, stabilised and transferred to the Regional Burns Centre in Wakefield. In addition to her broken leg and pelvis, and compression fracture to her vertebra, Kate had suffered 28 per cent burns to her body which would require extensive plastic surgery. She would remain in hospital for nearly three months.

Back in Hull, the surgeons stitched Ash's chin back in place and reset his teeth. Once on the ward, unaware of Kate's plight, light-headed relief at their survival gave way to darker thoughts. '*How badly injured was Kate? She was my responsibility. What had gone wrong? I'd now ejected twice in less than a year. Was it all my fault?* It was possibly an irrational notion, but, ultimately, I was responsible for all this. Whatever had happened, I was to blame.'

Seeing his distress, a nurse came over and held his hand. She told him it was his bravery that had saved Kate's life, and they had both been incredibly lucky. 'It just didn't feel that way at the time,' Ash says.[9]

He had to suffer a diet of liquids and softly mashed food for the next few months, but six weeks later – once his face had healed sufficiently to wear an oxygen mask and a good friend had boosted his dented confidence by taking him flying in a two-seat Harrier on a low-level sortie around Wales – he was back in the air. 'Flying was my life,' he says. 'I was initially somewhat reluctant, and it took a little persuasion, but I was determined to get back on the horse.'

For Kate Saunders, the journey was only just beginning.

CHAPTER SEVENTEEN

A DAY LIKE NO OTHER

Ash Stevenson and Kate Saunders had joined a growing list of aviators who, although injured, had had their lives saved by ejection seats. The risks involved in military flying were well known. Worldwide figures are elusive, but looking at the Royal Air Force's official statistics alone, the dangers become apparent. In just five years from the end of the Second World War to 1950, the RAF lost an astonishing 3,268 aircraft in training accidents and on operations, resulting in 2,158 fatalities.[1] Clearly some aspects of military flying were still in relative infancy, but the numbers are shocking.

In the thirty-five years from 1960 to 1995, the RAF recorded another 1,172 aircraft lost, with 897 fatalities. The combined losses from other nations' militaries must have been staggering. Obviously, these accidents include aircraft like transport and helicopters which were not equipped with ejection seats, but the figures still tell a story of the ever-present hazards involved in military aviation. The author attended a training course in the early 1990s where the participants were informed that the RAF *alone* 'statistically expected' to lose around twelve lives a year in flying training accidents.[2]

As has been seen, neck, back and leg injuries could still occur due to the huge forces needed to propel the comparatively frail human body high enough, clear and away from some 20 tons of fuel-filled metal about to impact the ground at high speed. Ejection seats were constantly evolving and improving, and, while injuries, some serious, certainly happened, the vast majority of ejectees were subsequently cleared to return to fast-jet flying.

Ejection from a combat jet – the final option to save a life – would always be a delicate balance between injured survival or certain death. And terrible accidents did still occur.

* * *

In November 2011 Sean Cunningham, a pilot on the RAF's Red Arrows display team, was strapped in and preparing for a flight when, stationary on the ground and without him pulling the handle, his ejection seat suddenly fired, blasting him from the cockpit. His parachute did not deploy and he was killed when, still strapped into the seat, it impacted the ground near his Hawk aircraft. The horrific incident had been caused by a sequence of largely unlinked, but ultimately fatal, events.

It later transpired that the seat's black and yellow ejection handle was probably not fully seated in its housing, and the safety pin not properly inserted to prevent the handle being moved. It was thought that the firing handle could have been dislodged by an incorrect strapping-in procedure on a previous sortie. The error had not been spotted and the seat had remained in a catastrophically dangerous position for some days. As Sean shifted around in the cockpit, completing his various pre-flight checks, it seems the ejection handle was moved enough to fire the seat.

However, those events on their own would not have caused his death. He was sitting on a Martin-Baker Mk10 'zero-zero' seat and, if the sequence had operated correctly, he would have been released from the seat at the top of its trajectory, with his parachute withdrawing and opening as normal. But a final, almost unbelievable, tragedy intervened. Although the drogue parachutes deployed, a small nut had been fractionally overtightened on the main parachute shackle system meaning it jammed and did not release Sean's personal parachute. His fate was sealed.[3]

Although it was not a fault of ejection seat design or manufacture, subsequent RAF engineering on the front line – the overtightening of a small bolt – was obviously deemed the key factor.[4] The later inquiry and court case raised countless questions surrounding the incident for both the RAF and Martin-Baker. The company was eventually prosecuted under Health and Safety legislation and fined £1.1 million for failing to inform its customer, the RAF, of the dangers of overtightening the bolt.[5] Many aviators at the time felt that the decision was unfair. The ruling made no difference to Sean's family, who gave an impact statement at the trial:

> Sean's interest in flying started when he was a teenager and he gained his private pilot's licence at the age of seventeen, having saved up to pay for his lessons. His first love was to fly. His two biggest fears in life were being ejected from an aircraft with the injuries that could be sustained, and dying at a young age. He was to tragically experience both of these. He was the perfect son and brother and the pride in him from the family is justifiably immense.[6]

The tragic incident, and the loss of one of the RAF's shining stars, was deemed 'entirely avoidable'.

The investigation into Sean Cunningham's accident had established that since 1959 there had been twenty-four successful Martin-Baker ejections – across all aircraft – under zero-zero conditions: the aeroplane on the ground and stationary. Sixteen had involved the identical mechanical shackle system used in Sean's seat. His death was the only such ejection where the main parachute did not deploy. The tragedy showed just how narrow the line between death and survival could be for military flyers. A misaligned cable, an incorrectly routed strap, a tiny bolt overtightened by a fraction of a turn could still prove fatal.

And it wasn't just engineering errors on the more modern seats that could cause problems. As aircraft and their ejection systems had evolved, older aircraft were constantly being withdrawn from military inventories. Some of those high-performance jets, and their elderly ejection seats, were increasingly passing into civilian hands.

NORTH WEALD AIRFIELD, EPPING, ESSEX
3 APRIL 1994

Tom Moloney had built a successful company supplying fellow aviators with flight equipment, clothing and survival gear. He had started the business in the bedroom of his home near Heathrow Airport, where his father worked as a baggage handler.[7] A keen pilot, Tom had dreamed of owning a jet for as long as he could remember and had gained his civilian Private Pilot Licence (PPL) in 1984. Now thirty-two, he was about to take his younger brother, Des, flying in the ex-RAF Jet Provost he had bought a few days earlier.

First flown in 1954, over 700 of the small twin-seat trainers had been built, and many thousands of aspiring RAF pilots had learned to fly in the iconic jet, affectionately known to all as the 'JP'. Heading for retirement in the 1990s, old JPs could be picked up for as little as £1,000. 'Flying restrictions back then were very different,' says Tom. 'Astonishingly, you could simply buy an ex-military jet, climb in and just fly it away.'

Even though he was already certified to fly a variety of civilian aircraft, Tom was too safety conscious to take any risks. Because he had no military aviation experience, he had hired a former RAF Qualified Flying Instructor who could show him the ropes and also supply him with technical information, the necessary checklists and a copy of the student manual. They had spent five hours together in the air as Tom got to grips with the jet's handling and learned its emergency procedures. 'It was a joy to fly. A solid, reliable little aircraft, and a real piece of aviation history.' Despite being powered by a jet engine, the 34-foot-long JP had manual flight controls and simple mechanical systems. Low maintenance and economical with fuel, it could climb to 31,000 feet and reach a top speed of over 400mph. Sitting on North Weald Airfield, north-east of London, with its red and white livery gleaming in the Easter Sunday sun-shine, Tom liked to think of it as his 'Ferrari of the skies'.

Unlike Tom, his brother Des was feeling slightly underwhelmed, even apprehensive, as they put on their flying suits and walked out to Tom's pride and joy. Both wore military-style flying helmets with tinted visors and boom microphones. Des was also sporting a leather flying jacket with the collar turned up for Tom's first flight in sole charge of his JP, and the first time they had gone up together. But he was no Biggles. 'I had never been interested in aviation the way Tom was,' he would admit to anyone who asked. 'I'm fairly adventurous, but high-speed flight wasn't something I was hugely interested in. Curiously, I remember a slight feeling of foreboding that morning. I couldn't put my finger on it, but I sensed this was going to be a day like no other.'[8]

Side by side in the Jet Provost's snug cockpit, Tom, in the left-hand seat, helped adjust and tighten Des's shoulder harness and lap and crotch straps. He then took him through a twenty-minute safety briefing. 'As I would with any passenger, I detailed all the systems on the aircraft, particularly the abandoning procedures. This was my baby brother, so I had to look after him and make sure he knew what he was doing.'

They would both be strapped to one of Martin-Baker's early ejection seats. Des had visited the company's stand with Tom at various air shows over the years, and they had both seen the accounts of aircrew shot down and ejecting during the Gulf War a few years earlier. 'I knew what an ejection seat *was*, and thought they were amazing devices,' says Des. 'But other than that, I didn't really give them a second thought. As far as I knew, I wasn't actually sitting on an "ejection seat" per se.'

In theory, he was right. The 1950s-developed Mk4 seat in Tom's Jet Provost had been stripped of all explosive charges and totally 'disarmed' before sale to its new civilian owner. Metal placards had even been fitted to the ejection seat headboxes, indicating they were 'inert'. Tom's RAF flying instructor had confirmed all this during his briefings. 'He told me: "Although your parachute is still stowed in the seat as normal, this is no longer an 'ejection seat'; it has been completely deactivated. If you have to abandon the jet, you need to do it manually." So as far as I was aware it was indeed "inert" and just my aircraft "seat". It contained my parachute, I strapped into it to fly, then unstrapped after we landed. Nothing else.'

If they did have to leave in an emergency, Tom would first unlock and jettison the cockpit canopy using the release lever. They would both then need to pull their individual seat 'manual separation' handle which would disconnect all their seat straps, but still leave them attached to their parachute harness. Then, they would have to stand up into the airflow and dive out over the side of the cockpit, towards the rear edge of the wing.[9]

'This all sounded very straightforward, but rather frightening,' Des recalled. But there was more to come.

'THEN . . . when you have eventually fallen clear of the aircraft, pull this,' Tom continued, tapping the metal D-ring on the harness just above his brother's waist with his index finger.

'You're kidding.'

'That's the ripcord.' Tom smiled. 'Pulling it opens your parachute.'

They were ready to go. Des was being dazzled by the sunshine, but was so tightly strapped in he couldn't find his sunglasses.

'Just pull down your helmet's dark visor instead, okay?' Tom slipped on his white flying gloves, cranked the canopy shut and fired up the engine. He gave a thumbs-up to Kay, his fiancée, who was there to wave them off. As they taxied out to the runway holding point, Tom noticed that his brother was looking twitchy. 'Tom,' Des asked, 'where's that D-ring for the parachute again?' Tom took his brother's hand and placed it firmly on the parachute-release ring. Des felt a lot happier.

After taking off, they flew eastwards, doing a few gentle turns. Gazing down at the rolling Essex countryside below, Des had to agree that the view from the cockpit was superb. 'I was actually beginning to enjoy the flight.'

Tom sensed his brother relaxing. 'Would you like to do a roll?'
'Perhaps just the one,' Des ventured.

Near Colchester, 15 miles from the Essex coast, Tom handed Des
his map and checklist for safekeeping and began a gentle roll to the
right at 3,000 feet. It had just gone 11.40, and the JP was travelling
at around 250mph. As the jet rotated, Des was sure he felt a jolt as
his seat moved slightly. 'It didn't seem quite right. But I didn't really
understand what it meant, so I said nothing.'

What neither Des nor Tom knew was that the ancient ejection
seat's 'top latch' – the locking mechanism designed to hold it firmly
on its rail – was both worn and not properly screwed in place. In
normal flight, the top latch prevented the seat from moving if the
aircraft was inverted. On a 'live' seat, when the ejection guns fired
the latch system disengaged, releasing its grip and allowing the seat
to rise, guided safely on its rail, blasting up and out of the cockpit.[7]
Unfortunately for Des, that first aircraft roll had dislodged the worn
locking mechanism sufficiently to render it inoperative. The basic
laws of gravity were now in control.

'Did you like that?' Tom asked as he righted the JP.

'It was fine,' Des lied.

Tom rolled the jet left into another rotation.

Des felt the seat jolt again. Now he was in no doubt. It was definitely moving.

The Jet Provost was almost on its back.

The latch had worked its way free. Now upside down, the whole
seat was sliding fast up its guide rail, towards the cockpit canopy.
And, in this case, sliding 'up' meant the seat – and its occupant –
were heading earthwards.

Des managed 'a single, instantaneous thought: *This seat is leaving the aeroplane and I am attached to it*. Everything then happened
in an instant and I wondered if I'd survive going through the canopy.
Then I was hit by a massive air blast.'

Out of the corner of his eye Tom saw what was happening and
instinctively shot out his arm. He got a hand to the seat, but the

Perspex canopy shattering told him it was too late. 'I watched the bottom of the seat and my brother's feet crash through the cockpit hood and simply disappear. I could not comprehend what had happened. Des had just gone and was now falling to his death.' He grabbed the stick to turn the jet the right way up, then, still flying at around 250mph, pulled a hard turn back to the point where Des had been thrown out. There was no sign of his little brother.

'I knew if Des didn't get out of the seat, he'd be killed. He would still be sitting in it, spinning over and over. How could he ever remember that he still had to pull the seat's manual separation handle to release its straps? Then push it away, before finally pulling his parachute D-ring?' It was almost the same procedure Jo Lancaster performed back in 1949 after ejecting from his experimental 'Flying Wing' aircraft. But Jo and the other ejectees using those early, manual seats had been trained to carry out the correct procedure. Des Moloney had not. Tom was bereft, smashing his hand on the cockpit panel in desperation. 'My God!' he screamed. 'He's just been killed.'

He pressed the transmit button on his radio and said, as loudly and clearly as his emotions would allow, 'MAYDAY, MAYDAY, MAYDAY . . .' He told North Weald Airfield what had happened, and that he was on his way back. He prayed he had been heard. The sound and force of the wind howling through the smashed canopy was strong enough to impair his vision and hearing, and had swept his map away. 'I was horrified, stunned, completely disorientated. I had to crouch forward to fly the jet because of the air blast in the cockpit. I was just desperate to land and get out of that aircraft.' The fifteen-minute flight back to the airfield was 'the longest, loneliest time of my life', Tom remembers. 'I knew in my heart that Des was dead.'

After an emergency landing, he took the first turning off the runway, pulled up and jumped out of the jet, falling to his knees, convulsing in shock. Kay, waiting to welcome the pair back to the airfield, rushed over to him. 'I kept repeating, "Des is gone." I just didn't know what had happened to us, or how to explain it.' His fiancée couldn't understand what he was trying to tell her. 'My brother's dead,' he screamed. 'He just fell out. Des is dead.'

But Des Moloney was very much alive. For the moment. 'I sensed I was falling and knew I was in big trouble. Part of me didn't want to accept what was happening. I remember thinking, *I don't want to deal with this, I'll just do nothing and see what happens*. That wasn't an option, so what could I do?'

An astonishing series of interconnected events now contrived to save him. Firstly, because his helmet's dark visor had been down, in place of his sunglasses, his face had been protected and not cut to pieces by the shards of Perspex as he smashed through the canopy, tumbling, still strapped into the seat. Then, a second piece of incredible luck. Unbeknown to the brothers, the ejection seat had not been properly deactivated. The 'barostatic time delay' device – which in a fully armed seat is mechanically activated as the seat rises up the rail, automatically releasing its occupant after 1.25 seconds – had actually been left connected on this 'inert' seat. It now did the job it had been built to do over forty years earlier, whirring into action, and, just over a second after Des's catastrophic departure through the cockpit hood, it activated the release mechanism, guillotining his straps and freeing him from the seat. Normally, an experienced aviator now knew they had to push away from the seat to properly break free. Des knew nothing about ejection seats, but lady luck intervened once more. As he fell out of the cockpit, he had unknowingly collided feet-first with the tailplane which sent him spinning, that motion now throwing him clear of the seat as his straps released.

Still totally unaware of what was going on around him, Des was now in free-fall. Did he have the capacity – and knowledge – to remember he needed to perform one final action if he was to survive?

'I was no longer tumbling and realised I was free from the seat, but had no idea how that happened. When I opened my eyes I could see green fields racing up to meet me. I suddenly remembered Tom's safety brief about the parachute-release D-ring. I presumed this would probably be a good time to pull it.' With his right hand, he felt for the piece of metal sitting to the left of his waist, threaded his fingers through the ring and pulled. A multicoloured canopy billowed above him and his earthward plunge ceased. He was now floating serenely downwards. Des suddenly felt calmer. 'The James Bond theme tune now played through my head. *I'm going to live.*'

But his ordeal was not yet over.

'As I descended, I mused, *Tom is really going to get into trouble for this.*' Then he glanced down towards his feet. 'All the straps were dangling around, which didn't look right.'

It was all very far from being 'right'. When he'd strapped into the seat, Des had not fitted his crotch straps correctly so the harness had never been secured beneath his torso and between his legs. It began to ride up and Des began to slip down and out of the harness. In a series of ominous jerks, the straps rose to his armpits. If they lifted over his head, he would drop away, back into free-fall. But this time with no parachute. Luckily, Des was both fit and strong. 'I braced my arms and elbows downwards and tight to my torso, trying to stop the parachute harness riding right up my body and off me. If that had happened, I would certainly have died. It took huge effort to stabilise myself as I floated down, but it seemed to be working.'

Not for long; worse was to come.

Now a thousand feet above the ground, the harness jerked upwards again and its round metal 'quick-release' box, which locked all the straps together, shot over his chest and rammed into his throat, sticking there like a limpet. 'It simply cut off my breathing and began strangling me. I was utterly terrified again.' While trying to ensure he didn't slip out of the harness, keeping his elbows jammed to his side, he tried to reach up with his hands and tug the harness box away from his neck. He could not shift it. 'I was being hanged by the parachute that had just saved me. For twenty seconds I couldn't breathe at all. I remember thinking, *I can't do this any more.* That's when I truly thought I was going to die.'

On the point of passing out, the ground finally rose up fast to meet him. Thirty seconds after pulling his ripcord, he slammed onto a grass verge outside a Sainsbury's supermarket on the outskirts of Colchester. His abandoned ejection seat crashed onto the roof of a nearby house, causing minor damage. It was a miracle nobody was injured.

Struggling groggily to his knees on the long grass, blood dripping from a gash to his mouth, Des wiggled his hands and feet to make sure he was okay. A young boy watching the drama unfold detached himself from his family's shopping trip and ran over. 'Hey, mister, can I have your parachute?'

'Yes,' Des replied, gulping in lungfuls of air. 'As long as you help me get out of the damn thing.' His young rescuer somehow knew how to twist and press the quick-release harness buckle, freeing him from the remaining straps.

Des was alive, but he still didn't understand what had happened, and now presumed there may have been some sort of collision, or the Jet Provost had somehow disintegrated. 'I heard a bystander say they had seen a dark black object falling from the sky. There was no parachute. If that was Tom's jet, there was only one possible conclusion. My older brother must be dead.'

Over an hour after he had landed, Tom and fiancée Kay were still at North Weald Airfield, shell-shocked and grief-stricken. Now sitting in the back of a police car, they did not know what to do next. At some point, Tom was going to have to call his parents to tell them their youngest son was dead. Then the police radio crackled into life. 'They had found *somebody* in a field,' Tom remembers. 'But that's all they knew. We presumed it was Des's dead body. It was really difficult to get information from the police because they were still rushing around trying to establish details.' Then another message. The body was being taken to Colchester General Hospital, a forty-minute drive from North Weald. 'We instantly set off with the police. I was convinced that I was going to identify my dead brother.'

Des, likewise, could not process the awfulness of losing Tom. Grief paralysed him. Then, as he was being stretchered into an ambulance, his neck now in a brace, a paramedic, who had been on his radio in the cab, came running up to his colleagues.

'Have you heard the news? The aircraft this man fell out of has landed safely.'

'Thank, God. Is he all right?' Des asked, desperate for news, still riding an emotional rollercoaster.

'The pilot's perfectly okay.'

Des Moloney's worst fears had been allayed. Tom's had not.

'As we raced to the hospital in the police car, we finally learned that the person was still alive, but their condition was unknown. I turned to Kay, saying, "He *can't* be alive?" How had he survived after hitting the ground in his seat? He must have been smashed to pieces. I was coping with it very badly. He was my younger brother and it was my responsibility to look after him.'

Early afternoon, two and a half hours after the incident, Tom and Kay were rushed along a corridor in Colchester General Hospital. Tom was dreading what sight would greet him at their destination. He prayed his courage would see him through. 'I was expecting to see my brother, his body shattered and barely alive,' Tom remembers. As they burst into the room, they found Des lying on a trolley in his neck brace, smiling up at them.

'Rumours of my death,' he said, 'have been greatly exaggerated.'

Emblazoned on his hospital wristband in red marker was the phrase: *Man who has fallen out of an aeroplane at 3,000 feet.*

'I couldn't believe it. We hugged and kissed. I don't think I had ever kissed Des before, and I probably never will again. The brother I thought was dead had survived. A sequence of totally unprecedented events surrounding that ejection seat had ensured his survival.'

'And I thought it was *you* who was dead,' Des told him.

Two days later he was released from hospital.

Nearly thirty years after the incident, Des still can't fully fathom quite how he survived, and he remembers every second of the incident in glowing technicolour. 'I was just so lucky that day,' he recalls. 'I wasn't seriously injured going through the canopy. Then that "inactive" time-release device activated and released me. Then I remembered to pull the parachute D-ring. Then I was nearly strangled to death by the harness. It certainly was a strange day.'

Amid many recommendations, the Air Accident Investigation Branch's report concluded that all occupants of civil aircraft equipped with ejection seats, whether they were disarmed or not, must 'receive appropriate training, information and certification regarding the use of the seats and the equipment installed on them'. It seems astonishing that this had not already been the case.

Tom remembers another change that came about because of his brother's near-death experience. 'In the aftermath, the powers that be in aviation realised that if ex-military jet aircraft – with inert ejection seats – were ending up in civilian hands, there needed to be better procedures. So, subsequently, any civilian operators could then *choose* to have the ejection seats in their jets fully functional, armed and serviced by trained personnel. Yes, it cost more money, but an active ejection seat would provide that second, life-preserving chance in the midst of a catastrophic emergency. All that said, Des was never allowed to fly in the Jet Provost with me again.'

'I often reflect on the fact that the ejection seat nearly killed me because it was not fixed in the aircraft properly,' says Des. 'But then it saved my life because it was not properly deactivated, so released me from its death-grip. I think some angels were watching over me that day and I quickly recovered from my injuries.'

* * *

Ex-naval gunner and Falklands veteran Neil Wilkinson's injuries were not physical. But they were serious and life-changing. Back in Leeds, he had left the Royal Navy in 1989 but could not get the conflict in the South Atlantic, and the Argentinian airman he thought he had shot down and killed, out of his head. He tried to shut out the haunting memories, but they were too deeply locked in. He avoided all discussion of the war, even refusing to watch any of its countless documentaries. 'I wasn't interested in watching those history programmes,' he says. 'Why would I want to relive those times? The ships sinking, the noise and the chaos. Deaths of friends. And enemies. My life took a dive and I suffered from a lot of mental stresses due to the conflict.' Try as he might to blank it out, he started having nightmares.

'I'd see images of HMS *Antelope* exploding. The aircraft bearing down on our ship. Bodies floating in the sea. I was replaying the film of my war constantly in my head and just couldn't work out why I was feeling so emotional, so distant from the rest of my life. I'd wake up some mornings dripping wet after a nightmare. I couldn't understand what was happening to me, and why.' He felt he couldn't talk to anyone about what he had witnessed. 'When you are in the military, you can talk more about your problems, the feelings you are experiencing. When you are a civilian, no one understands, so you bottle them all up.' Neil's bottle of emotions was close to exploding. He was losing control of his life. Apart from his marriage hitting the rocks, he never seemed to have any money, even though he was always in work.

In the early 1990s, he was eventually diagnosed as suffering from post-traumatic stress disorder (PTSD) and started seeing a specialist. It was a long, hard road back. 'The images didn't go, nor did my feelings, but with help I began to understand and accept how I was affected by *my* war. There was nothing to be ashamed of. If you have lost an arm in battle, you can see what is missing, understand why you have to do things in different ways. But an injury of the mind is very different. You can't see the damage, but it was certainly there.'

He was still having flashbacks of the Argentinian Skyhawk he had shot down on 27 May 1982, but his feelings were changing thanks to the coping strategies he had learned. 'I always presumed I had killed the Argentinian pilot, that there was no way he could have ejected from his jet. I still felt guilty about taking another man's life and I began to contemplate what his fate had been, who he was, how his death had affected his family and friends. I wondered how I might be able to find out.'[10]

Neil had no idea that Skyhawk pilot Mariano Velasco had survived both his gunfire and the subsequent ejection. He was alive and well, had returned to flying and had a distinguished career, becoming a Commodore in the Argentinian Air Force. Velasco still regretted having to eject from the Skyhawk he felt so bonded to, but there were no hard feelings about whoever had shot him down. 'There was nothing to forgive,' he said. 'Each of us had to fulfil our duties.'[11] Though he did wonder occasionally about the identity of the mystery British gunner who had come so close to killing him. What was *he* doing now? How was *his* life going?

* * *

Neil Wilkinson was not the only one trying to adjust to the past, treading that sometimes solitary path which occasionally affected those on the periphery of an ejection story. Mike Schaffner had still not come to terms with losing the father he had never known. He was just a baby when US exchange pilot Captain Bill Schaffner had crashed into the North Sea in his RAF Lightning in 1970; his ejection seat still in place but no trace of Bill.

Mike, still completely unaware of what had occurred that night, approached his twenty-ninth birthday with apprehension. 'I realised I would then have been alive longer than my dad, who had died three days short of his own. As a teenager, I remembered daydreaming that my father had somehow survived and was suffering from amnesia somewhere. Or perhaps he had been picked up by a Soviet submarine and was languishing in a gulag. As an adult, I put those thoughts away. I'd never known him beyond his military mementos that my mom kept in that old red box, which she had showed us as kids.'

Mike had trained as a lawyer and was now working in banking compliance, rooting out fraud in Chicago. Killing time in the office just before a Christmas break, he went online, idly looking for any references to his father. 'Dad was always on my mind but I presumed there was nothing more to be learned since those letters Mom had received after the accident in 1970.' The family had been promised the Board of Inquiry's accident report but had never received it. Mike had no idea it had been completed back in June 1972, and then restricted for security reasons. 'The internet was still in its infancy and it was the first time I had looked for any information. About ten pages into my search I saw the words *8 September 1970* and *RAF Binbrook*. It was the date my dad had died, and his base. I was astonished.'

He clicked on the link. Up popped a 1992 article in the *Grimsby Telegraph* – 'The Riddle of Foxtrot 94'. Next to the story was a photograph of a Lightning. It claimed that on the night of 8 September 1970 'a young USAF pilot, Captain Bill Schaffner, had been scrambled in his Lightning, along with six RAF fighters from other bases, three tankers and a Shackleton early warning aircraft to intercept a "mystery contact".' An unidentified flying object – UFO.[12]

He couldn't quite believe what he was reading. This was the first real information he had found about the accident. There was a whole clutch of stories about his dad, each one more astonishing than the last, all based on the sleuthing of UFO hunters.

Incredibly, one ran what it claimed was an official transcript of Bill's final radio calls with his flight controllers. Mike read through it, increasingly bemused:

Schaffner: *I have visual contact, repeat visual contact.*
Staxton: *Are your instruments functioning, 94?*
            Check compass.
Schaffner: *Affirmative. I'm alongside it now, maybe 600 feet.*
            It's a conical shape. Jeez, that's bright. It hurts my
            eyes to look at it for more than a few seconds.
Staxton: *How close are you now?*
Schaffner: *About 400 feet. He's still in my three o'clock.*
Staxton: *Is there any sign of occupation?*
Schaffner: *Negative, nothing.*

Apparently, the article claimed, Schaffner now saw an object *'the size of a large football, like it's made of glass, bobbing up and down'* at the back end of the mystery craft. *'It's not actually connected. Maybe a magnetic attraction. Could be the power source. It's within a haze of yellow light. There's no sign of ballistics.'*

Suddenly there was alarm in his voice. *'It's turning ... coming straight for me ... shit ... am taking evasive action.'*

His radio then went dead.

Time stood still as Mike read all the articles onscreen in his Chicago office. 'We had heard little about the accident other than Dad's ejection seat was still in the jet and he wasn't. And we had *nothing* about these reports and conspiracy theories. Nothing about UFOs. Now I was reading the supposed transcript of my dead father's final words. His radio calls that night talking about bright lights and strange shapes in the sky. It just didn't make sense, but I didn't know what to believe.'

Mike called his architect brother, Glennon. 'You really have to look at this. It seems almost unbelievable, but perhaps it's a way into discovering the real story about what had happened to our father?'

Over the coming days, the brothers made endless calls to people who had known their father, and to any contacts they had within the US Air Force who could offer any advice. And they discovered even more ludicrous stories, and TV interviews with so-called experts who delighted in promoting the UFO theory. Mike printed off the supposed transcript and sent it to one of his father's old USAF friends. The reply was prompt and precise. 'This is utter hokum. We didn't communicate like that. It's just a bunch of made-up junk.'

After his initial shock, Mike agreed. 'Clearly it was utter rubbish, but the fact that there was so much information out there about my dad spurred me on to investigate more.' He and his brother were also angered by the stories. They did their distinguished and decorated father a gross disservice. 'My dad was a pilot. I was horrified his name was linked to this UFO nonsense. I wrote to officials in the RAF and the USAF to find out the truth. I contacted old colleagues.' But, apparently, there was no further information available. Three decades since Bill Schaffner's disappearance, his son was getting nowhere.

'The Cold War was over. Any need for secrets about the job my father was doing on the east coast of England, staying alert for any kind of nuclear attack, was long gone. The world had changed. Surely it was time that the full facts were released? None of us could understand why the ejection seat was still in his jet, but my father was not. It became my mission to establish what had happened to my dad that night.'

CHAPTER EIGHTEEN

AGAINST ALL ODDS

RAF CONINGSBY, LINCOLNSHIRE
10 JANUARY 1996

Seasoned Tornado navigator instructor Ian Weaver had gone to bed looking forward to the next morning's cycle ride of 2 miles from his home in the Lincolnshire village of Tattershall to nearby RAF Coningsby, where he was scheduled to be part of an Air Combat Training (ACT) mission. Almost all aircrew delighted in these sorties, referring to their dogfighting training as 'the sport of kings'. Even official descriptions viewed them as 'undoubtedly the most satisfying and enjoyable exercise for fighter crews. But also the *least forgiving, demanding the highest skills and split-second decision-making.*'[1] What better for a man whose dream of flying fast jets was inspired by the first *Top Gun* movie? 'The air combat looked amazing,' he said. 'I joked to a friend that I really wanted to be like the character of Goose, the back-seater in Maverick's Tomcat. Just not the bit where he hit the canopy as he ejected. And was killed. Obviously ...'[2]

At 3am, Ian was woken by a cry from his son, Chris. He rushed to the bedroom, where the 8-year-old was having a nightmare. He quickly calmed Chris, tucked him up, then checked that his 10-year-old daughter, Jenni, was sound asleep before making his way back to bed. It is the last thing that Flight Lieutenant Ian Weaver of 56 Squadron remembers.

Wednesday, 10 January 1996 would see an event almost unparalleled in the ejection seat narrative. A day when its lifesaving capabilities would be tested to the limit. And beyond.

Sleek grey Tornado jets were lined up on the airfield outside 56 Squadron – the Operational Conversion Unit (OCU) responsible for all training of fast-jet crews ready for the Tornado F3, the interceptor version of the ground-attack Tornado. One of the RAF's oldest squadrons, its airmen enjoyed their morning marmalade and toast in a crew room crammed with photographs of historic aircraft, and the memorabilia of many of their First and Second World War fighter-pilot forebears.

Three crews – 'Rambo' formation – sipped from steaming mugs of tea as they prepared for a 'two-versus-one' air combat exercise. Two fighters, Rambo 1 and 2, patrolling as a pair defending UK territory, would be subject to a series of attacks by a single 'enemy' aggressor, Rambo 3.

Rambo 1, pilot Cliff Nichol and his navigator, would lead the mission. Rambo 2's crew had Ian Weaver in the back seat as the staff instructor with Stan, a retraining pilot, up front. Stan was the only one of the six who was not an OCU instructor. A highly experienced RAF pilot, he had just returned to the UK from an overseas exchange tour and was now learning to fly the Tornado F3.[3] The three-jet ACT sortie had been tailored specially for him. Rambo 3 would be the enemy or 'bounce' aircraft, crewed by pilot Mike 'Elvis' Costello and navigator Mark 'Skids' Richardson. Their job would be to put the defending pair to the test, repeatedly harassing them in order that Stan could gain experience in Tornado intercept and air combat tactics. 'It was a reasonably relaxed atmosphere with staff instructors and just one very experienced student,' says Cliff, who briefed the mission. 'It should have been truly routine; the type of exercise we flew regularly.'

Skids had enjoyed a glass of wine over dinner with friends the previous night, before driving to work that morning in his Golf GTI. 'It was a glorious winter's day, a bit of frost and the ground firm underfoot. It was going to be a great sortie to be involved in; the type all the instructors tried to get on.'

Though he had no recollection of doing so, Ian Weaver had woken up, enjoyed breakfast, kissed his family goodbye and then cycled the short distance to work as usual.

The three crews climbed into their Tornados after fully checking their ejection seats. Navigators Richardson and Weaver paid particular attention. Before joining the RAF, Skids had been a mechanical engineer and was fascinated by every aspect of the seat's design and function. 'The construction and complex engineering completely amazed me,' he says. 'What it did in a couple of seconds, after sitting dormant for years, was simply astonishing; a real marvel of engineering. I read everything I could about this lifesaving invention and completed every check forensically before strapping in. Of course, you never thought you'd need it.'

Ian Weaver had come within an inch of ejecting a few years earlier when his Tornado's nosewheel collapsed after landing. Seconds from disaster, his pilot had luckily managed to bring the jet safely to a halt. 'I took the ejection seat very seriously indeed. I know some people just thought of them as another bit of aircraft kit; always there if needed. But I was fanatical about doing every single pre-flight seat check. I viewed it as the one piece of kit that could one day save my life, and paid it the respect it deserved.'

And there was another reason for their attention to detail. Like the issues surrounding Red Arrows pilot Sean Cunningham's terrible death, a recent incident at RAF Coningsby had highlighted the vast range of dangers all aircrew faced. Ian and Skids had both been close friends with a fellow OCU navigator, Jesse, who had died of severe head injuries suffered during an ejection from a Tornado over the North Sea the previous year. The subsequent investigation suggested that he had probably been looking over his shoulder when the pilot command ejected them both as the aircraft went out of control. It appeared that as the canopy was blasted clear of the cockpit, it had struck Jesse's helmet, delivering a fatal blow. 'It was a massive shock to everyone across the Tornado force,' Ian recalled. 'It was absolutely not the fault of the ejection system, but it reminded us to ensure we assumed the correct sitting position if we ever had to pull the handle.' Skids was similarly punctilious. 'We were all truly devastated when Jesse was killed, and his death had emphasised the need to adopt the correct posture if you ever needed to pull the handle. The ejection seat was completely reliable, and we – obviously – trusted our lives to it. But other dangers still existed.'

Cliff Nichol, with nearly twenty years' flying experience, was similarly realistic. 'One tended not to dwell too much on the deaths of friends, which often happened during your RAF career,' he says. 'But what most of us certainly discussed were the vagaries of certain accidents, forming opinions on how we might react if confronted by similar emergencies. I suspect those crew-room chats saved quite a few lives over the years.' Like every crew on the Tornado force, they all flew with Command Eject activated.

* * *

To further assist the occupant during any ejection, their Martin-Baker Mk10 seats were equipped with the latest modification installed in the wake of some of those previous accidents. Aircrew like Neil Johnston, Ian McLean and Kate Saunders had all suffered severely broken legs after ejecting while their heavy personal survival packs were still attached to their backsides on landing. If an ejectee was knocked unconscious while blasting out of the cockpit, they would be unable to manually release the clips on the yellow, solid fibreglass box, allowing it to dangle down on its 15-foot tether. So instead of hitting the ground safely before them, the PSP added its own 10kg to the ejectee's body weight, making their landing even more brutal, and could also act like a battering ram against their legs, causing catastrophic injuries.

The solution to this serious problem had been, yet again, remarkably simple. Martin-Baker's designers had manufactured a red, cigarette box-sized explosive device – the Automatic Deployment

Unit – which could easily be retrofitted inside all PSP boxes. Now, by moving the PSP's seat cushion slightly, the occupant could check – and if required easily adjust – the device's 'AUTO' or 'MAN' settings.

The unit, enclosed within the fibreglass PSP, was connected by a short metal cable to the seat itself. After an ejection, if set to 'automatic', and once the ejectee and their PSP had been cut free, the cable attached to the falling seat would pull the firing pin of the explosive cartridge in the small deployment unit. After a short delay to ensure complete occupant–seat separation, gas from the triggered cartridge routed to a new locking piston on the airman's PSP's clips, unlocking the buckle so the fibreglass box fell free to dangle on its lanyard.

If set to 'manual', the device would not activate and the PSP would remain attached to the flyer after they separated from the ejection seat. The occupant needing, as before, to reach around and press the unlocking buckle themselves. It was envisaged that this might be useful when flying over heavily forested areas, where descending onto pointy trees after ejecting might cause unpleasant injury to the most delicate of aircrew body parts. The scenario was so unlikely compared to shattered limbs that almost all aircrew chose the 'automatic' option.

* * *

All checks complete, Rambo 3 – Elvis and Skids in the bounce jet – took off in a blaze of afterburner at 8.34, the second sortie that morning out of RAF Coningsby, home to the OCU and two frontline Tornado air defence squadrons. 'We turned south as we climbed,' Skids Richardson recalls, 'taking up our pre-planned position at the south of the exercise area west of King's Lynn, ready for the intercept and combat exercise.'

Rambo 1 and 2 taxied to the head of the runway and sat side by side, completing their final 'pre-take-off' checks. Even though it was an early winter morning the temperature out on the tarmac soared as orange-tinged flames spat out from the back of the Tornados. White strobe lights winked under their bellies and they rocked a little, like a sprinter going down into the blocks. Cleared for take-off, pilots Cliff and Stan pushed their twin throttles forward into 'combat' position, soaking their twin engines with fuel, igniting the afterburners and producing maximum available thrust. Releasing their brakes, the two jets catapulted down the runway together, wingtips just feet apart. The din they left behind rattled windows in the nearby buildings as they shot off the ground in a perfect pair's take-off at 8.39.

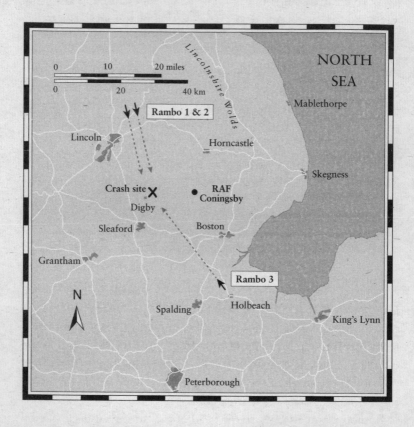

Rambo 1 and 2 flew to the north of the exercise area east of Lincoln, climbing to 15,000 feet. 'They soon came on the radio to say they were ready to start the training mission,' Skids Richardson in Rambo 3 says. All three Tornados were now logged onto the same radio channel so they could communicate with each other during the sortie. The combat exercise was live. The premise straightforward.

Rambo 1 and 2 would be patrolling the region to the north, detecting Rambo 3 trying to enter the area from the south. He would be confirmed hostile and they would head towards the 'enemy', initially using their air defence radars to track and try to destroy him with long-range Sky Flash radar-guided missiles. Then, as they merged – passed close by in the same piece of sky but separated laterally or by height – they would enter close-quarter combat and engage in visual ACT, dogfighting, trying to lock on their short-range, heat-seeking Sidewinder missiles or their guns.

'We were heading north-westerly with the coast to our right,' Skids remembers. 'It was all totally standard as we climbed up to 27,000 feet. To my left, the flat Lincolnshire countryside unfolded past Spalding and towards Grantham. Out on my right, I could see over the Wash, and past the Skegness coast. It was a stunningly clear day and the view really was beautiful.'

Sortie leader Cliff Nichol in Rambo 1 was heading south-east with Rambo 2 as his wingman. 'I was on the westerly side of our pair around 2 miles away from Stan and Ian and could see them clearly. The coast was on our left, the city sprawl of Lincoln on our right.'

Skids and Elvis were around 40 miles south of the interceptor pair. At their combined speed of some 1,200mph they were just two minutes' flying time apart, closing together at a mile every three seconds.

The Tornados' three navigators began to manipulate their radar systems to locate and track the opposition. Their displays told them where their particular 'enemy' was, and the height, speed and direction of travel. Swapping information over the radio, they were all satisfied they had good awareness of each other's positions.[4]

At 20 miles, around one minute apart, one of the interceptors locked their radar onto Rambo 3, so Elvis and Skids entered an aggressive spiral dive to 11,000 feet in an attempt to evade the simulated missile which would now be fired. Heading north again after the hard 360-degree turn, they squared back up like a boxer to Rambo 1 and 2. Now just 15 miles or forty-five seconds apart, Cliff and Stan widened their formation, ready to swoop down on the bounce from 16,000 feet.[5] They were now forty-five seconds from 'merging' with Rambo 3 and entering visual combat.

Cliff Nichol was completely happy. 'I could see Stan and Ian out to my left, we had radar contact with Rambo 3 heading towards us and I now picked him up visually, in front and below. It looked like he was positioning to attack my wingman so I manoeuvred further right, giving myself more space to slash in towards him and engage with my missiles after we merged.'

At around 5 miles – fifteen seconds – apart Rambo 3 was also happy. 'Elvis and I both had Stan and Ian in Rambo 2 visual above and to our right,' remembers Skids, 'and we then picked up Rambo 1 well out to our left. We were listening in to the radio calls between the crews of 1 and 2, and from what they were saying, we all thought Rambo 2 could physically see us.'

They were wrong.

Stan, flying Rambo 2, had indeed initially seen Elvis and Skids just over 5 miles away on his nose, but lost visual contact with them as they manoeuvred. In the rear cockpit, Ian Weaver was tracking events on his radar display, keeping his own pilot, and Cliff in Rambo 1, updated on positioning. In any event, the navigator's vision directly forward in the Tornado was severely limited because of various displays, and the back of the pilot's ejection seat.

Events now unfolded in fractions of a second.[6]

Cliff Nichol – the mission leader – saw Rambo 2 and 3 getting close to a merge. 'I was higher, about 3 miles away from the bounce jet, so I started to hook left into a descending turn towards them.'

Rambo 2 continued its descent towards the bounce. 'Stand by to mark,' Stan called over the radio. 'My nose 2 miles low.' He appeared to be ready to engage in visual combat and everyone presumed his radio call meant he could still 'see' Elvis and Skids. In fact, he merely knew from the Tornado's radar information presented in his display where they should be positioned in the sky.

With slightly decreased speeds, they were just eight seconds apart.

Elvis kept his jet steady at 11,000 feet, fired a simulated missile at Stan and Ian in Rambo 2, then turned away left.

'As we understood they could see us,' Skids says, 'there was no problem. Then, as we rolled back level, I glanced right and all I could see was a massive Tornado pointing at me. It completely blocked my field of vision. I was looking straight into its jet intakes. There was no doubt that we were going to collide and the nose of their jet was going to spear right through me. I resigned myself that I was going to die, simply muttering, *Holy shit*.'

In the front seat, Elvis had a similar thought. 'All I saw was an image of a Tornado filling my canopy. At first there was disbelief, then dawning shock. I didn't experience fear particularly, just an instantaneous reaction as I hauled on the controls, focusing on that giant Tornado fin about to come through my canopy.'

Stan had not actually seen Rambo 3 until he was half a second from impact – in simple terms, he had made a basic error in descending towards an unseen aircraft. Desperately, he pushed on his own stick trying to inch away and avoid the collision.

In Rambo 3, Elvis yelled a series of expletives as he rolled his jet over onto its side. The instant, captured on their cockpit voice recorder, is telling.

'Fuck. Fuck. Fuck. Shit. Fuuckkkk!'[7]

It was all too little, too late. The Tornados came together at a combined speed close to 1,000mph in a scream of tearing metal and roaring engines.

Cliff Nichol had initially not seen anything untoward and was still manoeuvring into position to simulate firing a Sidewinder missile at his foe. 'As the two jets merged, I expected to see my wingman Stan pull up high into a position ready to begin a visual combat reattack on Rambo 3. But what I actually saw was a huge

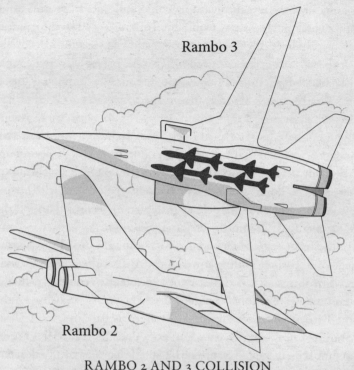

RAMBO 2 AND 3 COLLISION

amount of fuel suddenly venting from Elvis and Skids's jet. I still hadn't computed what was happening.'

With their jet now flying vertical, wings perpendicular to the horizon, Elvis and Skids's right wing had sliced off a large section of Stan and Ian's wing as though it was butter, severing fuel, hydraulic and control lines. Stan remembered a huge 'bang' then his aircraft spun totally out of control, rolling rapidly and plummeting earthwards as his wing disintegrated. One and a half seconds after the collision, although he has no recollection of doing so, he pulled his yellow and black handle initiating the Command Eject sequence. Exiting the cockpit, their seats were subjected to incredible forces caused by the pitch and roll of the wildly spinning, tumbling jet.

Rambo 3 was also in severe trouble. 'There was a brief shudder, just a jolt, nothing more,' says Skids. 'We were then rolling over and over, rotating incredibly fast. In the front seat, Elvis was desperately fighting to control the mortally wounded Tornado. 'I didn't feel any impact but clearly knew we were in severe trouble as we started rapid rolling. With everything now happening in fractions of a second, training kicked in as Skids and I tried to analyse exactly what had happened, and what, if anything, we could do to solve the problem.'

Elvis put out a rapid 'Mayday, Mayday' call over the radio then shouted at Skids, 'What have we lost?' His navigator glanced out to his right. 'I saw about 6 feet of wing was missing and realised we were going to have to eject. The very first thing that flashed through my mind was my fellow instructor Jesse being killed the year before while looking over his shoulder as his pilot command ejected them. So I instantly motored my seat lower, pulled down my visor, grabbed at my lap straps to tighten them, and sat squarely back against the seat, my head rigid against the headbox. I yelled at Elvis to do the same.'

Elvis was already on the case. 'As we continued rolling, and Skids was calling our heights as we descended, I was very conscious after Jesse's fatal incident of not ejecting as my nav's head was turned.'

Skids was now aware of a multitude of warning lights on the panel and the jet's increasing speed of rotation along its axis like a spinning drill bit. As the Tornado's nose began pointing down towards the Lincolnshire fields below, he called, *Fuck the jet!* 'I was already waiting in the textbook position, with my left hand on the ejection seat handle, my right gripping that wrist to steady it.'

It was 13.2 seconds after the collision. Plummeting through 7,750 feet at over 400mph, Elvis waited until the jet was rolling towards upright, canopy pointing away from the ground, and yelled, 'Eject! Eject!', pulling his handle and initiating the second Command Eject sequence of the morning.

'It all happened in the blink of an eye,' Skids remembers. 'The cockpit filled with smoke and fumes as the canopy was blasted free and I felt myself being thrown forward as the ejection guns fired.

Then, as the seat rose up and the rocket motor ignited, I was briefly looking down on my instruments as the aircraft moved away from me. I remember thinking, *I'm definitely not sitting where I normally do*. There was a blur as I shot away. I remember my face feeling cold and the whistling of the wind. Then two huge jerks.' In the front cockpit, Elvis felt the initial gun going off and his head was thrust to his knees, where it remained while the rocket pack ignited carrying him clear. He felt another yank as his parachute deployed.

Watching in dismay as events unfolded, Cliff Nichol saw the explosive flash as Elvis and Skids's canopy blew off, followed by two smaller ones when both ejection rocket packs fired. 'Everything was happening incredibly quickly. I saw their Tornado pitch rapidly nose-down, heading towards the ground.' He rolled his wings level and looked for his wingman, Rambo 2, Stan and Ian. He saw their Tornado also in a rapid, almost vertical rotating dive, plummeting earthwards. 'I immediately called for them to eject. But didn't get any reply.' Clearly, Cliff could have no idea that they had already ejected seconds before. 'Seeing the two jets tumbling, I now realised there had been a mid-air collision. We put out a Mayday radio call for Search and Rescue helicopters to be launched.'

He then spotted two parachutes floating gently earthwards. Elvis Costello and Skids Richardson had made it out. He put his Tornado into a gentle, descending turn, so he could monitor them going down and see where they landed. But in the scattered cloud he had lost sight of his wingman. What had happened to Stan and Ian? 'I could hear a number of emergency Personal Locator Beacons [PLBs] sounding their distress signals on the emergency radio frequency. They are automatically activated once you eject as the seat falls away, but I couldn't work out how many were sounding, so I had no idea if anyone had got out of Rambo 2.' Then he spotted shattered aircraft wreckage blocking the Sleaford to Lincoln railway line. 'We radioed a message in the hope that somebody could halt the trains and prevent a serious derailment. I still had no idea what had happened to Stan and Ian and was becoming increasingly concerned.'

*

Just after breakfast, 10-year-old Matt Phillips stormed out of his parents' farmhouse in the hamlet of Digby Fen, 10 miles south of Lincoln. 'I was in a fit of rage, following a cataclysmic bollocking from my mother,' he says. He was staring moodily to the west when he saw a Tornado overhead. It looked like it was falling. 'I dismissed it as an optical illusion. Still, I tracked the jet downwards until it disappeared from view behind the next farmhouse.' At which point he saw a burst of flame and smoke.

He ran back inside, shouting to his parents. 'I've just seen an aircraft crash.' His father John, a farmer and dog breeder, was still in his dressing gown. He and his wife Jenny hurried outdoors. All they could see was two puffs of smoke in a grey sky, dismissing them as low cloud. Then John glanced up. 'I saw two parachutes and the enormity of the situation suddenly hit me.'

He rushed indoors and in around thirty seconds got dressed, grabbed his keys and jumped into his 4x4. 'I started off across the fields, keeping an eye on the chutes through my sunroof. I could see that they would likely land about a quarter of a mile away.' After bouncing over two muddy fields, his progress was stalled by a dyke covered in barbed wire. He jumped out and took to his feet. 'I was not built for sprinting across the countryside,' he said, 'but I went as fast as I could.'[8]

When Skids's parachute had deployed, it jerked him upwards with such force that he wondered if the harness could hold him. 'I thought the parachute straps between my legs would simply snap and I'd fall out!' Luckily they didn't, but 'the crotch straps were yanked violently upwards, trapping my left bollock and causing the most excruciating pain I have ever experienced'. He then felt another large jolt as his PSP Automatic Deployment Unit fired, releasing the buckle and allowing the hefty yellow box to drop safely away on its lanyard.

Skids was successfully, if uncomfortably, under his parachute and Elvis was floating beneath him, close enough for each to shout and ask if the other was okay. The harsh ejection at high speed and under violent motion had dislodged Skids's oxygen mask and smashed his helmet visor downwards, cutting through his nose. Its fleshy tip was hanging off. 'I hadn't any pain, but did feel the blood flow down my face as it chilled in the freezing air.' He had also cut the bottom of his chin on the metal fastening of his life jacket when his head had been forced down as the rockets fired. One of the chains holding Elvis's oxygen mask to his helmet had been torn free during the ejection, and a leg pocket on his immersion suit ripped away. Of more immediate concern were the major roads, huge pylons and power lines Skids could see coming up fast towards him. 'I was thinking, "I can't believe I've survived a high-speed mid-air collision and I'm going to get killed on the front of an Eddie Stobart lorry on the A15!"'

Six minutes after ejecting, he and Elvis landed close by each other in a sugar beet field. 'Elvis appeared from around its corner, dragging his parachute like Will Smith in *Independence Day*! He looked at me and he said, "Are you all right, mate – your nose is a bit bloody?"' A driver had seen them, parked up and run over with a first-aid kit. 'Elvis clearly wasn't too concerned about my bleeding nose hanging off because he just asked the driver if he had a spare cigarette. I was just glad to be alive.' Then, in a truly surreal moment, Skids remembers a farmer appearing around a hedge. 'For some reason he clearly found the situation mildly amusing and shouted, *Are you the Germans?* Still in a bit of shock, I heard myself shouting back *Nein!*'

Reality kicked back in. 'As we waited to be picked up by a helicopter, Elvis told me he had seen another two chutes coming down a few miles away. Now we knew that Rambo 2 had also ejected. But what state were they in? We could only hope they were both okay.'

John Phillips had now sprinted across two more fields to reach the downed airmen. 'I was pretty winded when I reached the first one,' he recalled. It was Stan, the pilot of Rambo 2 who had just regained consciousness and was struggling to his feet. During the brutal high-speed ejection, his helmet had impacted the headbox of his seat and he had been knocked out. His oxygen mask had been torn apart and there was significant damage to his helmet. 'He said he was okay, even though he was bloodied and in shock,' Phillips remembered. In fact, Stan had suffered a severe neck injury and had lost feeling in his forearms. During the ejection his oxygen mask had been whipped from his face and swung around, lacerating and bruising his eyes, face and neck. Fuel from the rocket motors had burned his eyebrows. 'I told him to stay put, until help arrived,' John Phillips recalled. 'I had now seen a second airman 250 yards away. He was lying prone in the mud. Motionless.'

It was Ian Weaver, whose last memory that day had been his young son's early morning nightmare. Now, just a few hours later, his life was hanging by a very narrow thread.

Later analysis of the Accident Data Recorder (ADR; the so-called 'Black Box') would show that during their ejection at 10,950 feet, Stan and Ian's jet was so wildly out of control that it was rotating and flailing under 3.5 times the force of gravity while descending at 612mph. Because the ADR could only measure a maximum roll rate of 333 degrees per second, the exact parameters were unknown. But subsequent modelling of the incident suggested that the conditions were so violent, and the seats so unstable as they left the aircraft, that they could have been subjected to momentary motion of perhaps 1,000 degrees – almost three full rotations – every second. Outside the ejection seat's design capacity to cope, and an almost unsurvivable situation.

Because of those adverse aerodynamics, when Ian was ejected by the Command Eject system after Stan pulled the handle, as his head entered the violently oscillating airflow, the 600mph blast took hold under his helmet, shearing its thick nylon chinstrap and metal oxygen mask chains as though they were paper, and yanking it free. In doing so, it tore his head backwards, hyper-extending his neck, cleanly snapping the top of his spine and stretching his spinal cord and brainstem. It also tore apart the ligaments which hold the head onto the neck. Muscle had also been shredded so that, in effect, Ian's head was now only held onto the rest of his body by the skin around his neck and his spinal cord.[9] His skull was fractured. He had suffered a lesion to the frontal lobe of his brain due to the complex angular accelerations he had undergone during the ejection process, which caused localised brain damage. His scalp was lacerated, his left lung collapsed; he had broken six ribs on the right, and his left arm had been so severely extended that all the nerves controlling movement were ruptured.

Although the auto-lowering of their PSPs had prevented severe leg injuries to both unconscious crewmen, Ian Weaver was in a very bad way.

Reaching him, John Phillips recognised instantly that the navigator, still attached to his parachute, was severely injured. Against all the odds, his ejection seat had managed to save his life, but the whole process had triggered a journey into mortal danger. 'His face was in a molehill. He was gurgling and making horrible noises. I could see he couldn't breathe properly. But I remembered that moving an injured person might complicate an already dangerous situation.' John clearly had no idea what sort of neck or back injuries might be present, but, in his gut, he knew that if he did nothing the man would die. 'As gently as I could, I eased his face round so that I could clear his airway.' He noticed that one of Ian Weaver's bloodshot eyes was open. 'I thought he might be conscious, but unable to communicate.' John put his finger into Ian's hand and told him, 'If you can hear me, squeeze my finger.'

There was no response.

Supporting Ian's neck, John carefully rolled the navigator halfway across his knee to help his laboured breathing. He realised that the navigator wasn't wearing a helmet and blood was pouring out of the back of his head. Phillips then remembered the 'recovery position' he had seen demonstrated on a recent 999: What's Your Emergency? TV programme. He carefully moved Ian onto his left side. 'I held onto one of his straps to prevent him falling onto his back and put my coat under his head. Looking down at him, I noticed his wedding ring. Like me, he was probably married with children. Although I am not at all religious, I said a small prayer.'

As he called the emergency services on his mobile phone, Stan managed to hobble over to check on Ian. 'He was obviously in considerable pain himself but tried to give some comforting words to his crewman,' says John. 'He asked if I could call his wife to let her know he was safe.'

Weaver began to thrash around in the dirt and John was becoming seriously concerned for his wellbeing. He and Stan tried to steady and hold him as best they could while still trying to attract attention. 'I was now desperate for someone to come and help. I could see he was in a critical condition and that time was not on his side.' After what seemed like hours, he heard a helicopter and saw blue flashing lights speeding across the flat countryside towards the crash site.

Relieved beyond belief, Phillips watched the paramedics secure Weaver, still bleeding and unconscious, to a stretcher and load him into the helicopter. As he watched it speeding away, he presumed he would never see the airman again, but sincerely hoped that he had done the right thing and not caused further damage by moving him. And that his small prayer for survival would be answered.

* * *

Shrill, excited voices filled the dining room at Tattershall Primary School. It was a normal busy lunchtime and Ian's daughter Jenni was eating her packed lunch when one of the teachers approached. 'She was a middle-aged lady with a huge mane of permed hair and glasses,' she remembers. 'She was also very strict, so when she asked me to go with her to the staffroom, I was worried I was in trouble. I knew something was wrong.' Like the 'knock at the door' for their partners, it was a scene most aviators feared – their child being called out of a classroom to be given horrendous, life-changing news. As Jenni walked through the hall, every pupil stopped eating and talking, and followed her with their eyes. 'My heart was racing and my legs were a bit wobbly. I felt quite numb, as I simply had no idea what was happening. The school bully jeered at me, "Oh look, Jenni's in trouble!"'[10]

The frightened 10-year-old arrived at the staffroom door. 'Whenever I'd been there before, it would only open slightly as the face of an annoyed teacher disturbed from their tea break appeared.' Now it was swung fully open and she was ushered inside. 'None of this was helping my worry.' She sat at a desk and was handed a phone. 'They told me my mum was on the other end. She said that Dad had been in an accident at work, had hurt his head and was in hospital. I knew Dad flew Tornados but that's about it, really. I remember him being away for long periods of time and missing birthdays, but there was nothing exceptional about what he did. It was just a job, the same as any of my friends' dads, whether they were a mechanic or a builder.' Looking back on that day, and now a mum herself, Jenni realises her mother was incredibly upset and trying to keep her composure to reassure her daughter. 'She told me somebody would pick us up from school and take us to a friend's house. I wasn't really able to process everything or fully understand the severity of it all.'

She went outside to find her brother. Chris was standing in the playground by a large tree. It fell to her to tell him their beloved father was seriously injured. 'We stood under the tree, talking about Dad's accident. When lunch finished, all the other kids raced out and started crowding around us, asking what had happened. I just remember trying to shield Chris and protect him from all the shouted questions. He was upset and I felt really isolated, with my back against a fence, trying to fend the crowd away.'

Several hours after his high-speed collision and near-death ejection, Ian Weaver was still unconscious and now in intensive care. His wife was being given a stark, truly horrifying summation by a specialist at Queen's Medical Centre in Nottingham. They gently explained that he would either die in the next twenty-four hours from the head and neck injuries, or would be quadriplegic and on a ventilator for the rest of his life. He had suffered what medics considered the worst spinal injury possible: a clean separation of vertebrae at the very top of the neck. The so-called 'hangman's break'. Needless to say, the injury was mostly fatal.

Luckily, even though it went against most conventional medical advice, there was no doubt that by clearing his airway and putting him in the recovery position, John Phillips had saved Ian Weaver's life at the scene. But would he survive the night? The doctors and nurses kept him immobilised on a spinal board, put a breathing tube down his throat, kept his lungs ventilated, and hoped.

He did make it, but was paralysed from the neck down. While he was still unconscious, a huge weight hanging from a cradle was screwed into his skull to immobilise and extend his neck. The combination of his head injury and morphine triggered hallucinations. 'My first "awareness" of the coma was a shining white light. Was this heaven? Then an image formed of a young boy's head stuck on a tray travelling above me. The boy would stare at me with big blue eyes and laugh. Then he would disappear.' The hallucinations were so real that even to this day Ian can't sort out the facts of his treatment from the fiction of his dreams. 'I became aware of night-time on the ward as the lights were dimmed. In the dark, I would be loaded onto a trolley and pushed across the fields to the banks of the River Trent where I was left among the dark shadows, the rats, the stench. Each morning the nurses would arrive to see if I'd survived the night. Finding me still alive they would wheel me back to intensive care for another day's treatment.'

On 16 January the surgeons operated to repair the spinal damage. 'They basically used screws to reattach my head back onto the top of my neck with a small part of my pelvic bone thrown in to help long-term binding,' Ian says, reading from his medical notes. 'Thankfully, it seemed that I had a very slack spinal column and the cord, although damaged, was unbroken.' Not that he knew any of this at the time.

*

A week after the collision, Elvis Costello and Skids Richardson returned to the spot where their Tornado had crashed. The stench of jet fuel still filled the air. Both were wearing neck braces for the strain injuries caused during the ejection. What greeted them was a scene straight from a First World War battlefield. Their jet had punched a crater in the heavy clay soil 15 feet deep, similar to the effects of a 1,000-pound bomb. 'What had been an incredibly complex flying machine was now thousands of tiny pieces,' Skids recalls. And 10 feet beneath the bottom of the crater, thousands more fragments had been buried by the impact. Huge swathes of debris had been blown across the fields. 'Looking into that massive hole really brought home just how close we had been to dying, and our remains being down there with the shattered jet. It put the whole experience into stark reality.' They didn't share many words. No need. The silence spoke volumes.

Rambo 2 had punched a similarly dramatic hole in farmland 5 miles away. It would take more than sixty RAF personnel another two weeks to fingertip-search 14 square miles to recover all the wreckage from the two Tornados. Contaminated soil from the two fields would have to be removed and replaced layer by identical layer. Not forgetting 18,000 sugar beets that needed digging up before work could begin.

Twelve days after the collision, Ian Weaver regained full consciousness. 'I simply didn't know where I was or what had happened. I was totally immobile. My first thought was that I had somehow been killed in an accident and perhaps this was now hell? I later discovered I was paralysed from the neck down, had a 5kg weight hanging from my head and a machine breathing for me.' Moments later intense pain smacked into him for the first time. 'I then prayed that I wasn't dead, as I couldn't possibly endure that level of pain for minutes, let alone eternity. The only pain relief I had initially was morphine, but as the pain was neurological, caused by the damage to the spinal cord and the brain, the drugs had little or no effect. It was truly horrific – the worst agony you can imagine, magnified tenfold. And it was constant. Like living a never-ending nightmare.'

Unable to speak thanks to the tube down his throat, Ian couldn't explain his agony so it went untreated. If nurses or visitors patted his arm or stroked his head it caused unbearable surges of burning pain. As it sank in that he was in hospital, he presumed he had been in a car crash. 'I worried about my wife and kids. Had they been involved? Were they injured or worse?'

He still had no recollections. He was desperate to communicate. He could hear and see, but still had the breathing tube in his windpipe. His eyes would have to do the talking. Twenty-four hours after he came around, his father visited, and they managed to communicate by blinking. One blink for yes, two for no. 'He asked me if I knew what had happened. I blinked twice. So he told me about the mid-air collision, the four ejections, and that no one else was badly injured.' At least that was some relief.

Two weeks after he regained consciousness, and still in dire pain, Ian's children were allowed to visit him in hospital. Since his accident, Jenni and Chris had mostly been looked after by family and friends while their mother spent hours at the hospital. The television news was always turned off so they wouldn't see the endless reports about the crash. They knew their father was seriously injured, but had been spared the worst details. 'I found out more information at school because friends had seen news reports,' Jenni says. 'They told me details of what had happened to my dad, about his aircraft crash, that he was in hospital and badly injured. It was all quite upsetting. Everyone was always kind to us, but as a 10-year-old I felt quite alone.'

Entering the hospital, the children were taken to a family room where their granddad treated them to chocolate bars from the large purple Cadbury vending machine. Their relations were all putting on brave smiles. A nurse took them through the heavy double doors into intensive care. 'I can still remember how they slammed firmly shut behind us. It seemed so final. We trailed behind her, holding hands, between the two rows of curtained off bays. It was deathly quiet, apart from the bleeps from banks of medical machinery.'

And then they saw their father.

'I hardly recognised him,' Jenni says. 'He was in a huge metal bed with all sorts of controls on it. There were wires attached to every bit of skin. He had a big metal cage screwed into his head. Under the bed there was a pool of blood congealing on the floor. Being as little as we were, the blood was a lot easier for us to see than the adults. There was a smell of wee. Chris and I stood there in absolute shock, just staring . . .' Jenni's voice breaks as, twenty-five years later, she recalls the events. 'I can still see the whole image clearly today.'

Ian, immobile and still intubated, was unable to speak to his children. 'It was very emotional for me and incredibly tough for them,' he says. 'They just wanted to climb onto the bed with me, but I couldn't hold or hug them.'

Jenni couldn't keep her eyes off the bloodied floor. 'It was so shocking, I couldn't think of anything else. I held Dad's hand. It was really difficult trying to speak to someone who couldn't answer, although Mum did her best to speak to us on his behalf.' When it was time to go, Jenni and Chris were lifted up to give their father a kiss. She was fearful of snagging some of the many wires attached to him. 'I kissed him, but it didn't feel like my dad. It was so confusing. Normally he would kiss and hug us back. It was like kissing an inanimate object. Then we were ushered away. I didn't know if I would ever see him again.'

But Ian was getting better. Slowly, painfully, over the coming weeks some movement started to return to his legs. His left arm remained paralysed and he still had the breathing tube inserted thanks to the massive chest trauma he had suffered. Still unable to speak, he could not tell the medics about the unbearable agony he was suffering. Every part of his body seemed to resonate with shooting pain. Made worse by the well-meaning staff trying to keep him alive. 'Every couple of hours a physiotherapist would push a tube up my nose and into my chest, then proceed to pump out the liquid collecting in my lungs, by compressing my chest.' With broken ribs, and the intense pain of being touched, it was easier said than done. 'The whole procedure caused astonishing levels of pain, and I remember starting to cry whenever I saw a physiotherapist moving around the ward.'

Eventually, his breathing tube was removed. 'As soon as it came out, I could start to growl out the word "PAIN". They brought in an expert who put me on a massive cocktail of drugs which helped only slightly. The pain in my arm and shoulder was still so intense that, over the following days, I regularly asked to be allowed back into a coma, or at the very least to have the left arm amputated.' Despite the hardships, Ian gradually improved – to the amazement of the doctors and his family – and a month after he had regained consciousness, he was moved from intensive care to the high-dependency ward.

Although out of immediate danger, his recovery was still in its infancy.

Three months after the collision and neck injuries healed, Skids Richardson was approaching the point when he could go flying again, having completed all his post-ejection medical checks. He was in the 56 Squadron crew room, where one of the windows faced out onto the flight line. The sleek grey jets were all lined up, ready for their crews. As he was brewing a coffee, he glanced out. 'Holy shit!' he gasped. The mug and spoon fell out of his hands onto the bench. 'One of the Tornados was pointing into the window, and presented the exact same picture I had seen in the seconds before the collision. I had been immediately transported back into the cockpit that day, with a vision of my life unravelling before my eyes.'

The first time he got back into the jet he gave his ejection seat an extra check over. 'I drew comfort from the fact that I now knew it was a lifesaver. It was always going to be there as a final option to survive in the most acute moment of need. Another chance at life.'

Elvis Costello too had suffered flashbacks in the wake of the incident. 'For months afterwards, every time I closed my eyes, all I could see was the image of a Tornado filling my canopy. Then, if I blinked, I could see its huge fin, just inches from impacting my face.' Things did not improve when he returned to flying.

'For my first half-dozen trips back in the air, I was simply terrified strapping back into the jet. I couldn't rid myself of the image of the impact, or get over the thought that my nav might eject me. I didn't talk about it to anyone and just tried to work through it myself. But I loved the job so much, I just had to force myself back into the cockpit.'

Elvis, Skids and Stan all continued successful careers in the RAF.

Four months after the accident, and during intense rehabilitation, Ian was allowed his first home visit. His neck was in a brace, his useless left arm was strapped across his body and he had difficulty balancing. But he had managed to convince the medics he could walk unaided.

Top of his list of things to do was to see his mates. Friday night was always 'happy hour' in the RAF Coningsby Officers' Mess so, despite advice to the contrary, he persuaded his wife to help him into his flying suit. 'I stumbled into the bar and it was the first time in my life that I received a spontaneous round of applause simply walking into a room. It was really emotional and not a minute passed without someone wanting to shake my hand. I felt incredibly lucky to be alive. Two of my three fellow ejectees, Elvis and Skids, were there and it was a special moment when they tried to hug the life out of me. I didn't have the heart to tell them it hurt like hell.'

The next day, Ian was driven out to Digby Fen. When the farmhouse door opened, he said, 'Hi, I'm Ian Weaver, and this is the second time I have dropped in on you unannounced.'

John Phillips couldn't believe it. The two men hugged one another. 'I was so grateful to finally meet the man who had saved my life. John had always worried that he had done the wrong thing in moving me after the crash. But seeing me on his doorstep was living proof that he had done the right thing. I can't say enough about him. He's my hero.'

'Those events on the day of the crash only lasted around forty-five minutes,' John recalled. 'But it seemed like days. You just do what you have to, but have no idea if you are helping.' And, ever modest, he told Ian, 'Like that Martin-Baker ejection seat, I was just a rung in the ladder of your journey to recovery. It's you who is the hero and it gave me great satisfaction to be able to help such nice people.'

Ian didn't feel particularly heroic. Moved to a military rehabilitation centre, his therapy started in earnest. 'I was subjected to every kind of challenge you can imagine, both physical and mental. Month after month I was pulled, bent and stretched until I begged for relief. I was prodded, probed and tested at every stage. They put me through maths, logic, IQ and psychological tests. I even had the amount of muscle wastage on my tongue measured by a speech therapist.'

His walking got better and his family was overjoyed at his progress. But his levels of pain were still high and, because of their neurological nature, there was little relief. There were many down days. 'We always concentrated on the positives of Dad's recovery,' Jenni remembers. 'But one day we were walking through a supermarket car park, his arm still strapped to his body, and he said suddenly, "I wish they'd just cut the damn thing off." For a child, that was truly shocking.'

Ian was finally discharged home in December of 1996. He had surgery to repair the major nerve junction in the arm doctors had initially thought he would never use again. The physio would spend years with him, teaching him to build up the few muscles that still functioned until he achieved better movement. 'I still don't have the deltoids operating, but today my skinny little shoulder does okay, even to the extent of playing a half-decent game of golf.' Despite all the hard work, the serious extent of the injuries meant it was decided Ian could never fly again. 'I loved the RAF and could have taken a uniformed ground role, but I wasn't able to fly, to do the job I loved. In December 1997 I decided to make a clean break from the RAF and start the next phase of my life.'

* * *

Around a year after his accident, Ian visited the Martin-Baker factory. 'I toured the facilities, thanking all those involved in saving my life, from the ladies in the sewing room to the senior executives. What they design and produce is an astonishing piece of engineering, which has saved so many thousands of lives.'

He and his fellow ejectees were Martin-Baker's 6,492nd, 93rd, 94th and 95th successful ejections. In the two weeks after his own accident, another seven airmen's lives would be saved by Martin-Baker seats alone. Ian was presented with the same commemorative ejection tie received by all ejectees – the dark-blue material emblazoned with the bright red warning triangle signalling the danger of explosive seats. 'He was incredibly proud of that Martin-Baker ejection tie,' Jenni says. 'It still comes out for every family occasion.'

Ian puts his remarkable survival from a spinal injury that invariably kills down to 'sheer bloody-mindedness and a massive stubborn streak inherited from my mother'. And, of course, to John Phillips.

He still has a lot of pain in his left arm. His worst problem is a severe burning sensation in his feet and legs, the backs of his hands and around his kidneys. He accepts it is just something he has to live with. 'Another manifestation of this is a reversal of hot and cold sensations,' he says. 'Cold water burns, and hot water feels pretty neutral. It makes swimming bloody painful and meant that when I first went home the kids had to check the bathwater for me, rather than the other way around.'

Ian is one of many thousands eternally grateful to have been offered another chance at life. 'Events after my ejection caused great suffering to both me and my loved ones. But that seat performed against all the odds to save my life, and that's the only thing that matters.'

'Without the ejection seat my dad would have been killed,' says Jenni, now in her mid-thirties. 'I'm lucky to still have a dad, and my two children have a wonderful granddad. There are many others like me who still have a loved one because of those ejection seats.' As an adult, Jenni became a Community First Responder with Yorkshire Ambulance Service. 'I was often first on the scene of an emergency and involved in helping to save lives. That's all as a direct result of Dad's accident, ejection and John saving his life. I hope that I can be someone else's "John Phillips", somebody who makes a difference, perhaps saves a life and gives a child more time with a father.'

'If I had known what was going to happen that day in 1996,' Ian says, looking back on the incident that nearly claimed his life, 'I would have called in sick! But reality doesn't work like that. You can't go back and change anything, and so many good things occurred in the wake of that traumatic incident. Yes, it was the worst time of my life, but nothing that I have today would have come about without that ejection seat.'

CHAPTER NINETEEN

MANHUNT

HOLLOMAN AIR FORCE BASE, NEW MEXICO
MARCH 1999

Stealth fighter pilot Dale Zelko was heading to war for the third time in nine years. His wife Lauren was also a serving US Air Force officer, and as resilient as they come, but even she could not hold back the tears. After twenty successful missions in his F-117 during Operation Desert Storm in 1991, Dale had returned to Iraq seven years later to fly more bombing raids against Saddam Hussein's regime, which was refusing access to UN inspectors on the lookout for weapons of mass destruction.

Now he was leaving home yet again. And Lauren knew all too well that this latest conflict, in the former Yugoslavia, a region of Europe riven by unrest since time immemorial, was a very different prospect. 'As soon as they announced an American-led NATO operation against the Serb government of Slobodan Milošević, I just knew Dale would be deployed. Before he and the other pilots left for their operating base in Italy, Holloman Air Force Base staff organised a dinner for their wives to get together. We were told that if something happened to our husbands they would come to our homes and tell us in person. I knew he was going to be flying very dangerous missions and there was a good chance he could be shot down.'

Tall, square-jawed and athletically built, Dale was less concerned on his own account about deploying again. He would be the only F-117 Nighthawk pilot with combat experience in theatre. 'This was my job, what I'd trained to do as a professional aviator. It was an enormous honour to be chosen as a Stealth fighter pilot, it was considered a special duty. I didn't really think about potential dangers. I was just focused on doing the best job possible.'[1]

He was concerned, nonetheless, about its impact on his family. 'My mother was permanently worried about me being in danger. As far back as graduating from flying school in the eighties she asked me, in all seriousness, if I could now give up being a pilot. And being military herself, Lauren knew just how quickly this could all descend into chaos. Having your loved ones live in fear because of your job is a great responsibility, and this really played on my mind as I headed for Europe.'

On the non-stop fourteen-hour flight across the Atlantic, Dale Zelko did his best to persuade himself he wouldn't be flying his Nighthawk into a hornet's nest. 'Just like I had in the run-up to Desert Storm, in 1990, I presumed we would not get into a shooting war, and that Milošević would come back to the negotiating table. The threat we presented to him was so great I reckoned he'd back down.'

After the death of its last dictator, Tito, in 1980, the socialist nation of Yugoslavia, a satellite of Soviet Russia, had been broken up along ethnic lines into a federation of states. When the Soviet Union collapsed in the early 1990s, Yugoslavia began to fragment. Slovenia, Croatia and Bosnia opted for complete independence. Serbia and Montenegro created a new Federal Republic of Yugoslavia, headed by the authoritarian nationalist, Slobodan Milošević. In the late 1990s, President Milošević, obsessed with creating a greater Serbia, unleashed his forces on neighbouring Kosovo, claiming the country and driving more than 300,000 ethnic Albanians from their homes.[2] When Serb troops crossed into Albania itself, NATO, fearing more massacres, issued an ultimatum. Get out of Kosovo by 24 March

or face the wrath of twenty NATO and four other nations. The planners reckoned two or three days of limited airstrikes would convince the Serbian leadership to change its rogue-like behaviour.[3]
Milošević didn't flinch, and the operation was green-lit.

Touching down at Aviano Air Base near Venice in northern Italy, Dale Zelko's F-117 was one of 550 US and 650 allied combat and support aircraft assigned to strike key targets across Serbia. It was the first time in its fifty-year history that NATO had launched hostilities against a sovereign nation. On 24 March, Zelko and his fellow Stealth pilots prepared to repeat the success they had enjoyed in the Gulf. That night, NATO launched 214 strike aircraft, in wave after relentless wave, from bases in Italy, Germany, the UK and the United States. In the opening three days of the air campaign, the alliance suffered no casualties, while five enemy MiG-29 fighters and key military installations and infrastructure targets were destroyed. Every night the Belgrade sky turned bright white and orange as its citizens ran for underground shelters to escape. But Milošević and his commanders refused to follow the NATO script and throw in the towel.[4]

Colonel Zoltán Dani, who commanded the 3rd Battalion of Serbia's 250th Air Defence Missile Brigade, was determined to defend his *dear Yugoslavia*. 'Surrounded by so many hostile attacking countries, we were up against impossible odds,' recalls Zoltán, a meticulous, dark-haired man who had devoted his adult life to the army. 'But our leaders decided we must defend ourselves. We had high resolve and were morally motivated; fighting for our country, for our existence.'

Zoltán's wife Irena, mother of their three young children, Atila, Tibor and Iren, remembers the incessant wailing of air-raid sirens and TV images of skies glowing with fire. 'I had to wear a smile during the day for the children.' While Zoltán was defending Belgrade and the country they both loved, she told herself, 'The kids mustn't see me crying. I did chores and the ironing, trying to stay calm and wondering how we would get through it.'[5]

Standards had to be maintained. She was always smartly dressed, her blonde hair stylishly cropped. But the strain was already beginning to tell. As is so often the case, the innocent civilians on the ground were suffering the most. One eyewitness recalled the terror of waiting for the inevitable, when being attacked by a high-tech adversary that could land a missile on a pinhead.

I do not remember having any feelings. Just sitting in the dark, underground, praying that the bombs wouldn't hit us. We did not sleep. The electricity was cut off. There was no running water. Everything that we had taken for granted just a day before was gone. We were thrown into a warzone in a European capital, and NATO showed no mercy.

I crashed on the mattress and fell asleep, but then the rollercoaster started up again after just a few hours. The sound of air-raid sirens woke me up with noise that cuts through the bones and chills to the core. A paralysing fear washed over me and for a moment I could not catch air. The next series of explosions was headed for us. Someone, somewhere – who does not know me, who I do not know – was deciding if I was going to live or die. The feeling of hopelessness overwhelmed me. I could not envision the future. I found myself with the rest of my family and the nation at a dead end, knowing that some will live and some will die. This time, the floor beneath us started to shake as the bombs fell down on our neighbourhood. People started to scream. Babies and children were crying.[6]

Irena Dani barely saw her missile-commander husband during the first three days of the NATO blitz. Then, 'on 27 March, Zoltán suddenly came home'. Her relief was short-lived. 'We were having a normal meal when he stood up, saying, "I have to go now."' What could it mean? The family walked out of their neat, apple-green detached house surrounded by a white balustrade in the small town of Kovin, 30 miles east of Belgrade. There were hugs and kisses as they waved Zoltán off. Then Irena clicked the ornate wrought-iron

gates shut and watched him drive away. 'Once more we waited for
the phone to ring. We never had any idea where he was. I was terrified he wouldn't live to see the next morning.' Irena dreaded being
given the news that she was a widow. 'I told the children he would
be coming home safely. I shared a bed with them and they'd hug me
so tight I felt I was suffocating. They didn't want to lose me when
their dad wasn't there.'

* * *

On the night of 27 March the weather was so bad NATO had cancelled all airstrikes, except one. Eight F-117 Nighthawks, America's
supposedly invisible, invincible Stealth aircraft, were to take out a
high-value, heavily defended military facility near Belgrade. Giant
B-2 Stealth bombers would follow them. 'The only show in town
was us stealth guys,' Dale recalls. 'My objective had been targeted
previously, unsuccessfully. Unlike in Desert Storm, we had not really
degraded their air defence system sufficiently. We had been picking
at it rather than destroying it.'

Before that evening's brief, Dale ate his customary king-size bowl
of cereal with dried cranberries, like a marathon runner loading
carbs before a race. He put four chewy power bars in his pocket,
'just in case', then went to kit up. 'Preparing for the worst scenario –
ejecting – I put on four layers of clothes: flight suit, vest, t-shirt, my
winter flight jacket, as well as three pairs of socks, two thin cotton
and one high-quality wool. I tucked an extra set into my suit.'

He completed a final mission check with his 20-year-old targeting officer who had planned the sortie. 'She gave me her American
flag to take with me. It was not unusual for folk to give us stuff; a
physical memento they could say had been on a stealth mission.' He
put the Stars and Stripes in his helmet bag, but then changed his
mind. 'If I had to eject, I didn't want to leave it in the cockpit. So I
unzipped all my layers and stuffed it inside my flight suit, right on
top of my heart.'

His Nighthawk, call sign Vega 31, was Bible-black and Darth

Vader-menacing. Although he loved the jet, walking out for his second combat sortie over Belgrade, Dale was uneasy. 'As on all my previous sorties over hostile territory, the thought of my family's anguish, suffering and sorrow if something bad happened was terrifying. But there was a job to do, a mission to be flown and I had to do it.' The last thing the Nighthawk pilots did before boarding for a combat sortie was to give their security passes to the Crew Chief for safekeeping. 'He took hold of mine, but for some reason I couldn't let go. He was tugging at it. I could see him wondering why I was acting so strangely. Eventually I gave it up and climbed in.' That night's mission was unusual. The atrocious weather prevented support from aircraft designed to jam, or attack, enemy surface-to-air missile systems. More worryingly, the route they were about to take had already been flown multiple times. No routes had been used consecutively over Iraq. NATO's air superiority might already have led to complacency. And it was later discovered that the Serbs had spotters monitoring aircraft movements at Italian bases.

Dale strapped himself into his ACES II 'Advanced Concept Ejection Seat'. Over 5,000 of these lightweight rocket seats would be fitted into a range of US military aircraft. During over 600 live ejections, it automatically sensed airspeed and altitude, and selected the appropriate drogue and main parachute deployments for the occupant. Another leap forward in design, the propulsion system adjusted to varying aircrew weights, so an 8-stone female pilot and a 15-stone male would benefit from a similar rate of acceleration. Just 1 per cent of ejections resulted in back injuries, compared with up to 40 per cent in some other systems.[7]

Although totally confident in his jet's escape systems, when he took off from Aviano at around 6.30pm, Dale was consumed by one thought: *If there was ever a mission for an F-117 to get shot down, this is the night, and my mission is the one.*

Missile battery commander Zoltán Dani had left his family's dining table that day in great haste after receiving an Attention Order Number One. 'It's the highest warning level in the Serbian air defence missile units. We were told, "Targets are heading into our airspace."'

'They say that the F-117 is invisible,' Zoltán had told his men, 'but it's a marketing trick. No aircraft is completely invisible. One of my pilot friends says it's a plane with a smaller radar cross-section, but it still has one. It's just a case of finding it.' The nine men of Zoltán's unit were now huddled around their screens in a command vehicle near the village of Šimanovci, 15 miles west of Belgrade. Their missiles were elderly Russian SA-3s dating from the 1960s. Twenty feet long, weighing just under a ton, the missile was powered by a solid propellant rocket motor and carried a 32-pound 'proximity' fused warhead. It did not have to physically strike a target to bring it down, just get close enough to explode. Variations of the missile system had been used successfully in Iraq, Angola, Egypt and Syria, but could an elderly, bog-standard Soviet surface-to-air missile really pose a threat to a state-of-the-art Nighthawk, bristling with anti-detection technology? Zoltán was confident. 'A Nighthawk's flight ceiling is 49,000 feet,' he said. 'With our system, that is very reachable, and we have an operational range of 22 miles. They claim it is outdated technology, but let me tell you, we don't have anything better in our defence, old or not. And, to remain covert, we used no radio communications. We ran a telephone landline cable to the last house in the village so we could keep in contact with HQ.'

Now it was just a question of waiting.

Loaded up with two 2,000lb precision-guided bombs, Dale Zelko flew east from Aviano, into Slovenia. 'As I went over the top of the capital, Ljubljana, I rocked my wings as an acknowledgement of where my paternal grandparents had lived before emigrating to the United States in the aftermath of the First World War. It was a little wave to my heritage and a reminder of my own family back home. I wondered if there were still people down there I was related to.' Dale's thoughts strayed to his family in America and his daughter Gina, from his first marriage. She would celebrate her tenth birthday the next day. 'I was so looking forward to calling her to wish her happy birthday. I had planned it all out: which phone I'd use on the base, and what time. Nothing was going to prevent me connecting with her. It was really important to me.'

His route took him through Hungary then down along the Romanian border towards northern Yugoslavia. He prepared to stealth up. Just like Zoltán Dani, he knew his F-117 was neither invisible nor invincible. Before entering Serbian airspace, he cut all lights, brought in the aircraft's antennae and turned off every transmitter that might give away his position. As he approached the border, the familiar wave of pre-combat fear washed over him. 'Not for myself, but for my loved ones waiting back home. I waited until the last moment to switch off my radio. I was desperate for a message that said, "We've figured it out peacefully. You can return to base." I didn't get that call.'

Without his external antennae deployed, he was operating alone, effectively blind. Ready for combat, he turned a sharp right into Serbia and hostile territory, heading east towards Belgrade.

At 8.30 that evening Colonel Dani and his SAM crew got warning that the intruders were heading their way. They turned their radar systems east towards Belgrade, covering the Nighthawks' suspected flight path. The radar operator switched on his system. There was no sign. Zoltán told him to switch off, wait a few seconds and then turn it on again. Still nothing. Well drilled at setting up ambushes, it looked as though he was going to miss out on attacking those who were hellbent on destroying his homeland. Could he risk another burst on the radar, with the associated risk of being detected and destroyed by a NATO anti-radiation missile? His instincts took over. Zoltán had taken a keen interest in the F-117 and how it could be detected. 'The Americans entered the war a bit overconfident. They thought they could crush us without real resistance.' He had a passion for electronics and had had his radar equipment modified so it would operate beyond the usual wavelengths. 'I ordered my team to try one final time, but to set the system on its absolute lowest frequency, and therefore its largest bandwidth. I guessed that was the one that would most probably resonate with Stealth aircraft.'[8]

It was at that very moment, 5 miles from his target, that Dale Zelko opened the doors of his weapons bay, greatly reducing his aircraft's stealth properties. Zoltán was ecstatic. 'Its distinctive profile appeared onscreen, accompanied by a series of insistent bleeps. I saw him coming straight at us, entering our kill range, our zone of no escape. I called our operations centre for permission to shoot it down.'

Dale Zelko had no idea that he had just been captured on the SAM's radar. Having unleashed his weapons, he took an aggressive turn north-west, heading out of the hostile zone and back to Italy.

Four miles below him, Zoltán had clearance. 'The only thing for me to say to my guys was, "Launch!" At 8.42 the first missile streaked away from the battery. Five seconds later number two followed.'

As Zelko rolled out at just over 23,000 feet, he looked left and right for potential threats. Nothing. 'Then suddenly, down in my right four o'clock, there they were, two missiles punching out of the bottom layer of cloud. As soon as I saw them coming at me at three times the speed of sound, I thought, *They got me. My mission is over.* I had a front-row seat for the entire event.'

Zoltán's first missile just missed, passing right over the top of the Nighthawk. It buffeted the Stealth fighter but not closely enough to detonate. The second came within 140 feet. Its proximity fuse activated. Zoltán and his crew were ecstatic. 'We watched it exploding on the screen as the warhead broke into 30,000 shards of jagged steel, ripping into the jet. The target had been neutralised.'

'I closed my eyes and turned my head, anticipating the impact,' Dale recalls. 'I knew there would be a fireball and didn't want to be blinded. The blast slammed into the aircraft, like being hit by a train. A huge flash of light engulfed my jet, the left wing was blown off, sending the aircraft into a violent roll. Tumbling earthwards,

I tried moving the throttle and stick to see if I had any control. I didn't.'

Thoughts of his family flooded Dale's mind as the Nighthawk plummeted, now totally out of control, 10,000 feet in a matter of seconds. As it gyrated, the g-forces on his body were enormous and, even though he was tightly strapped into his ejection seat, he was now sliding out from underneath its lap straps. 'I was forced against the top of the shoulder straps, with my butt way out of the seat. The back of my head was pinned up against the canopy and my torso was doubled over. The worst possible position for an ejection.' Immobilised by the negative g-force – which Dale estimated was in excess of -7g – of his falling, doomed jet, he was desperately trying to get his hands down to the ejection handle on the side of his seat. 'I was thinking, *This is really, really, really bad. Chances are I'm going to break my neck, have massive back injuries, if I even live.*'

All he could think of doing was to push with the back of his head against the top of the canopy, which would perhaps straighten his spine once it blew, and before the seat went up the rails. At high school he had wrestled and was often pinned on his back; he had bridged his body to get out of the hold. Maybe, just maybe, that move would work again? 'Mentally and emotionally I felt very calm. I realised I was unlikely to survive, but I didn't have any fear. There was little I could do about it, so no point in worrying.' Dale has no memory of reaching for and pulling his ejection handle. 'The next thing I recall was being out of the aircraft, in the seat and looking down. I saw the cockpit falling away from me with all the red and yellow alarm lights flashing. The environment in which I had felt extremely comfortable, which I absolutely loved and was where I belonged, was no longer there for me.'

Zoltán was thrilled. 'When it hit, it felt very, very good. Like scoring the winning goal in a football match.' The first F-117 to be shot down in combat, it was one of the most spectacular US Air Force losses. David knocking out Goliath. He didn't need telling that taking down a supposedly 'invincible' Nighthawk was a major propaganda coup for his beleaguered nation. The only thing that would trump it was capturing the pilot. 'We had seen him eject and put out an alert,' Zoltán says. 'I had every confidence that our 72nd Paratroop Brigade would soon pick him up.'

From pulling his ejection handle to being under a fully inflated parachute would take Dale around two seconds. It seemed like hours. Tumbling through the air, random thoughts raced through his mind. 'I had a mental image of myself landing, kicking the dirt and saying, "Nuts, isn't this inconvenient? My mom's not going to be happy with me and I won't be able to call my daughter tomorrow on her birthday."

'Then *BLAM*! At around 7,000 feet, I was out of the seat and my parachute bloomed. In just a couple of seconds, I had survived the missile strike, the jet going out of control, the force of the ejection and finally escaping from the seat. I was now entering the next phase of the ordeal – the battle for survival.'

Apart from his own personal survival, Dale knew how important it was to deny the enemy the propaganda coup of capturing a senior F-117 pilot. 'To those who shot me down, I was the ultimate prize.' He vividly remembered Larry Slade and the other allied aircrew who had been shot down during the Gulf War being paraded on Iraqi TV, battered, tortured and abused. He was determined that was not going to happen to him, even though he knew his chances of avoiding capture were slim. 'I had no idea how this was going to play out, but I was rapidly going through the various stages of combat aviation and was about to transition from ejectee to evader. I had to start thinking and acting like one.'

Dale disconnected his oxygen mask and threw it away. He started to orientate himself. The clouds had parted and there was a near full moon. Looking right, he saw Belgrade, fully lit. To his south-west there were two fires burning – all that remained of his Nighthawk. After the violence of his ejection he was floating in near silence. All he could hear was the gentle swishing of the life raft and survival kit hanging on a lanyard 25 feet beneath his feet, and the oscillation of his parachute. He looked up. 'I said out loud, "You've gotta be kidding me!" My chute was made of orange and white panels. In the moonlight it was glowing like a Chinese lantern.'

If he were to benefit from Combat Search and Rescue 'it was vital to make good two-way contact with a friendly as fast as possible'. A minute after his parachute opened, he reached into his g-suit pocket and pulled out his mini flashlight, complete with red lens cap. He shielded the beam with his body as best he could and did a quick visual confirmation that he had the correct settings on his survival radio. He started transmitting on the emergency 'Guard' frequency.

'Mayday, Mayday, Mayday. Vega 31.'

He made several calls, giving the last position of his aircraft.

'Out of the aircraft. Vega 31 is going down.'

'I just wanted to let somebody, anybody, know, "Hey, I am out and alive."'

His radio crackled into life. 'Vega 31, this is Magic 86 on Guard.'

It was a British Flight Lieutenant on board a NATO AWACS (Airborne Warning and Control System) aircraft, in a high-altitude orbit not far away. Sensing the Serbs would be listening in, Dale did not respond. At least someone knew he was down.

The AWACS crew began frantically digging through reams of data. Who was Vega 31, and what had been his mission?[9]

At about 2,000 feet, Dale passed through a band of cloud. 'That gave me around two minutes under the canopy to take stock of where I was going to land and make some initial plans.' There was a stiff breeze and he was drifting down, south of a town, towards flat, open farmland. 'There were lots of vehicles on a two-lane road and a major highway beyond. I was convinced somebody would spot me.' He steered his parachute towards a freshly ploughed field not far from the road. Landing softly, he collapsed the parachute and lay motionless on his stomach. Waiting. Expecting to hear cars screech to a halt. The traffic kept flowing.

There was precious little cover, but Dale spotted an irrigation ditch bordered by scrubby undergrowth with a single cherry tree,

250 yards away. As somebody who had excelled at combat survival training, he knew it was a far from ideal hiding place, especially under the bright moon, which kept breaking through the clouds like a spotlight. 'I had never been in this situation before, or imagined it happening, but I was beginning to think like a man on the run.' He wasn't scared but he knew he was in real danger. 'I had to get away from my landing spot fast, and travel small and light.' He clambered to his feet, placed his parachute and helmet into a deep furrow, put his life raft on top and packed earth around it so the wind would not expose them. He grabbed his survival kit and walked carefully towards the cherry tree, sticking to the bottom of the furrows rather than crossing them, leaving as few clues as possible.

It was still afternoon in New Mexico, eight hours behind Belgrade, and Lauren Zelko was lying on her couch with the profound sense that Dale was in severe danger, even though she had heard no news. 'I just started crying, crying out to God, "Please let him live, bring him home." I loved him so much, and I just wanted him to be okay.'

Soon afterwards her phone rang. 'We've lost a plane,' a high-ranking officer told her. 'Dale's plane.'

She was dumbstruck.

'The good news is that we know he ejected safely and he's made a radio call.'

'Where is he? Please tell me he's okay.'

A pause. 'We have no way of knowing at this point, ma'am.'

'Is he captured? Is he injured?' The questions kept tumbling out.

'We are not clear on his status. I'm sorry.'

He did his best to reassure her that every effort would be made to rescue Dale. How could they if they didn't even know where he was?

Belgrade TV station RTS had suspended its programming to show Dale Zelko's wrecked jet blazing on the ground, with close-ups of its US insignia. 'Immediately after it was hit, the F-117 hovered in the air for thirty seconds,' the Yugoslav state news agency reported, 'then it turned into a large blazing fireball and fell to the ground. There is no information as yet about the pilot of the aggressor aircraft.'

Lauren had switched on her TV and watched the drama unfold via CNN. Jubilant Serbs had come out of their cellars and gathered around the flames, increasing her sense of cold horror. Next day she saw pictures of a Serb teenager sitting in her husband's abandoned ejection seat, giving a 'V for Victory' with his right hand and brandishing a piece of Dale's jet, like it was a sports trophy, with his left. 'That was so hard to watch,' she says. Even now, the memory makes her very emotional. Dale's mother, a devout Christian, came out of church, got into her car and switched on the radio. A newsflash. 'She just knew it was me,' Dale says. 'She went right back into church and spent several hours praying.'

Zoltán Dani's wife, Irena, was as forlorn as Lauren Zelko, but for different reasons. She had no idea that it was her own husband who had given the order to take out Dale's Nighthawk. 'But when I heard that an F-117 had been shot down, a fear came over me. *Now we'll be bombed non-stop.*'

Dale had pushed his way through bushes and slid down into the ditch. He scooped up dark, moist Serbian soil and smeared his face, hands and neck. In the darkness, he felt for his survival and evasion equipment. Radio, flares, strobe light. He took out his 9mm Beretta pistol and hunting knife and laid them at his feet. The ground shuddered. He remembered that their B-2 bombers were scheduled to hit targets near Belgrade. 'Although I was a safe distance away, I felt the shock waves hammering over me. Now I was experiencing a little of what it was like to be on the receiving end of the NATO attacks.' He shrugged it off. He had more pressing things to think about. Like where the Serb search parties were. 'When the moon came out, I felt very vulnerable, but I wasn't afraid. I just needed to deny the Serbs their ultimate prize. Me.' All he could do was wait, and hope that friendly forces would find him first. Once well hidden, he took out his handheld military-grade satnav,[10] but, hunkered down, he couldn't get a signal. 'I had to crawl up the side of the ditch so I could hold out my arm to get the best line of sight with the horizon. It was very risky but I needed accurate information to pass on to any rescuers.'

Apart from checking his radio, taking little peeks to check for enemy searchers and to answer calls of nature, he kept his movement to a minimum and hoped. The night was damp and cold. He was thankful for his four layers of clothing. Later, he managed to make contact with a friendly aircraft, giving his position to a Lockheed EC-130 command and control aircraft. Now all he could do was wait. He dared not dwell on how long. He was sure that the Serbs would already be searching, and knew that in that sparse landscape he was incredibly vulnerable. 'I was motivated by what would be at stake if I was captured . . .'

As Dale suspected, all Serbian military assets in the area had been fully mobilised, and a huge manhunt was already underway. Hundreds of troops, local civilians and dog-led search parties were combing the area, desperate to catch the American pilot who had attacked their country.

But a major rescue plan was in place to get to Dale first. Six A-10 Warthog attack aircraft based at Aviano had been scrambled into the area. If they located Dale, they were to guide in and protect three Combat Search and Rescue helicopters, the same assets that had plucked Devon Jones from the sands of Iraq during Operation Desert Storm in 1991. As the various rescue teams organised and headed towards hostile territory, all Dale had to do was remain undetected until they arrived.

He knew none of this, of course, and was already facing a threat much closer to his hiding place. Around midnight he heard something moving across the field. He turned to see a visitor nearing the far end of his ditch. 'Clearly silhouetted against the moon was some sort of massive hunting dog. Weighing maybe 80lb, it was walking very deliberately and purposefully, like it was looking for something.' Or someone? Was it just a farmyard hound on a nightly prowl, or a military attack dog?

Dale reached down slowly, feeling for his survival knife. The dog stopped about 20 yards away, exactly where Dale had last been using his satnav. It sniffed at the ground. 'I figured he had my scent . . .' He crouched lower, his heart pounding. 'He still hadn't seen me. As he pawed at the undergrowth I was thinking, *How will I kill this dog if it starts barking?*' That wasn't something he'd been taught during survival training. Using his pistol wasn't an option. Too noisy. 'Would I, could I, stab it? Maybe it would be better to wrestle it to the ground and cut its throat? I had no idea. And even if I had managed, its death throes would probably give my position away.'

Dale stayed motionless, barely breathing. 'The dog kept looking around. His gaze swept across me a couple of times.' After all he had endured, to be unmasked or savaged by a dog was beyond cruel. Finally, the beast lost interest and moved on. Dale allowed himself to breathe once more.

The first A-10s eventually arrived overhead and made radio contact. Despite some confusion over precise positioning and timing, the rescue was well underway. 'Speaking to those friendly voices brought joy to my heart,' Dale recalls. 'The sense of comfort in knowing that my brothers-in-arms were there, looking out for me, was huge. They were my direct line to a rescue and return home.'

Before calling in the rescue helicopters, which were now heading to the area, the lead A-10 pilot had to be sure that he was really talking to Vega 31, or was it Serbian spoofing, laying on an ambush?

'Vega 31,' the pilot radioed, 'I understand that the colour of your motorcycle is black. Do you copy?'

All aircrew who risked being stranded behind enemy lines filed a detailed pre-mission Isolated Personnel Report (ISOPREP). It included a list of personal questions that only they would know the answer to. Authentication would be needed before Dale could be picked up. If Dale answered yes, it meant that he had a gun to his head. 'No,' he replied. 'It's red.'

The pilot asked him if he had a hobby.

'I'm a numismatist.' There was no way that captors could know that Dale had been an avid collector of coins since the age of nine.

In his flight-suit pocket he always carried his favourite, a ten-cent piece first minted in 1916, known as the 'Mercury Dime'. On one side was the winged capped head of Liberty, symbolising freedom, and on the reverse a bundle of sticks bound together, representing strength through unity.

'Okay, copy that,' the pilot said. 'Let's get this thing going.'

Finding Dale was one thing, getting him out posed a completely different challenge. In the early hours of Sunday morning the moon had set and a mantle of darkness enveloped the area. Thick, overcast weather rolled in, and heavy rain started to fall. 'That made everything more difficult for the Combat Search and Rescue force,' he says. 'Hopefully it would work against the Serbs as well.'

As Dale stayed in touch with the A-10 pilot, the three helicopters were hurtling towards him just 50 feet off the ground, their crews scanning the murk for trees, power lines and tall buildings. As they got closer, the rescuers received intelligence that eighty Serbian troops and police were closing in on Dale's position. They had found boot prints. The helicopters pressed on, determined to get their man. The A-10 pilot asked Dale further authentication questions from his ISOPREP form. They had to be sure that Vega 31 had not now been captured. Two miles away, the helicopters managed to make contact with Dale, but still couldn't see him. They told him to activate his infrared strobe light. He climbed to the lip of the ditch and switched it on.

'Vega 31, we're not getting your strobe.'

Dale slithered back down and examined the device. 'I couldn't believe it. It wasn't working. Could anything else go wrong? Was my luck about to run out as we neared the final moments?' He finally spotted the helicopters. 'You guys are north of me a couple of miles. Head south.' Then he spotted a searchlight. A Serbian helicopter was also heading his way. 'This was now a real *you gotta be kidding me* moment. What the hell was going on? Was I about to be rescued or captured?'

'Hey, Vega 31,' the rescue helicopter co-pilot told him. 'If we are that close, just go overt.'

Dale had two flares in his survival-vest pocket. He pulled one out and crawled back over the top of the ditch. 'Lying flat on the field, I held it above my head and just popped that sucker. It went up like a massive roman candle, blazing across half of Serbia.' He let it burn for two seconds then snuffed it out in the soil. Back into darkness.

Then rotor blades started pounding the air a matter of feet above his head. The noise was deafening. One of the helicopters landed close by, scarcely more than a rotor arc away from him. 'It was so dark I couldn't see them until they settled and the very top of the helicopter became barely illuminated by static electricity generated from dust hitting the rotors. That sparkling light, the crews call it the "fairy dust", was an unimaginably beautiful sight. Here they were at last.'

The two other helicopters hovered just above the ground, providing cover. As Dale emerged in a low crouch, their rotary cannons, each capable of firing 4,000 rounds a minute, were ready, in case he had been captured and was being used as ambush bait. Two indistinct shapes materialised out of this blackness. The pararescue men (also called PJs) had jumped out of the lead helicopter. 'I didn't see these guys until they were 10 feet away. With their helmets, night-vision devices and weapons, they looked like aliens.' Then, a terrifying thought. *I have no sense of who they are. Serbs or friendlies?* One of the PJs grabbed Dale by his upper arm and yanked him in close. 'I could feel his breath on my face. Then I realised. He was doing a visual identification. The final authentication that it was me and not a trap.'

'How're you feeling, sir?' the PJ yelled.

'Great!' Dale bellowed back. 'Let's get out of here.'

Shepherding the downed pilot towards the waiting helicopter, he said: 'Your PJs are here to take you home.' As he relives that moment, Dale comes close to tears.

The rescue helicopter, which had been on the ground just forty-five seconds, raced away unmolested by the searching Serbian troops, leaving hostile airspace at 3.46 in the morning.

Dale had been recovered almost eight hours after he had pulled his ejection handle. 'It had been a truly unnerving, almost unreal experience, and waves of relief washed over me as we finally crossed the border.' He would be able to wish his daughter happy birthday and tell Lauren that he was okay. He knew she would have been to hell and back waiting to hear his voice.

A quick medical check revealed severe lacerations to his hand and pronounced bruising on his right leg, around his eyes and on his backside, all courtesy of his ejection seat. Back in Aviano, his young targeting officer was waiting expectantly, with a large crowd. Dale stepped forward smartly, gave her a wide grin, unzipped his flight suit and, saluting with his bandaged right hand, returned her Stars and Stripes. She threw her arms around him to cheers and whoops of delight. 'I had also brought it back for all those who had been involved in my rescue,' Dale says.

He was then submerged in a mass of hugs from comrades who had feared they would never see him again. He met the six A-10 pilots who had found and watched over him as fifteen helicopter crewmen sped in, supported by many other aircraft and ground-based intelligence personnel. As one of the Warthog pilots said of the rescue, 'We went in with thirty-seven Americans, and we came out with thirty-eight.'

Dale told them, 'Boy, that Combat Search and Rescue was beautiful, and it's wonderful to be home, and God bless everybody for being with me all night.' Highly unusually, as he doesn't drink, the Nighthawk ace even joined them for a beer. 'The cheer for that was even louder than when I had got off the plane.'

After debriefing, Dale was handed a phone. When he heard Lauren's voice, 6,000 miles away at Holloman Air Force Base, all he could manage was, 'Hi Lauren . . .' and then the floodgates opened. 'I was still pumped with adrenaline. I had been shot down, ejected, evaded, feared capture and death, been rescued and had the joy of liberation. All these emotions were suppressed and rolled up into a small ball inside me. Lauren had been through the agony of not knowing if I was alive. Being strong, being brave, waiting minute by minute for news.' They both broke down, weeping uncontrollably. 'This was the release, that dam of pent-up, buried emotions bursting at the relief and realisation that it was finally all over.'

The day after his rescue, Dale received a phone call from General Wesley Clark, the Supreme Allied Commander Europe, who urged Dale to take part in a media event celebrating his rescue.

He refused. 'Actually, sir, I'd like to do just the opposite,' Dale told him. 'I don't want my name released at all. If you do that, it will preclude me from flying any more Nighthawk missions.' For security reasons, the identities of serving F-117 pilots were never revealed. And no way was Dale Zelko ready to be an ex-Nighthawk pilot. Clark sounded annoyed at the loss of a major PR opportunity. 'I guess it's your call,' he said and hung up.

Hours later, the phone rang again. 'Are you the pilot?' a voice asked.

'Affirmative.'

'Please stand by for the President.'

Bill Clinton told Dale that he was 'deeply relieved' that he had ejected and managed to evade capture. The President emphasised 'what it would have meant had you been captured'. If he had been paraded on camera it would have been a strategic and political crisis of the highest order for the United States and its NATO allies. 'When I told him I wanted a low media profile, he said, "You have full White House support."'

The Serbs held out for seventy-eight days before withdrawing troops from Kosovo. Slobodan Milošević was eventually brought before a court in The Hague, found guilty of crimes against humanity and died in jail.

The shooting down of Vega 31 was the highlight of Colonel Zoltán Dani's career. 'In my military life, I never had a team as good as that. That motivation and our moral victory versus the enemy against all odds makes me proud I was the commander of people who did such good work. Even now, the hairs on the back of my neck stand up when I think about that moment.'

After serving so many years in the army, Zoltán, now in his early forties, had come to a fork in the road. He could stay in the military, a highly respected and decorated officer, famed for shooting down an F-117, and very lucky to be alive after surviving the bombardment that had brought his country to its knees. But he was in demand internationally; nations hostile to the USA were offering vast sums for his expertise in combating stealth technology. 'A country in the Middle East was offering millions of dollars for my services,' he says. Or should he do something more peaceful, devoting the rest of his life to Irena and the children, who would get to see more of their father? The choice was easy. 'All the money in the world means nothing when you can't be with your loved ones.' He eventually decided to leave the military and become a baker. Irena was delighted. Holding his hand, she told him, 'When you decided to come home to us, I just thought, *At last!*'

All Dale Zelko wanted was to continue flying and, thanks to his ejection seat, he had survived to fight another day. He would file his lucky escape and rescue under past history and put it all behind him.

Or so he thought.

CHAPTER TWENTY

THE HUMAN CANNONBALL TRICK

FRANCESTOWN, NEW HAMPSHIRE
2009

Dale Zelko had wanted nothing more than to get straight back in the air after his adventures around Belgrade. 'I really didn't want to leave my colleagues behind in the combat zone, but senior officers insisted I was sent back to the USA for my family's sake. The risks of another "incident" were deemed too great.' Dale had reluctantly accepted the decision. 'It was wonderful to be back with my loved ones, to give them that peace of mind, but I still felt I had left a job unfinished. Although I was physically back home, a part of me was still in the theatre of ops, waiting to complete another mission.'

Lauren Zelko's greatest fear had been that if Dale had gone back to Serbia he risked having to eject again, risked being captured. But having her husband home was not easy. 'It quickly became apparent he wasn't ready. It became really difficult. We were in two entirely different places. He was so unprepared for normality. I really needed to be there for him, but mentally he was still in Aviano, waiting to complete another mission.'[1]

Dale eventually left the US Air Force after twenty-five years' service. He, Lauren and their four young children moved to a red and white clapboard house in New Hampshire, set on a large leafy plot, with the obligatory basketball hoop out back, and a Stars and

Stripes hanging over the garage. Nearly a decade since he had been blasted out of his Nighthawk and narrowly evaded capture, he revelled in the quality time he could now spend with his family. 'I was far more relaxed. I didn't really give much thought to my adventures over Serbia. Or my brush with death.'

Then, out of the blue, his past came rushing back to meet him. 'A Serbian documentary maker wanted to make a film about the day I was shot down. He wanted me to return to Belgrade and meet the man who had been responsible so he could film the reunion. I learned that the person who had taken me down was a surface-to-air missile commander, Colonel Zoltán Dani. I couldn't quite believe what I was reading. It seemed an astonishing proposition.'

But something sparked in Dale. Like many former enemies who had faced each other across the battlefield, he became consumed by the idea of meeting Zoltán, not as an adversary but as a fellow officer. 'I needed to meet him and the Serbian people under very different circumstances, even if it meant leaving my beautiful home and blessed family. I needed to explore the possibilities of reconciliation. To meet those I had been at war with, those I had tried to kill, those who had tried to kill me. It quickly became a passion. But the first time I was in Serbia, I was dropping bombs. How would I be received now?'

Dale's wartime survival mechanism had been to think of his heavily defended targets as no more than concrete and steel. But there had been thousands of casualties, including many civilians, and hostility to NATO was still rife. 'Memories were still fresh. Would some people in Serbia take a second chance to get at me? It was a real possibility. I was fearful for my safety.' So much so, in fact, that he decided it would be foolish to go. 'I emailed the filmmaker to tell him my decision, explaining that it was not that long ago that I had been over their country dropping bombs.'

He received a message from Zoltán Dani himself by return. *Please tell Dale that this time when he visits, he is invited, and there will be no missiles.*

'My anxieties were allayed. I decided, *Okay, I'm going*. I really needed to make that journey of redemption, to revisit the worst day of my life.'

* * *

The ejection seat that gave Dale the chance to embark on his own journey had come a long way and saved many tens of thousands of lives around the world since the Second World War era in which it had been conceived.

Just seven decades earlier, Jo Lancaster's prototype 'Flying Wing' had begun to break up in mid-air. But instead of plunging him into the Warwickshire countryside, his Martin-Baker Mk1 propelled him into the history books, as the first of over 7,600 aviators who owe their lives to that single company's increasingly sophisticated devices. Modern pilots like Dale Zelko would barely recognise the Mk1. How on earth did it save anybody's life? Sceptics regarded those cumbersome, explosively charged 'bang seats' with suspicion and fear. As Jo himself declared after his first test on the static rig, 'It was a bloody dangerous invention!'

Unimaginable by today's standards, Jo's whole ejection process had been completely manual. To save his life, he first had to pull a toggle to release the cockpit canopy, then haul down the face-blind handle to trigger the seat.

Blasting out of the aircraft, risking serious back injury, he then had to wait for the contraption to stabilise before unstrapping. Feeling across his sternum for its release buckle, desperate not to unfasten his personal parachute harness which sat beneath it. Plunging earthwards all the while. Finally, after pushing clear of the seat and into free-fall, he had to locate and pull his ripcord. The whole process took Jo around thirty seconds.

And since then, Martin-Baker and their fellow designers across the globe had one simple guiding principle: 'Striving to reduce that delay between pulling the handle to parachute opening to the shortest time possible.'[2]

The technology has developed at breath-taking speed thanks to gifted engineers and, in the early days, human guinea pigs prepared to risk life and limb. It is still being constantly refined. New devices, utilising either airbags or a metal frame, now force the head forward on ejection rather than back. The brace position, chin down and tucked into the chest, prevents the rear of the cranium slamming back against the headbox, a major cause of whiplash injuries. Seat stabilisation and parachute extraction times have also reduced the entire sequence to a mere one and a half seconds – beyond the imagination of Jo Lancaster and his fellow pioneers.

The smartest of current seats no longer merely 'waits' idly for its occupant to pull the handle. They can now be digitally linked into the aircraft's computer systems, constantly monitoring airborne parameters. The very latest have the capability to adjust the escapee's trajectory to suit the host aircraft's height, speed and attitude, greatly increasing survivability in situations previously way beyond ejection limits.

The most astonishing, but perhaps most logical, development is to give the ejection seat autonomy over the pilot. The Lockheed Martin F-35 Lightning II Joint Strike Fighter's 'short take-off and vertical landing' capability allows it to hover in the same way the Harrier once did. But now, the Martin-Baker Mk16's *auto-eject* facility is enabled with the jet hovering. In the event of life-threatening power failure, presenting imminent danger to the pilot, the ejection system can trigger its electronic sequencer more swiftly than human reaction time. The seat, not the pilot, takes the decision to eject. And in just fractions of a second. The first the aircrew might know about this is when finding themselves leaving the cockpit.[3]

* * *

Dale Zelko was not the only veteran on tenterhooks about embarking on a journey to meet a former enemy. In 2011, ex-Royal Navy gunner Neil Wilkinson landed in Argentina to meet the aviator he had always believed he had killed.

A few years earlier, still suffering from complex PTSD, he had flicked onto a TV documentary about the Falklands conflict – the first he had been able to endure. He had been transfixed listening to the pin-sharp detail of an Argentinian aviator's description of the events of 27 May 1982. He and his wingman had slipped into Falkland Sound two days after he had sunk HMS *Coventry*, to mount another raid. The story that followed was uncannily familiar. Neil's eyes began to fill with tears. There was only one Argentinian pilot shot down that day. 'I was totally gobsmacked. The man I thought I had killed twenty-five years earlier was alive and now had a name: Mariano Velasco. Incredibly, he had actually ejected from his flaming Skyhawk and survived. A huge weight was lifted from my shoulders. I was determined to meet Mariano and tell him how happy I was he hadn't died that day. It would be a way of putting my demons to bed, to finish off my own story.'[4]

After nearly a year's research, Neil finally tracked him down. Velasco responded to his email, describing how he had ejected and been rescued by Argentinian soldiers. He had stayed in uniform, becoming Assistant Commander-in-Chief of the Argentinian Air Force. Now he was enjoying quiet retirement in a comfortable, rustic farmhouse with his family in the mountains near Córdoba, in central Argentina, about 435 miles north-west of Buenos Aires. He finished by saying, 'I hope that one day we can meet in person. The doors of my home are open, so I can welcome you with an affectionate hug.'[5]

Like Dale, Neil was conflicted. 'I was really unsure how things might unfold. How would we feel about each other? Thirty years before, I had been trying to kill him and probably killed one of his friends. He had certainly killed some of my colleagues. I was very nervous about going to Argentina. They still claimed the Falklands as their own and had lost countless lives trying to make that a reality. Would they resent me? Would there be any hostility? All of these thoughts were playing heavily on my mind.'

In November 2011, pushing his fears aside, Neil was on his way, courtesy of a documentary crew who were captivated by his story. Driving north through picturesque, fertile valleys, he felt increasingly apprehensive as the road twisted and turned into the mountains of Córdoba. 'I had no idea what to expect. My mind was awash with possibilities about how the meeting might go. What do you say to a man you thought you had killed?' Argentinian flags flew proudly above public buildings, increasing his anxiety. When the road became a narrow track, he got out and walked through the wooded garden that led to Mariano Velasco's farmhouse.

His host was standing outside, a smile of greeting on his face. The once jet-black moustache and hair were now white and he had put on a little weight, but he did not look so different from the Skyhawk pilot who had flown missions to drive the British from the islands he called Las Malvinas. He walked briskly towards Neil and threw his arms around his shoulders.

'Hello, Neil, welcome to my house.'

'Thank you, sir. It's an honour.'

'Are you good?' Mariano asked.

'Yes, very . . .' Neil faltered. 'A few tears, maybe.' He had waited a long time for this moment, and the relief was overwhelming. 'It's too massive to put into words,' he said. 'I'm just so happy to be here.'

'These are the important things in life, when you go through such bad times, to meet, and to forgive each other,' Mariano replied. 'But there's nothing really to forgive. Both of us had our duties to fulfil.'

After a family meal, they sat together on the terrace, warmed by friendship and the late afternoon sunshine. It was hard to imagine the pair as bitter enemies. Mariano motioned towards the clock he had plucked from the wreckage of his Skyhawk. 'See . . .' He pointed to the hands, which had stopped at 4.50pm, the precise moment his jet impacted the ground after he ejected.

'I am very happy that he has come to see me.' Mariano addressed his words to the camera. 'My family feel the same. It doesn't matter that we were in battle against one another. The horrible times should be put to one side now.'

'He welcomed me with open arms,' Neil said. 'The only reason Mariano was alive, that I could now shake his hand, was because of his ejection seat. That device, which I knew so little about, had saved his life, and given him and his family a second chance. His family had a husband, father, grandfather. And it gave me the opportunity to meet the man I thought I'd killed in war. It is difficult to explain, but I felt something like euphoria. My journey was coming full circle, and I was getting the closure I realised I had been searching for. I still relive the Falklands every day of my life and see Mariano's jet in my gunsight as clear as day. The images will never go, but being able to meet my former enemy really helped put my demons behind me.'

Neil and Mariano did not discuss the rights and wrongs of their war. Thirty years had passed, and most wounds had healed. As Mariano says, 'Good soldiers should be able to forgive each other and, afterwards, why can't they be good friends?'

* * *

The ejection seat used by Mariano Velasco was not dramatically different from those developed in the early 1950s, but many escape systems pioneered since wouldn't have been out of place in *Mission: Impossible*. Such as totally enclosed capsules, and seats that transformed into basic gyrocopters allowing ejectees to fly clear of a combat zone.

The Russian Ka-52 Alligator attack helicopter boasts an ejection system that prevents its polymer rotor blades transforming themselves into a giant food processor. They are blown away from the spinning shaft by explosive charges when ejection is initiated, before the canopy over the twin-seat cockpit is jettisoned. Then the seats – and pilots – are hauled out by a rocket extraction system similar to the Yankee device fitted to American Skyraiders during the Vietnam War.

More than 35,000 light civilian aircraft have been fitted with an even more ambitious device. On activation, a huge 1,000-square-foot parachute – large enough to entertain 100 guests if it were a marquee – is extracted by rocket from a specially designed compartment behind the cockpit and attached by cables fore and aft. When fully inflated, it lowers the entire machine gently to the ground. As the manufacturer points out: 'You don't want to leave an unmanned 7,000lb bomb to fly into a residential neighbourhood when you can bring everyone down safely, with the airplane intact, for about one-tenth the price of ejection seats.'[6]

* * *

However sophisticated the ejection, there is little to protect the escapee from the trauma of the journey that may follow. Fast-jet pilot Ash Stevenson and his university cadet passenger Kate Saunders are still haunted by theirs thirty years on. After Ash rescued Kate from their blazing Harrier, brought down by something as innocuous as a small bird, she was rushed to the specialist burns unit in Wakefield. Over the next few months, he regularly made the five-hour round trip from his base at RAF Wittering near Peterborough to sit by her bed. The first time he saw her, surrounded by fresh flowers, cuddly toys and get-well cards stuck on the walls, he had done his best to hide his shock. Kate managed only a brief smile; two-thirds of her face was covered in vivid red flash burns. She was bandaged head to foot and unable to move or feed herself. Almost a third of her body had been severely burned. She had a metal plate in her shattered leg and her pelvis was broken in two places.[7]

'Driving home from those visits, I was often in bits, and I had to park up so I could cry; let the emotions out and try to compose myself,' Ash says. 'Seeing Kate lying seriously injured and in pain was incredibly difficult. The feelings of guilt surged back. They still do today. She had nearly died and I was the pilot and senior officer. I was responsible.'

After ten weeks, Kate was released, barely able to walk and unable to sit with any comfort or bend her knees. More operations and grafts followed but she eventually completed her degree and joined the RAF to start the pilot training she had always dreamed of. Sadly, the severity of her injuries meant she couldn't continue. Following a medical discharge, Kate began a career as a teacher, and is now a trained counsellor. 'I think my own experiences of ejecting, facing death and being badly injured have helped me relate to other people who encounter difficult circumstances in their lives,' she says. 'I'm a practical person; I just get on with things.'

But the scars are still there. While discussing the incident with the author, Kate holds up a hand, upon which they are still vivid. She still can't fully flex one of her fingers and gripping things can be a challenge. 'My back is still very badly scarred,' she says, 'as are my backside, the front of my ankles and parts of my arms. I had another operation on my ankle only a few weeks ago to release some of the tension in the scars. I was twenty-one when it happened, and it was tough having such visible injuries. If I went to a party you wouldn't see me in a sleeveless dress or a t-shirt. I always covered up. I still wish I didn't have to choose clothes that completely cover my arms and legs. I'd love to wear a short skirt or a strapless dress. But I can't. Bad things happen in life and sometimes we are tested by the circumstances we are thrust into. I'd rather not have ejected, been badly injured, badly burned, but I was. And when we are tested, we begin to learn what we are made of.'

Ash Stevenson received the Queen's Commendation for Brave Conduct in July 1992. 'There's nothing I can say to fully acknowledge how I feel about Ashley and what he did for me,' Kate says. 'I simply would not be alive if he had not been ready to risk his own life to save mine. I was probably outside the survivable parameters when I ejected, but I'm alive today because of that ejection seat. The accident defined my life and I wouldn't want to change that.'

Ash often thinks how different his two ejections, separated by mere months, really were. 'The first time, I simply hit the ground, gathered up my parachute and headed off to the bar to celebrate. That's the reality for the vast majority of ejections. In the second, I was faced with a catastrophic, almost unimaginable set of circumstances no one could be prepared for and I can still see Kate trapped in the flames, her hands and back burning.'

It is this image which, thirty years on, still causes Ash such distress. 'A few months ago I was at a business dinner with new

colleagues. They discovered I was a former fast-jet pilot and, as so
often happens, asked me if I'd ever ejected. When they heard I'd
ejected twice, they wanted a bit of detail. I began telling them about
the incident with Kate, but as I described seeing the wall of flames
in the field, finding her on fire, I simply broke down in tears. I was
in the middle of dinner, sobbing my heart out, struggling to draw
breath in front of new business partners I had only just met that
night! I was mortified. I eventually managed to regain my composure and carry on, but they must have thought I was crazy.'

Ash is far from alone in discovering that the post-ejection experience could be as traumatic as the event itself. 'The ejection seat
had saved our lives,' Ash says. 'But it really was only the beginning
of our shared story. We were both so lucky to survive the whole
incident; the bird strike, the ejection, the inferno. In spite of all the
suffering, we are just lucky to be alive.'

* * *

Lightning pilot Bill Schaffner had not been similarly blessed. His
sons Mike and Glennon had given up all hope of ever finding
out what had happened that pitch-black night in 1970 when his
Lightning jet had crashed into the North Sea off Flamborough
Head. Cockpit firmly closed, ejection seat still in place, harness
unstrapped and no sign of their father. All the brothers had heard
were the ridiculous stories from UFO hunters and conspiracy
theorists.

Then, out of the blue, thirty-two years later, Mike's phone rang
in his Chicago home. 'It was someone from the BBC telling me they
were going to investigate the incident and establish the truth once
and for all. And they would fly us to England to follow the story;
here at last was the chance to get some answers.' Bill's widow Linda
was pleased that the boys were going back to see where the tragedy
had unfolded. 'It was a chance to walk on the same ground as their
father and see if they could find closure, especially for Mike who
only knew Bill through pictures and occasional stories.'[8]

The BBC was putting pressure on the Ministry of Defence to release the accident report, but were initially told it had been shredded. While journalists pushed hard at the MOD, the TV crew took the brothers to their father's old base at RAF Binbrook in Lincolnshire. They stood outside their home on Cambridge Crescent. 'It's weird, it's what I remember dreaming about ...' Glennon gazed up at what had once been his bedroom window. 'I can still picture the backlit front door, my mother opening it and the two officers standing there. Then her collapsing as she was told our dad was missing.'

The airbase looked very different from when their father had been stationed there. Abandoned by the RAF a decade earlier, it was used briefly to film scenes for the war film *Memphis Belle* and then left to rot. The Schaffners picked their way through waist-high grass. Gutted, derelict buildings, some reduced to burned out shells, were piled high with elderly computers and TV sets. But Glennon still felt 'an intense connection to our father's life and our own past. We were going back in time to the world where he had lived and worked.'

A silver Lightning jet fighter towered above the brothers at a nearby aviation museum. 'Holy mackerel,' Glennon gasped. 'It's huge,' Mike echoed. He climbed the ladder and lowered himself into the cockpit, as his father had done so many times. He took a couple of slow breaths to steady himself. 'It was incredibly moving to sit on that ejection seat, looking at the controls, and to finally feel that connection with the father I never knew but still mourned for.'

Glennon had also spent years trying to make those connections. 'I closed the cockpit and let my mind roam free.' His hands flitted around the aircraft controls, touching the control column and throttles the way his father would have done. 'I looked at the yellow and black ejection handle and felt what it was like for him to sit in that cramped space. I tried to work out what went through his mind on that fateful mission. When did he realise he was in mortal danger? When had it all been over for him? Why hadn't he ejected? It was intensely emotional.'

Every detail was precious as they tried to assemble a portrait of their father's world.

The brothers then headed with some trepidation to London and Whitehall. 'Our cab driver asked where we were going,' Mike says. 'We explained why we were off to a meeting at the Ministry of Defence; to find out what had happened to our father. He said, "YOUR father was THAT Lightning pilot?" He had read all the stories and wanted to check that we understood how much the UFO stuff was nonsense. A British taxi driver had known more about our American father than we did when we were growing up.'

Thanks to pressure from the BBC, the MOD had managed to find the supposedly shredded accident report, including the real transcript of their father's final conversation with ground controllers, and the photographs of the aircraft and the empty cockpit. The other 'UFO' transcript that had been doing the rounds for all those years, with its utterly manufactured conversations and ludicrous descriptions of floating lights, was finally exposed for what it was: a total fake.

After years of official secrecy, the RAF's official Aircraft Accident Report was handed to his sons by an RAF Wing Commander. 'Tragic though it was,' he told them, 'it was an accident. There is no other logical explanation.'[9]

In the austere and imposing confines of the MOD, a stone's throw from Downing Street, they read that a string of operational and procedural errors had led to their father's flight to oblivion. There was no single root cause. The report said:

> His lack of training led to a situation where he failed to monitor the height of his aircraft while slowing down and acquiring his target. And he had inadvertently flown his aircraft into the sea.[10]

A battle-hardened combat pilot had made a simple error while flying a new aircraft on a complex night sortie and it had cost him his life. He would not be the first, or the last. But why had Bill not ejected?

Why was his seat in the cockpit while he was not? The explanation was tragically simple.

The inquiry had revealed that the 'lower ejection handle had been pulled to the full extent'. But a tiny, misaligned screw in the separate hood-firing mechanism meant that its explosive charge failed to detonate. To avoid fatal head injuries, the seat was designed not to eject if the canopy was still in place. There was no problem with the design of the seat itself, but it was never going to blast Bill to safety. A simple engineering error, not dissimilar to the one that had prevented Red Arrows pilot Sean Cunningham's seat from operating successfully, resulted in the death of a husband and father.

'Crashing into the sea at 180mph would probably have resulted in severe injuries,' Bill's friend and former colleague Chris Coville says. 'He had obviously tried to eject at some point and failed. But had then clearly unstrapped from his ejection seat, managed to open the cockpit canopy and struggle out as the jet initially floated on the water.' Schaffner had also unclipped himself from his personal survival pack, which would still have been secured into his ejection seat. Its lanyard was dangling over the side of the fuselage when the wreckage was raised to the surface.

So why was the jet later discovered with the canopy fully closed? Didn't *that* point to some outside interference? No. After Schaffner had escaped from his cockpit the aircraft had begun to sink. The increasing water pressure and diminishing hydraulic supply had forced the canopy shut again as it descended.

According to Chris Coville, there was only one answer to the mystery of his friend's disappearance. 'Incapacitated and without his survival dinghy, Bill almost certainly drowned in the cold North Sea. It was sheer bad luck that several elements combined on that fateful night to take him to his death. In spite of his bravery and determination to survive, Captain William Schaffner had finally run out of luck.'

For Glennon Schaffner, 'discovering that my father's ejection seat hadn't worked because of a servicing error was a shock. The tiniest detail, the smallest piece of equipment – a damaged screw thread – meant that the canopy jettison unit had been incorrectly aligned and therefore failed to fire. If it had been serviced correctly, my father might still be alive today, but I don't blame any single person.'

No culprit was ever identified but, as the report says, the MOD acted quickly.

> The servicing procedures for the inspection, re-arming and servicing of cockpit canopy firing units have been amended. All ejection firing units of a type that prevented ejection in this accident have been inspected for signs of damage.

Just as well. Glennon later discovered that 'other Lightnings were found to have exactly the same issue, which was hurriedly rectified. There is no doubt that our father's death, and the discovery of the fault, meant other lives were saved. At least that's something we can hold onto.'

He and Mike smiled as they leafed through the report. 'It's such a relief to finally see this, and the pictures of our dad's aircraft, and to know what happened.' They finally had most of the answers they had craved for so long.

They called their mother to tell her the news.

'I look back on my life with Bill with joy and sadness, but mostly pride,' she says. 'Losing my husband in such tragic circumstances was truly terrible, but our lives together in the military had been wonderful. I still think about him, and wonder what might have been if he had lived. Where would we be now? Where would I be? I loved being in that military family, and I was proud to belong. They gave us the best of times, but, in the end, the very worst of times. I wouldn't wish my personal experiences on anyone, but I don't think I'd change them either. You can't change the past, can you?'

Mike and Glennon had one final pilgrimage to make. Travelling north, they stood side by side on the Yorkshire coast, staring out over the calm, grey-green sea off Flamborough Head. The place their father had died, and where his remains almost certainly still rested. Gentle waves lapped over a beach of polished rocks. It was hard to imagine anybody coming to grief there. 'There was a large boat a mile or so out to sea,' Glennon said. 'It climbed and dipped in the swell, disappearing from sight, then rising to appear again. That image really made me think of my dad, injured, unstrapping from his ejection seat and climbing from his Lightning out there on a stormy night. What must he have had to endure? He really didn't stand a chance.' He brushed away a tear with the back of his hand.

Mike grimaced, holding back his own tears. The brothers walked up to the rugged cliffs and sat down on the grass not far from the gleaming white Flamborough Lighthouse, looking out over the North Sea. 'I've waited my whole life for this.' Mike's long-held emotions poured out. 'I don't have any memory of my dad and I never will. This is as close as I'm ever going to get. But at least I finally made it. It was an ignominious way to die, and as a child, because Dad was rarely spoken of, I felt that he had been forgotten after his death, just a footnote in military history.'

A lone seagull soared over the slow rhythm of the waves, only a beat away from the dark night tide that had captured their father. 'It was all quite sad; somewhere out there, my father's life had ended,' Mike reflected. 'But it was also uplifting, a chance to say a few words, a sense of saying the "goodbye" I'd never had as a child. And it marked the end of our journey of discovery; our father had been a fairy tale, now he was real.'

Back home in Chicago, Mike describes the conclusion of their thirty-two-year quest to the author. 'Fifty-two years after he died, I still have a Dad-shaped hole in my life, but now I know why. I still dream about him, but because I never knew him, in my dreams he never moves or talks. He is just a shadow. I still miss him every day. He died doing what he loved, and I am grateful for that. I have told my wife that when I die, my final wish is that she takes my ashes to Flamborough Head in England and throws them into the North Sea. I want to be back with my dad, in his final resting place.'

* * *

On Tuesday, 31 May 2011, Dale Zelko was heading for a very different kind of redemption. In high spirits, he had woken the children to say goodbye. Four bags were sitting in the hallway, zipped up and ready to go. He checked his passport and tickets one more time. His destination: Belgrade. At last, he would meet Zoltán Dani, the man who had blasted him out of the sky.

'He's very excited.' Lauren smiled nervously. 'We're all a little "not" excited. We don't like it when he goes on trips. We miss him.' She had never forgotten being told he had been forced to eject, and the hours not knowing whether he was alive or dead. She could still see those TV pictures of jubilant Serbs dancing around the wreckage of Dale's crashed jet. And now he was going back there. Lauren knew it was something he had to do. But she was fearful.

'I am driven to go,' Dale said. 'I *need* to meet Zoltán under entirely different circumstances.'

This time he would come to Serbia bearing gifts and good wishes rather than precision bombs. He wanted to close one chapter and hopefully open another with the man who had tried to kill him. Only they could really understand what had brought them together. They were two sides of the same coin.

Zoltán now ran a successful bakery business in the village where he grew up, a few miles from his home in Kovin, east of Belgrade. He had a talent for making paper-thin filo for apple strudel alongside the loaves and chocolate cakes. Belgrade was a bustling city, but here in the countryside donkeys still clopped along the cobblestones, carrying farmers' produce to market.

Zoltán and his family had tidied their house in readiness for Dale's visit. Not a speck of dust to be seen. His son, Atila, had given up his bedroom for their special guest. Zoltán climbed a ladder to sand and paint the basketball hoop and backboard mounted on the garage, in case his American visitor wanted a game.

But what kind of a man was Dale Zelko?

As Zoltán rubbed away flaking paint, images of the moment he had fired his missiles flooded back. He knew they were probably not the same men they had been that cold March night in 1999. When the documentary had first been suggested, he'd had some misgivings. Did he really want to dig up the past, a world he had turned his back on? But Atila had won him round. His son realised how important it was that the two men met to try to lay their past to rest.

'Are we near Kovin?' Dale asked as they sped along the motorway. 'I'm starting to get this swell of anxiety in my chest.'

'Yes, we're near,' his driver replied. 'Very near.'

'Oh boy, is he really going to be there, in his bakery?'

Right on cue, the sign to Kovin came into view, and the car headed off towards Zoltán's bakery, on a tree-lined, residential road, with the occasional tractor parked alongside the neat verge.

To calm his nerves, Dale broke into a somewhat tuneless version of 'Oh What a Beautiful Morning'.

'Fifty metres,' the driver said, flicking on his indicator.

They came to a halt in front of a beige brick building with a red-tiled roof.

'Really? I'm really here?'

The driver confirmed that indeed he was.

This was it. Dale hauled himself out, patted down his red sweatshirt and blue jeans, and walked through a side gate into the goods yard. He felt a smile creasing his face as he peered around an open door.

'Hello, Colonel,' he said, saluting the man in chef's whites and a 'Bekeri Dani' t-shirt.

Zoltán saluted right back. 'Hello, Zelko.'

The two former officers shook hands formally, then threw their arms around each other.

'So, this is my bakery,' Zoltán said.

'I can see that; you've got flour on your face.'

Zoltán laughed as Dale brushed it off his nose with a finger, then gave his guest a quick lesson in filo making. Dale watched the expert handle the dough like it was fine silk. He himself had never baked anything in his life. But under the master's careful tutelage and now kitted out in a matching white cap and apron, he was soon rolling and cutting with confidence.

'Very good touch,' Zoltán told him. 'This is baker Zelko and baker Dani.'

'Yeah!' Dale said. 'I would like that.'

Later, the two went in search of the place Dale had spent the dark hours hoping to be rescued before the Serb military hunted him down. Clutching a map, he checked the landmarks. 'We're close, very close . . .'

Zoltán pulled over and the two men clambered up the embankment. Dale pointed across a field. 'I came down close to that railroad track, right over there.'

They slid down into a field of young maize. Everything was coming back into focus. He recognised the solitary cherry tree where he had initially hidden, now in full leaf. At the field's edge he led the way, pushing through shrubs and branches, and peered into a drainage ditch. 'This is it,' he said triumphantly. 'Right here. I can't believe it. This is where I stayed for the rest of the night.'

'You stay here *seven hours*?' Zoltán couldn't hide his astonishment. 'Are you sure this is where it was?'

'This is absolutely the place. How can I forget? It was the seven most intense hours of my life.'

Next stop was the Belgrade Museum of Aviation. Walking in, Dale did a double-take. 'Hey!' he said in mock outrage. 'That's my stuff!'

Right in front of him, in pride of place, was what was left of his multimillion-dollar Nighthawk Stealth jet.

In among a wing, sections of wrecked fuselage and a smashed cockpit canopy was the helmet he had buried in the field after he landed. 'Seeing my helmet, such an incredibly personal item which had protected me in flight for so many years, sitting amid the debris, was really emotional.' Even more poignant was the battered ejection seat. 'A great piece of technology that had saved my life. I truly felt grateful to be looking at that seat and helmet alongside Zoltán. It was another one of those moments where I realised how blessed, how lucky I was to be alive.'

He pointed to one of the many shrapnel holes in his F-117. 'Very nice shot, Zoltán. That was well done.'

Exploring the city NATO had bombed, Dale told his former enemy, 'I'd like to thank you for bringing me here. It is very important for me that I connect as closely and deeply as possible with the Serbian people.'

Zoltán Dani, the missile battery commander turned master baker, agreed. 'Both you and I were given a second chance and created a new road for friendship.'

When they arrived at Zoltán's home for a lunch Irena had been preparing all morning, Dale stopped by the house number to the right of the front door and traced the figures 1-1-7 with his fingertip.

Zoltán grinned. 'We should paint an "F" next to it.'

Inside, Irena and Atila were waiting to greet him. In a smart skirt and top, she had been to the hairdresser's and was wearing her best pearls. 'Welcome in our house,' she said simply. Although she only had a few words of English there was no doubting her warmth. Any animosity towards American pilots was a thing of the distant past. She could see how much this meeting meant to both men.

Dale put his arm around her and kissed her cheek.

After eating, Dale gave them the gifts he had brought from America. Zoltán convulsed with laughter when ceremoniously presented with a scale model of an F-117 Nighthawk.

Dale grinned. 'Now try not to blow this one up, okay?'

Reaching into his bag again, he turned to Irena. 'I have something for you from my wife.' He unfurled a beautiful 5-foot-square quilt, a series of concentric red squares on a cream background with a border of flowers and a red star at its centre, then read from Lauren's letter.

When Dale travelling to Serbia to meet Zoltán became a reality, I decided I would like to make this for you as a gesture of friendship, to a woman I've never met, halfway across the world, but who is connected to me through the shoot-down of my husband's aircraft.

Lauren had spent months on the quilt, often stitching the fabric into the early hours.

The star on your quilt represents the night Dale was evading on your soil. Many people were searching for him and converging on him. There were Serbian military, citizens and United States rescue services.

Dale had to pause to compose himself.

The star on your quilt also reflects a sense of patriotism. Dale and I both served in our nation's Air Force because we love our country. I know that Zoltán and your family will also feel pride in your nation. From this military event Dale and Zoltán have become well known, so the star also represents them across the lines of war in which a friendship has been born. Hence, your quilt's name: Many Roads Lead to Friendship. I hope you will accept my offer of friendship and peace and that the quilt will warm you not only physically, but in your heart.[11]

Irena was visibly overcome, as was Zoltán. She took Dale's hand. 'Please give my best regards to your wife. What she has woven into this quilt is what I believed in those difficult moments. We are always beside you. I'm beside my husband and your wife is beside you. Let this be the beginning of friendship and peace.'

Dale Zelko returned to Serbia in 2012 for the premiere of the film documentary *The Second Meeting*. He hoped that all those attending would take away the conviction he now held dear. 'It may sound trite, but if only there were a way for all the religious, cultural and ethnic groups of the world to meet and get to know one another in a meaningful way – the way Zoltán and I have – how could we ever go to war again?'

Unusually booted and suited, the two comrades sat together during the screening, witnessing their dramatic life stories collide before an astonished audience. At one point, they held hands. 'It was a really strange experience,' Dale says. 'The story we watched was about one of the key moments of my life. A time I had faced deadly danger in the truest sense of the word, and an event I am still surprised, thanks to the ejection seat, that I survived.'

During the Q&A that followed, one woman in the audience told Dale, 'When you were shot down, I celebrated. I cheered with my friends. You had attacked our country and we were upset that you were not killed. We thought you deserved to die.'

'You can imagine the hush in the audience,' Dale remembers. 'Then she said, "But now that we have got to know you, I'm so glad that you are still here. That you survived your ordeal and were able to come back to visit us."' Dale wept. He is close to doing so again as he tells the story. 'It was one of the most astonishing, life-affirming moments of my life. I had been given that second chance at life, and her sense of humanity, of forgiveness, of togetherness, was so wonderful to hear.'

* * *

In his iconic book *The Right Stuff*, detailing early test-flying of high-speed aircraft, Tom Wolfe wrote: 'In the era of jet fighters, ejection meant being exploded out of the cockpit by a nitro-glycerine charge like a human cannonball, and hoping that the goddamned human cannonball trick works.'[12]

The ejection seat – a 'curious contraption' so many early flyers were deeply suspicious of – has more than proved its value to many thousands of grateful aviators and their loved ones, often marking the beginning of an incredible journey of discovery and redemption which echoes across the generations.

Speaking at ejection seat pioneer Sir James Martin's memorial service, test pilot Bill Bedford said, 'As one of those whose lives have been saved by his genius, I can say from my very heart: never in the history of aviation has the safety of aircrew and the happiness of their families owed so much to one man. Thank you, Jimmy.'

Bill spoke for all aircrew – especially the author – whose lives had been saved by that incredible 'goddamned human cannonball trick'.

Thank you, Jimmy.

ENDNOTES

NB: JN = John Nichol

Foreword

1 With limited information from the USA, and no information from Russia or China, this figure is merely an estimate. However, in 2018, the combined US military air forces alone were over twenty times the size of the RAF.

Prologue: 'It Will Never Happen to Me'

1 Jo Lancaster's account constructed from JN interview; and 'Setting the Record Straight', Jo Lancaster, *Aeroplane* magazine, October 2006. Some quotes edited for clarity.
2 'Time Machine: Too Far, Too Fast', *Airscape* magazine, 18 December 2014.
3 *Flight* magazine, 14 January 1948.
4 'Test Pilots', Tim Bracey, RAF Museum website.
5 'This Day in Aviation', 27 September 1946: https://www. thisdayinaviation.com/
6 David Gunby, *First Out in Earnest: The Remarkable Life of Jo Lancaster DFC* (Fighting High Ltd, 2016).

7 The description of Jo's seat is compiled from: Sarah Sharman, *Sir James Martin: The Authorised Biography of the Martin-Baker Ejection Seat Pioneer* (Patrick Stephens, 1996); John Jewell, *Engineering for Life: The Story of Martin-Baker* (Martin-Baker Ltd, 1979) and Martin-Baker Mk1 factsheet. There are some discrepancies in the various accounts. Jo himself described a 'pre-production Mk1 seat'.
8 In fact, there had been a couple of previous ejections in other countries, but in the UK nothing was known about these at the time.
9 'The rocket-powered rise of the ejector seat', Paul Marks, 21 May 2015: https://www.bbc.com/ future/article/20150521-the-rocket-powered-life-saving-seat

Chapter One: A Shaky Do

1 Jo Lancaster personal account and JN interview. Gunby, *First Out in Earnest*, op. cit.
2 Jo Lancaster Pilot's Flying Logbook 6/7/37–15/9/43.

3 'Neutralise the Threat', *Britain at War* magazine, April 2012.

4 Chaz Bowyer, *Wellington at War* (Ian Allan Ltd, 1982).

5 Jo was carrying a second pilot that day.

6 Gunby, *First Out in Earnest*, op. cit.

7 The *London Gazette*, 24 July 1917: www.thegazette.co.uk/london/issue/30204/supplement/7622

8 Valentine Henry Baker: www.rainydaygallery.co.uk/valentinebaker.html; 'The Llanfairfechan pilot who taught a future king to fly', 9 January 2015: https://www.dailypost.co.uk/news/north-wales-news/llanfairfechan-pilot-who-taught-future-8409808

9 Jewell, *Engineering for Life*, op. cit.

10 Sharman, *Sir James Martin*, op. cit.

11 Jewell, *Engineering for Life*, op. cit.

12 Sharman, *Sir James Martin*, op. cit.

13 'Punching Out', Robert E. van Patten, *Air Force* magazine, March 1995. There are a number of historical references to 'the first' escape system. This is just one.

14 https://www.nytimes.com/2012/12/13/opinion/100-75-50-years-ago.html

15 Schenk's account is constructed from a number of sources: Bryan Philpott, *Eject! Eject!* (Motorbooks International, 1989); Jim Tuttle, *Eject!: The Complete History of US Aircraft Escape Systems* (Motorbooks International, 2002); 'Ejection Seat Works, Pilot Elated', *Wired* magazine, January 2001. There are some discrepancies regarding certain actions and events; resolved here as far as possible.

16 The Germans experimented with several types of ejection seat – or *Schleudersitzapparat*, which translates as 'seat catapult device'. The one Schenk used was activated by compressed gas, another relied on a spring-operated mechanism, and a third used a propellant charge.

17 Tuttle, *Eject!*, op. cit.

18 Philpott, *Eject! Eject!*, op. cit.; Doddy Hay, *The Man in the Hot Seat* (Collins, 1969).

19 Sharman, *Sir James Martin*, op. cit.; William Douglas Bow Symington Davie, Mearns History Group: www.mearnshistory.org.uk; 'The rocket-powered rise of the ejector seat', Paul Marks, op. cit.

20 Hay, *The Man in the Hot Seat*, op. cit.

21 Sharman, *Sir James Martin*, op. cit.

22 Jewell, *Engineering for Life*, op. cit.; Sharman, ibid.

23 Sharman, ibid.

24 Jeanne Cavelos, *The Science of Star Wars* (St Martin's Press, 1999); *Scientific American* magazine, 11 August 2008; 'Acceleration in Aviation: G-Force', Federal Aviation Administration website: www.faa.gov/pilots/safety/pilotsafetybrochures/media/acceleration.pdf

25 Hay, *The Man in the Hot Seat*, op. cit.

26 Tuttle, *Eject!*, op. cit.; Operation LUSTY factsheet, National Museum of the United States Air Force: www.nationalmuseum.af.mil

27 Data provided by Air Commodore Rick Peacock-Edwards, a former officer who commanded the RAF's Flight Safety organisation.

Chapter Two: A Curious Contraption

1 Lynch's account is constructed from a number of sources: Sharman, *Sir James Martin*, op. cit.; Philpott, *Eject! Eject!*, op. cit.; Hay, *The Man in the Hot Seat*, op. cit. There are some discrepancies regarding certain actions and events; resolved here as far as possible.

2 Royal Aeronautical Society
Heathrow Branch November
2016 Lecture, 'The History
and Development of Martin-
Baker' by Tony Gaunt, Business
Development Executive at
Martin-Baker Aircraft Company
Ltd: https://www.youtube.com/
watch?v=XFzWAM5uWxQ&ab_
channel=RAeSHeathrowBranch

3 Described by Andrew Martin
in: https://www.youtube.com/
watch?v=X1MBluWD7NI

4 Quoted in Hay, *The Man in the Hot
Seat*, op. cit.

5 Early test seats also had another
parachute fitted to aid in the
recovery of the seat itself for
examination. Once in production,
only the drogue parachute was
fitted.

6 Martin, op. cit.

7 Philpott, *Eject! Eject!*, op. cit.

8 'Ejection Seats Improving the Breed',
J.C. Edwards, *Aerospace Safety*
magazine, January 1968.

9 Sharman, *Sir James Martin*, op. cit.

10 Tuttle, *Eject!*, op. cit.; Sharman,
ibid.

11 Sharman, ibid.

12 Ibid.

13 Ibid.; Hay, *The Man in the Hot
Seat*, op. cit.; Philpott, *Eject! Eject!*,
op. cit.; Peter Page *Guardian*
obituary by Barry Page: www.
theguardian.com/world/2014/feb/16/
peter-page-obituary

14 There is no further information on
Mr Keyes or his background.

15 Sharman, *Sir James Martin*, op. cit.

16 Jo Lancaster, JN interview.

17 Royal Aeronautical Society
Heathrow Branch November 2016
Lecture, op. cit. Although some of
the seat-testers like Benny Lynch had
the process down to around eight
seconds, they were experienced and
had rehearsed the sequence many
times.

18 'The "Flying Wing" Crashes', Alan
Griffin, Our Warwickshire website:
https://www.ourwarwickshire.
org.uk/content/article/
flying-wing-crashes

19 It is difficult to see how the ejection
seat had not already fallen below
Jo – but this is his recollection of the
incident.

Chapter Three: All or Nothing

1 'Two Chute Out at 555mph, Tell
About It!', *Daily News* (New York),
18 July 1949.

2 *Aerospace Safety* magazine, January
1968.

3 There are some conflicting reports
about the first American to eject in
an emergency. Farley's story is taken
from 'Ejected to Safety', *Flying
Safety* magazine, November 1949,
which claims he was the first.

4 Hawker P1081: www.baesystems.
com/en-uk/heritage/hawker-p1081

5 'Medals of Battle of Britain Spitfire
pilot set to fetch £70,000 at auction',
Daily Mail, 26 November 2021.

6 Accident Report – Hawker P1081
VX279, 3 April 1951: https://
aviation-safety.net; 'Cause of
Ringmer Plane Crash Still a Mystery
Inquest Report', *Sussex Agricultural
Express*, 11 May 1951.

7 Figures provided in document
'Military Aircraft Accident
Statistics' supplied by Air
Commodore Rick Peacock Edwards.

8 Sharman, *Sir James Martin*, op. cit.

9 *Swift to Destroy: An Illustrated
History of the 77 Squadron RAAF
1942–2012*, No 77 Squadron RAAF
Association Inc.

10 Ron Guthrie's story based on
JN interview; Col King and Ron
Guthrie, *Escape from North Korea:
The Ron Guthrie Story* (Meteor,
2002); interview with Ron Guthrie

at the Australian War Memorial, June 2001: www.awm.gov.au. Some quotes edited for clarity.

11 Igor Seidov, *Red Devils Over the Yalu: A Chronicle of Soviet Aerial Operations in the Korean War 1950–53* (Helion & Co, 2013). Some quotes edited for clarity.

12 www.biography.com/personality/casey-jones

13 Seidov, *Red Devils Over the Yalu*, op. cit.

Chapter Four: Simple Solutions

1 Tony Blackman and Anthony Wright, *Valiant Boys: True Stories from the Operators of the UK's First Four-Jet Bomber* (Grub Street, 2014) – some quotes edited for clarity. And Gabe 'Jock' Bryce obituary, *The Scotsman*, 2 June 2014.

2 *Christchurch Times*, 18 January 1952: http://www.royhodges.co.uk/Bomber%20Crash%20Bransgore%201952.pdf

3 Sharman, *Sir James Martin*, op. cit.

4 Ibid.

5 Ibid.

6 Tuttle, *Eject!*, op. cit.

7 King and Guthrie, *Escape from North Korea*, op. cit.; Ron Guthrie, JN interview. Quotes edited for brevity and clarity.

8 Sharman, *Sir James Martin*, op. cit.; 'Ben' Gunn – Test pilot for Boulton Paul at Defford: https://deffordairfieldheritagegroup.wordpress.com/

9 Sharman, ibid.

10 Philpott, *Eject! Eject!*, op. cit.

11 Royal Australian Air Force Association, New South Wales Division newsletter, October 2021. Edited for clarity.

12 Sharman, *Sir James Martin*, op. cit.; Tuttle, *Eject!*, op. cit.

13 'I Made the First Jump', Colonel Arthur M. (Chic) Henderson, *Popular Mechanics*, March 1955; Philpott, *Eject! Eject!*, op. cit.

14 Quote edited for clarity.

15 Sharman, *Sir James Martin*, op. cit.

16 Jewell, *Engineering for Life*, op. cit.

17 Sharman, *Sir James Martin*, op. cit.

18 'George Smith becomes 1st person to survive supersonic ejection', Super Sabre Society, 2022; 'This Day in Aviation': www.thisdayinaviation.com/26-february-1955/; Tuttle, *Eject!*, op. cit.

19 There are a number of incidents described as 'the first supersonic ejection'.

Chapter Five: A Second Chance at Life

1 Tony Blackman, *Vulcan Test Pilot: My Experiences in the Cockpit of a Cold War Icon* (Grub Street, 2019); Tony Blackman, *Victor Boys: True Stories from Forty Memorable Years of the Last V Bomber* (Grub Street, 2009); Inside the Cockpit of an Avro Vulcan: https://www.youtube.com/watch?v=NFGkZPPNrkA; Jet Age Museum (University of Gloucestershire) Avro Vulcan Video Cockpit Experience: https://www.youtube.com/watch?v=lz7fqk9YiG8

2 Aviation Safety Network, Flight Safety Foundation: www.flightsafety.org; Vulcan Air Crash, British Pathé: www.youtube.com/watch?v=m16xTQmIzCM&ab_channel=BritishPath%C3%A9; Vulcan Aircraft Crash (Report) Hansard, 20 December 1956; Hay, *The Man in the Hot Seat*, op. cit.

3 The debate surrounding rear ejection seats is constructed from: 'The V-Bomber Ejector Seat Story': https://h2g2.com/edited_entry/

A49097307; and the thoughts of
Doddy Hay in his autobiography,
and Sharman, *Sir James Martin*, op.
cit.

4 By the date of the Heathrow Vulcan
crash, Martin-Baker seats alone had
saved 160 lives.

5 Sharman, *Sir James Martin*, op. cit.;
Hay, *The Man in the Hot Seat*, op.
cit.

6 Hay, ibid. Some quotes edited
for brevity. Not all of Doddy's
recollections marry exactly with
some other accounts.

7 Sharman, *Sir James Martin*, op. cit.

8 Hay, *The Man in the Hot Seat*, op.
cit.

9 Aviation Safety Network Handley
Page Victor B2 XM714, 20 March
1963: https://aviation-safety.net; 'V-
Bomber Crash at Barnack': http://
stamfordlocalhistorysociety.org.
uk/v-bomber-crash-barnack

10 Quoted in 'The V-Bomber Ejector
Seat Story', op. cit., edited here for
clarity. The discussions surrounding
rear-crew seats went on for a
number of years. This is a brief
synopsis.

11 Hay, *The Man in the Hot Seat*, op.
cit.

12 The Martin-Baker company has
a very accurate database of all its
ejections, but there are limited
details concerning the multitude of
different US systems developed, and
almost no details from the likes of
the Soviet Union, China and beyond.

13 These conversations are reported in
Doddy Hay's autobiography. Some
sequences have been condensed for
brevity.

14 Sharman, *Sir James Martin*, op. cit.;
Hay, *The Man in the Hot Seat*, op. cit.

15 These are Doddy's own figures
which sometimes differ from other
sources.

16 Sharman, *Sir James Martin*, op. cit.;
Hay, *The Man in the Hot Seat*, op. cit.

17 Sharman, ibid.

18 'Convair B-58 HUSTLER
supersonic ejection tests':
https://www.youtube.com/
watch?v=-KLnqorLgDM&ab_
channel=AVhistorybuff

19 Guy Gruters' account compiled from
JN interview; and his book, *Locked
Up with God* (Guy Dennis Gruters,
2012).

20 https://www.mirror.co.uk/news/
world-news/bears-were-fired-out-of-
us-supersonic-1501169

21 Sharman, *Sir James Martin*, op. cit.;
Hay, *The Man in the Hot Seat*, op.
cit.

Chapter Six: Eject in Time

1 Guy Gruters' account is compiled
from JN interview; and Gruters,
Locked Up with God, op. cit.;
Malcolm McConnell, *Into the
Mouth of the Cat: The Story of
Lance Sijan, Hero of Vietnam* (W.W.
Norton & Company, 1985); 'A
Day in the Life of the Misty FACs',
Rick Newman and Don Shepperd,
Air Force magazine, 2006; 'The
Courage of Lance Sijan', John T.
Correll, *Air Force* magazine, July
2004.

2 Sandy Gruters, JN interview.

3 Mari's full name is Dawn-Marie; she
now goes by 'Mari'.

4 Bert Kinzey and Rock Roszak, *F-
100 Super Sabre in Detail & Scale*
(Detail & Scale, 2020).

5 'Trends in US Air Force Aircraft
Mishap Rates 1950–2018', RAND
Corporation: https://www.rand.org/
content/dam/rand/pubs/research_
reports/RRA200/RRA257-1/
RAND_RRA257-1.pdf (Those
figures do not include F-100s flown
by the US Navy and Marines.)

6 https://www.youtube.com/
watch?v=QbgeUNjsenA

7 B-58A Escape Capsule: www.
   ejectionsite.com; National Museum
   of the United States Air Force:
   www.nationalmuseum.af.mil/
   Visit/Museum-Exhibits/Fact-
   Sheets/Display/Article/196439/
   convair-b-58a-hustler/
8 Vietnam War: Causes, Facts &
   Impact: https://www.history.
   com/topics/vietnam-war/
   vietnam-war-history
9 'The Inventor', *Redlands Daily
   Facts*, 10 December 2007:
   https://www.redlandsdailyfacts.
   com/2007/12/10/the-inventor/;
   'Survival Package Deal', W.C.
   Thomas, Naval Air Systems
   Command, *Naval Aviation
   News*, February 1968, pp. 26–7;
   Tuttle, *Eject!*, op. cit.; General
   Dynamics F-111 Aardvark Crew
   Module Escape Ejection System
   8044: https://www.youtube.com/
   watch?v=F7O5lL8gPwo&ab_
   channel=PeriscopeFilm; Philpott,
   Eject! Eject!, op. cit.; https://www.
   key.aero/article/f-111-capsule-
   ejection-system-explained
10 http://f-111.net/Page-9.html
11 https://www.warhistoryonline.com/
   war-articles/he-managed-to-convey-
   the-truth-by-blinking-torture-in-
   morse-code-during-the-vietnamwar.
   html
12 Charlie Neel, JN interview.
13 https://www.archives.gov/
   research/military/vietnam-war/
   casualty-statistics
14 https://www.presidency.ucsb.edu/
   documents/annual-message-the-
   congress-the-state-the-union-28
15 Dawn-Marie 'Mari' Williams (Guy
   Gruters' daughter), JN interview.
16 Proclamation 3819 – Thanksgiving
   Day, 1967.
17 https://www.stripes.com/
   from-the-archives-thanksgiving-in-
   vietnam-1967-1.254830

Chapter Seven: The Heart of Darkness

1 Guy Gruters' account compiled from
   JN interview; and his book, *Locked
   Up with God*, op. cit.
2 US National Archives Military
   Records, Vietnam War US Military
   Fatal Casualty Statistics: https://
   www.archives.gov/research/military/
   vietnam-war/casualty-statistics
3 Ejection Seat Alternative –
   The Yankee Escape System:
   https://www.youtube.com/
   watch?v=8Yw8g1Soigk&ab_
   channel=WhatYouHaven%27tSeen –
   edited for clarity.
4 Bill Bagwell, *Plagued by Good Luck*
   (Naked Fanny Publishing, 2011);
   Tuttle, *Eject!*, op. cit.
5 Quote from another user of the
   system.
6 The Code of the US Fighting Force
   is a code of conduct that is an
   ethics guide and a United States
   Department of Defense directive
   consisting of six articles to members
   of the United States Armed Forces,
   addressing how they should act
   in combat when they must evade
   capture, resist while a prisoner or
   escape from the enemy: https://
   www.army.mil/values/soldiers.html
7 Lance Sijan's story, along with that
   of Bob Craner, compiled from: Guy
   Gruters, JN interview; Gruters,
   Locked Up with God, op. cit.;
   McConnell, *Into the Mouth of the
   Cat*, op. cit.; 'The Courage of Lance
   Sijan', Correll, op. cit.

Chapter Eight: There But for the Grace of God

1 Charlie Neel's account based on JN
   interview.
2 Alan White, *Lightning Up: The
   Career of Air Vice-Marshal Alan*

White (Pen & Sword Aviation, 2009).

3 'What Was the Cold War?', *National Geographic*, 22 March 2019 https://www.nationalgeographic.com/culture/article/cold-war

4 English Electric Lightning Supersonic interceptor and jet fighter capable of unrivalled capability during the Cold War era: www.baesystems.com

5 Sir Christopher Coville, JN interview; and his autobiography, *Fighter Pilot: From Cold War Jets to Spitfires – The Extraordinary Memoirs of a Battle of Britain Memorial Flight Pilot* (Air World, 2021).

6 Armament depended on the variant: https://www.baesystems.com/en-uk/heritage/english-electric-lightning

7 Sharman, *Sir James Martin*, op. cit.

8 10kg (22lb) is an approximate weight and depended on the equipment contained and the mark of ejection seat.

9 This is an approximate figure based on Lightning aircrew recall of costs in the absence of officially published figures.

10 'Out of Harm's Way', Michael G. McDonell, *Naval Aviation News*, October 1970.

11 'The Complicated Mechanics Behind the Ejection Seat', Kevin V. Brown, *Popular Mechanics* magazine, June 2021; 'Ballon + Chute = Ballute', Douglas Garr, *Popular Science* magazine, April 1972; and Philpott, *Eject! Eject!*, op. cit.

12 www.pow-miafamilies.org

13 Walter J. Cronkite was anchorman for *CBS Evening News* for nineteen years, from 1962 to 1981.

14 Letter from Charlie Neel to William Schaffner, 31 October 1969, Schaffner family archive. Some letters edited for clarity.

15 Linda Schaffner, now Linda Preston, JN interview. And 'Foxtrot 94', Ian Black and Mike Schaffner, *Flypast* magazine, July 2014.

Chapter Nine: The 'Knock at the Door'

1 William Schaffner story based on JN interviews with his widow, Linda Preston, and his sons, Mike and Glennon Schaffner. Sir Christopher and Lady Irene Coville, JN interview; and Chris's autobiography, *Fighter Pilot*, op. cit. And 'Foxtrot 94', Black and Schaffner, op. cit.

2 White, *Lightning Up*, op. cit.

3 Ibid.

4 https://www.raffca.org.uk/art_ControllingTheLightning.php

5 Transcript taken from RAF Aircraft Accident Report Lightning F6 XS894 5 Squadron, 8 September 1970. Sequence and content of radio calls edited for brevity and clarity. Note: In the report all timings are given in military Zulu (or GMT). The timings here have been converted to local time in September.

6 Letter from the Shackleton pilot to Mike Schaffner, Schaffner family archive.

7 'Military Aircraft Accident Statistics', supplied by Air Commodore Rick Peacock-Edwards.

8 Sharman, *Sir James Martin*, op. cit.

9 It is impossible to establish an accurate number for worldwide ejections – there is limited information from America and no data whatsoever from the likes of Russia, China or other major military powers. These estimations are merely extrapolated from the number of aircraft the UK had. In 2018, the US military alone was

over twenty times the size of the RAF, with around 14,000 aircraft as opposed to the RAF's near 600.

10 Figures vary depending on the source.

11 This conversation was reported in Chris Coville's autobiography, *Fighter Pilot*, op. cit.

12 Sandy Gruters, JN interview.

13 Letters from Schaffner family archive, edited for clarity.

Chapter Ten: 'My Daddy's Coming Home!'

1 William Schaffner story based on JN interviews with his widow, Linda Preston, and his son, Glennon Schaffner.

2 B-52 Stratofortress: www.af.mil; www.boeing.com/defense/b-52-bomber

3 'The B-52 gunners', *Air Force* magazine, 1 January 2012: https://www.airforcemag.com/article/0112gunners/

4 Terry Gruters' account based on interview in Guy Gruters' book, *Locked Up with God*, op. cit.

5 'Ending the Vietnam War, 1969–1973', US Department of State Office of the Historian: https://web.archive.org/web/20070202082232/http://www.axpow.org/stories-february2003.htm

6 Guy Gruters' release and homecoming based on JN interviews with Guy Gruters, Sandy Gruters, Dawn-Marie 'Mari' Williams (Guy and Sandy Gruters' daughter) and Charlie Neel. And Gruters, *Locked Up with God*, op. cit.

7 National Museum of United States Air Force: https://www.nationalmuseum.af.mil/Visit/Museum-Exhibits/Fact-Sheets/Display/Article/197496/operation-homecoming/ – Operation

Homecoming returned 591 POWs: 325 Air Force personnel, 77 Army, 138 Navy, 26 Marines and 25 civilians.

8 US National Archives Military Records, Vietnam War US Military Fatal Casualty Statistics https://www.archives.gov/research/military/vietnam-war/casualty-statistics; Defense POW/MIA Accounting Agency.

9 Vietnam War US Military Fatal Casualty Statistics, op. cit.; 'Vietnam War 1954–1975', Ronald H. Spector, Britannica: www.britannica.com/event/Vietnam-War; List of Aircraft Losses in Vietnam War: https://en.wikipedia.org/wiki/List_of_aircraft_losses_of_the_Vietnam_War; United States Fixed Wing Aircraft Losses of the Vietnam War 1962–1973: https://www.nhahistoricalsociety.org/wp-content/uploads/2017/08/aircraftlossesofthevietnamwarnhahs.pdf; Earl H. Tilford, *Search and Rescue in Southeast Asia* (Centre for Airforce History, 1985). p. 155.

10 'The Courage of Lance Sijan', Correll, op. cit.

Chapter Eleven: 'It Could Have Been Me ...'

1 'Oceans Ventured: Winning the Cold War at Sea', Center for Strategic and International Studies (CSIS), Washington DC; 'US and Allied Navies Starting Major Test Today – Military Analysis', *New York Times*, 1 August 1981.

2 Rodolfo Castro Fox's story is constructed from: 'I Was a Naval Pilot' (*Yo Fui Piloto Aviador Naval*), quoted from: www.castrofox.blogspot.com, April 2011. Quotes have been translated from Spanish and edited for clarification.

And Rowland White, *Harrier 809: Britain's Legendary Jump Jet and the Untold Story of the Falklands War* (Bantam Books, 2020).

3   Although he uses the term 'rocket' in his account, this was not a seat with a rocket pack; it was a standard seat with three cartridges.

4   'Skyhawk A-4: Courage in the Air', Eduardo C. Gerding, February 2018: http://nottinghammalvinas. blogspot.com

5   Sharman, *Sir James Martin*, op. cit.

6   'Falklands War: A chronology of events', The History Press: www. thehistorypress.co.uk; 'Scrap dealer who accidentally set off the Falklands War', Daniel Schweimler, BBC News, Buenos Aires, 3 April 2010.

7   Neil Wilkinson's account based on JN interview and his letters home during the conflict (some edited for clarity).

8   'A Brief History of the Falklands War', *Smithsonian* magazine, November 2020.

9   General Leopoldo Galtieri obituary, *The Guardian*, 13 January 2003.

10 'The Falklands War – day by day and blow by blow', Forces Net, June 2021: https://www.forces.net/news/ remembering-falklands-day-day

11 https://military-history.fandom. com/wiki/Bofors_40_mm_gun

12 White, *Harrier 809*, op. cit.; 'Sink the Belgrano!', Patrick S. Barker, March 2019: www.historynet.com

13 White, *Harrier 809*, op. cit.

14 Marcelo Larraquy, journalist and historian, 23 May 2021: https://www.infobae.com/ sociedad/2021/05/23/el-ultimo- vuelo-del-capitan-zubizarreta-el- piloto-que-murio-tras-regresar-de- una-mision-en-malvinas/

15 Argentinian Air Forces in the Falklands War: https://military- history.fandom.com/wiki/ Argentine_air_forces_in_the_ Falklands_War#cite_note-8

16 White, *Harrier 809*, op. cit.

17 This section is constructed from: Santiago Rivas, *Skyhawks Over the South Atlantic: Argentine Skyhawks in the Malvinas/Falklands War* (Helion & Co, 2019); and White, *Harrier 809*, op. cit. And the memoirs of the Captain of HMS *Intrepid*, Rear Admiral Peter Dingemans, *My Incredible Journey from Cadet to Command* (Brewin Books, 2016).

18 Rivas, *Skyhawks Over the South Atlantic*, op. cit.

19 https://weaponsandwarfare. com/2018/11/page/2/ – edited for clarity. Website no longer active.

20 There is some confusion about this aircraft and others claimed it was a Mirage.

21 Rivas, *Skyhawks Over the South Atlantic*, op. cit.; White, *Harrier 809*, op. cit.; www.castrofox. blogspot.com, April 2011.

Chapter Twelve: Friendly Fire

1   Steve Griggs's story based on his personal account and JN interview; and Ian Black's account of the incident, *Hunting Big Cats*.

2   Paddy Mullen's story based on JN interview.

3   https://en.wikipedia.org/wiki/ Inner_German_border

4   Michael Napier, *Blue Diamonds: The Exploits of 14 Squadron RAF 1945–2015* (Pen and Sword Aviation, 2015).

5   Data from Tony Gaunt at Martin-Baker Aircraft Company Ltd – timescales obviously vary depending on the type of seat.

6   'Pilot Equipment and Associated Systems', *Jaguar GR Mk1 and T Mk2 Aircrew Manual*, Ministry of Defence.

7 Martin-Baker website: https://
   martin-baker.com/products/mk9-
   ejection-seat/; Sharman, *Sir James
   Martin*, op. cit.; Hay, *The Man in
   the Hot Seat*, op. cit.

8 Squadron Leader Richard
   Groombridge giving evidence at
   the court martial of Lawrence
   and Inverarity quoted in the *Daily
   Telegraph*, 14 January 1983.

9 Daily reports of court martial in
   the *Daily Telegraph*, 11–15 January
   1983.

10 Royal Air Force Aircraft Accident
   Report Jaguar GR1 XX963, 25 May
   1982, Inspectorate of Flight Safety
   (RAF) MOD; daily reports of court
   martial in the *Daily Telegraph*,
   11–15 January 1983.

11 Daily reports of court martial in
   the *Daily Telegraph*, 11–15 January
   1983.

12 'How Sidewinder Missiles Work',
   Tom Harris: https://science.
   howstuffworks.com/sidewinder8.
   htm

13 'The Sidewinder Story: The
   Evolution of the AIM-9 Missile',
   Carlo Kopp, *Australian Aviation*,
   April 1994.

14 Daily reports of court martial in
   the *Daily Telegraph*, 11–15 January
   1983.

15 Sharman, *Sir James Martin*, op. cit.

16 *Daily Mirror*, 20 December 1982.

17 Napier, *Blue Diamonds*, op. cit.

18 Ibid.

Chapter Thirteen: 'Long Live the Homeland'

1 Ricardo Lucero story and some
   quotes are based on JN interview
   with Lucero's friend and former
   AAF pilot, Enrique Lippi. Also,
   Enrique Lippi social media posts
   translated from the Spanish and
   edited for clarity. And Rivas,

   Skyhawks Over the South Atlantic,
   op. cit.; Rick Jolly, *Doctor for
   Friend and Foe* (Osprey, 2019);
   https://www.youtube.com/
   watch?v=RI5M9lX9nKY&t=169s
   and his interview in *The Times*, 29
   May 1982.

2 https://second.wiki/wiki/
   jorge_osvaldo_garcc3ada

3 Rivas's investigations state the Jason
   Islands. Some reports say his body
   was found on Golding Island, south
   of Pebble Island.

4 Jolly, *Doctor for Friend and Foe*, op.
   cit.

5 Rivas, *Skyhawks Over the
   South Atlantic*, op. cit.; https://
   weaponsandwarfare.com/2018/11/
   page/2/ (website no longer active);
   HMS *Coventry* sinking anniversary,
   'War we weren't prepared for', BBC
   News, May 2017.

6 Falkland Islands documentary,
   Inside Out, BBC One (Yorks and
   Lincs), 16 January 2012.

7 https://www.facebook.com/
   watch/?v=597465150413766

8 Lucero interview, *The Times*, 29
   May 1982.

9 'Skyhawk A-4: Courage in the Air',
   op. cit.

10 Rivas, *Skyhawks Over the South
   Atlantic*, op. cit.

11 Jolly, *Doctor for Friend and Foe*, op.
   cit.

12 Velasco interview, *Inside Out*, BBC
   One (Yorks and Lincs), 16 January
   2012.

13 Ibid.

14 Sadly, Ricardo Lucero would be
   killed in a civilian flying accident in
   March 2010.

15 Naval History homepage: www.
   naval-history.net/F63-Falklands-
   British_aircraft_lost.htm

16 Royal Air Force Aircraft Accident
   Report Jaguar GR1 XX760, 6
   June 1983, Inspectorate of Flight
   Safety (RAF) MOD; Steve Griggs,

JN interview; Sharman, *Sir James Martin*, op. cit.

17 RAF Board of Inquiry – some terms paraphrased for ease of understanding.

18 Laurence Griggs, JN interview.

Chapter Fourteen: Command Eject

1 Account supplied by Dave Gledhill; some quotes edited for brevity.

2 At that time it was a purely male force – females would not be flying fast jets for a number of years.

3 Tornado Aircraft Safety, British Aerospace Defence Systems and Services.

4 Ian Hall, *Tornado Boys* (Grub Street, 2016).

5 Although information surrounding this incident is in the public arena, the pilot's family requested that the surname not be used. The author was happy to oblige.

6 Gordon Moulds's account based on JN interview and RAF Proceedings of a Board of Inquiry into an Aircraft Accident involving Phantom FGR2 XT908, 9 January 1989.

7 'Cold Water Survival', United States Search and Rescue Task Force.

8 This is a very basic explanation of a very complex system and many other actions occur in the sequence.

9 JN interview with the pilot.

10 Accident to Royal Air Force GR1 ZA603, 8 November 1984, Ministry of Defence Military Aircraft Accident Summaries published January 1986; Hall, *Tornado Boys*, op. cit.

11 The accounts of Dim Jones, Ian McLean and Neil Johnston and Ivy Johnston are based on JN interviews; RAF Accident Report Jaguar GR1 XX114 tactical training sortie, 19 September 1983 (Ian McLean's first ejection);

Ministry of Defence Military Aircraft Accident Summaries, Accident to RAF Jaguar GR1A XX393 and Tornado GR1 ZA408, 12 July 1984 (Dim Jones's first collision and ejection); Royal Air Force Proceedings of a Board of Inquiry into the mid-air collision involving Jaguar GR1A XZ108 and Tornado ZA394 on 9 January 1990.

12 It is difficult to establish what each witness actually saw and how it related to each aircraft and the crash, but these are their personal recollections.

13 Reports, *The Journal*, 10 January 1990. Some quotes edited for brevity.

14 Ivy Johnston, JN interview.

15 PSP weights differed slightly depending on the aircraft and the survival equipment they contained.

Chapter Fifteen: Storm in the Desert

1 Larry Slade's account based on JN interview.

2 'What Was the Gulf War?', Imperial War Museum website: https://www.iwm.org.uk/history/what-was-the-gulf-war; History.com Editors, November 2009, updated January 2017: www.history.com/topics/middle-east/persian-gulf-war; 'The Operations Room: Desert Storm – The Air War, Day 1', 2020: https://www.youtube.com/watch?v=zxRgfBXn6Mg

3 Desert Storm aircraft losses from *Gulf Air War Debrief* (Airtime Publishing, 1992).

4 'The Operations Room', op. cit.

5 SA-2 Surface-to-Air Missile, National Museum of the United States Air Force: www.nationalmuseum.af.mil/Visit/Museum-Exhibits/

Fact-Sheets/Display/Article/196037/
sa-2-surface-to-air-missile/

6 Devon Jones's ejection and rescue
based on: 'Down in Iraq, Part 1:
Survival in the Desert', *Gulf Air
War: Debrief* (Aerospace Publishing,
1991) – some quotes edited for
brevity; 'Desert Rescue', Captain
Paul Johnson (A-10 pilot), *From
the Line in the Sand: Accounts of
USAF Company Grade Officers in
Support of Desert Shield/Desert
Storm* (Air University Press, March
1994); William L. Smallwood,
*Warthog: Flying the A-10 in the
Gulf War* (Brassey's, 1993); Darrel
D. Whitcomb, *Combat Search
and Rescue in Desert Storm* (Air
University Press, September 2006);
'Hog Heaven', Walter J. Boyne, *Air
Force* magazine, December 2010.

7 GPS – global positioning system.

8 In his account, Devon uses the terms
'1,000 feet' but also '1,000 yards'.

9 Smallwood, *Warthog*, op. cit.

10 Sandy 57, Captain Paul Johnson,
was awarded the Air Force Cross;
Captain Paul Goff, Sandy 58, was
awarded the Distinguished Flying
Cross. The crew of Moccasin 5
were: Captain Tom Trask, Major
Mike Homan, Master Sergeant
Timothy Hadrych, Tech Sergeant
Gregory Vanhyning, Tech Sergeant
James A. Peterson Jr, Staff Sergeant
Craig Dock, Sergeant Thomas
Bedard. They would be awarded the
'Mackay Trophy' for 'extraordinary
heroism and self-sacrifice during
the rescue'. The trophy is awarded
yearly by the United States Air Force
for the 'most meritorious flight of
the year' by an Air Force person,
persons, or organisation. The
trophy is housed in the Smithsonian
Institution's National Air and Space
Museum.

11 Jeff Tice's account based on *Air
Force* magazine, January 2016

(some quotes edited for clarity) and
discussions with JN.

12 'US POWs Tell of Mental, Physical
Abuse by Iraqis', Melissa Healy, *Los
Angeles Times*, 15 March 1991.

13 *Financial Times*, January 1991.

14 'War in the Gulf: Prisoners; Iraqi
TV broadcasts; Interviews with 7
Identified as Allied Pilots', *New
York Times*, 21 January 1991.

15 Dale Zelko's account based on JN
interview.

16 https://www.airforce-technology.
com/projects/f117

17 *Gulf Air War Debrief*, op. cit.;
and 'Remembering the Gulf War:
Key Facts and Figures About the
Conflict': https://www.forces.net/
news February 2021

18 For obvious reasons, the number
of Iraqis killed varies according to
sources.

Chapter Sixteen: 'Oh My Gosh, You're a Girl!'

1 Ashley Stevenson's personal account
and JN interview.

2 Aviation Safety Network BAe
Harrier II GR.5 1 Sqn Royal Air
Force, 17 October 1990: https://
aviation-safety.net/wikibase/55530

3 Kate Saunders' account (now Kate
Morris) based on JN interview.

4 www.lindashaffer-vanaria.com/
home and interview with JN.
Inspirational Business Women Series
featuring Linda Shaffer-Vanaria,
2014. Some quotes edited for
brevity. And: https://www.youtube.
com/watch?v=9fMABfWvsoc&ab_
channel=VirginiaParsons

5 Sharman, *Sir James Martin*,
op. cit.; https://www.
wombat-womenincombat.com/
blog/2019/8/7/former-lt-commander
-linda-maloney; https://www.
ripleys.com/weird-news/

meet-brave-women-ejection-club;
'Meet the First Woman Martin-
Baker Ejectee': www.bremont.com.
blogs

6   Much of this section is taken from
    Kathleen Sherit, *Women on the
    Front Line: British Servicewomen's
    Path to Combat* (Amberley
    Publishing, 2020). The author
    is grateful to Kathleen for her
    research.

7   Julie Gibson, the RAF's first female
    pilot.

8   Joanna 'Jo' Salter joined the RAF at
    eighteen with the intention of
    becoming an engineering officer
    but went on to train as a pilot after
    the British government announced
    that women would be allowed to fly
    combat jets in 1992.

9   Ejection and rescue material
    additional to the accounts of Ashley
    Stevenson and Kate Saunders, based
    on Royal Air Force Proceedings
    of a Board of Inquiry into an
    Aircraft Accident involving Harrier
    T4A XZ147 at Great Driffield, 25
    September 1991.

Chapter Seventeen: A Day Like No Other

1   RAF accident figures supplied by Air
    Commodore Rick Peacock-Edwards,
    the RAF's former Inspector of Flight
    Safety.

2   Flying Authorisers' Course, 1994.

3   This is a very basic account of a
    truly complex but ultimately tragic
    accident using the two main sources
    below. There are many differing
    views about who was ultimately
    culpable and this brief summary
    is not designed to examine the full
    story. Many commentators feel
    Martin-Baker was unfairly treated.

4   https://www.gov.uk/
    government/publications/
    service-inquiry-into-the-accident-
    involving-hawk-tmk1-xx177

5   https://www.judiciary.uk/
    wp-content/uploads/2018/02/
    sentencing-remarks-mrs-j-carr-r-v-
    martin-baker-lincoln-crown-court-
    23022018.pdf

6   This is an edited section of the full
    statement.

7   Tom Moloney's account based
    on JN interview; Air Accidents
    Investigation Branch (AAIB)
    Bulletin No. 7/94.

8   Des Moloney's account based
    on JN interview; Air Accidents
    Investigation Branch (AAIB)
    Bulletin No. 7/94.

9   This is a very basic description of
    the system.

10 Neil Wilkinson, JN interview.

11 Velasco interview, *Inside Out*, BBC
    One (Yorks and Lincs), 16 January
    2012.

12 Mike Schaffner, JN interview.
    'The Loss of Lightning F6 XS894',
    English Electric Lightning
    website, June 2012, Archive Story
    reproducing articles from *Grimsby
    Evening Telegraph*: www.lightning.
    org.uk/jun12sotm.html; 'Did UFOs
    bring down fighter jet off UK coast
    killing American pilot "who saw
    flying cone in sky"?', Mirror Online:
    https://www.mirror.co.uk/news/
    uk-news/ufos-bring-down-fighter-
    jet-11472396; 'The mysterious death
    of a Lincolnshire RAF pilot that
    continues to fascinate UFO hunters':
    www.lincolnshirelive.co.uk/news/
    local-news/mysterious-death-
    lincolnshire-raf-pilot-733359

Chapter Eighteen: Against All Odds

1   Air Commodore Rick Peacock-
    Edwards, Inspector of Flight Safety,
    RAF Aircraft Accident Report
    Tornado F3s ZE166/ZE862, 10

January 1996. Edited for clarity.
Author's italics.

2　The events described in this chapter
are based on JN interviews with
Ian Weaver and his own personal
account, and those of Mark 'Skids'
Richardson, Cliff Nichol and Mike
'Elvis' Costello. And the RAF's
Board of Inquiry into the collision
between Tornados ZE166 and
ZE862.

3　The name of the pilot has been
changed as the author could not
contact him to discuss the incident.

4　This is a very basic explanation of
how the full intercept occurred.

5　Some of the distances the
participants recalled are slightly
different, as are those in the official
accident report. At closing speeds of
around a mile every three seconds,
and speeds changing rapidly during
manoeuvres, this is understandable.

6　This, and the following description,
is a very basic explanation of the
complex and varied reasons for the
accident. The complete Board of
Inquiry report is available via the
Ministry of Defence for any aviator
who wants to examine the full story.

7　Because of the noise on the tape, it is
difficult to establish everything that
was said in the final seconds. This is
as described by Skids Richardson.

8　John and Matt Phillips'
contributions posted on Ian
Weaver's blog 2011 and 2012,
and reports in *Lincolnshire Echo*,
Daily Telegraph, *The Guardian*,
Horncastle News, 10–24 January
1996. And 'First Aid Is Everyone's
Business', a first-aid magazine article
provided by Ian Weaver – source
unknown and some quotes edited
for clarity.

9　This is a very simplified explanation
of a long and complex list of medical
injuries detailed in the report.

10　Jenni Weaver's account based on JN
interview and her contribution to
Ian Weaver's blog.

Chapter Nineteen: Manhunt

1　Dale Zelko's account based on
JN interview, and interview with
Revista Força Aérea magazine,
Brazil 2010; TV documentary *No
Man Left Behind*, RAW TV.

2　Kosovo Air Campaign (March–June
1999): https://www.nato.int/cps/en/
natohq/topics_49602.htm

3　'Milošević: Serbia's fallen
strongman', Tim Judah, BBC News
Channel, 30 March 2001; 'Manned
Aircraft Losses over the Former
Yugoslavia, 1994–1999', Daniel
L. Haulman, Air Force Historical
Research Agency, 5 October 2009;
Darrel D. Whitcomb and Forrest
L. Marion, *Air Power History* Vol.
61, No. 3, Air Force Historical
Foundation, 2014; Kosovo Air
Campaign (March–June 1999), op.
cit.

4　Haulman, op. cit.

5　Zoltán and Irena Dani's accounts
based on Zoltan's lecture:
https://www.youtube.com/
watch?v=hvgyiFCoGoU; and
documentary, *Second Meeting*,
Director-producer Željko Mirković,
Optimistic Films: https://www.pbs.
org/video/second-meeting-zoxrdw/ –
quotes edited for clarity. There are
some discrepancies in the various
accounts. Resolved here as far as
possible.

6　Ksenija Pavlovic McAteer, 24 March
2022: https://thepavlovictoday.com/
nato-bombing-of-serbia-i-was-there-
i-remember – edited account.

7　'The ACES II Seat: Tech Info',
The Ejection Site: http://www.
ejectionsite.com/acesiitech.htm
https://wikipedia.org/wiki/ACES_II

8　An in-depth analysis of how Serbs

were able to shoot down an F-117 Stealth fighter during Operation Allied Force. Dario Leone, March 2020: https://theaviationgeekclub.com

9 Dale Zelko's rescue based on JN interview, and interview with *Revista Força Aérea* magazine, Brazil 2010; 'Team Sport, Combat Search and Rescue over Serbia, 1999', Darrel D. Whitcomb and Forrest L. Marion, *Air Power History*, op. cit.; 'Stealth Fighter Down: Rescue of Vega 31', Colonel Jim Cardoso, *Air Commando* journal, September 2017.

10 GPS – global positioning system.

Chapter Twenty: The Human Cannonball Trick

1 Dale Zelko's account based on JN interview, and interview with *Revista Força Aérea* magazine, Brazil 2010; *Reader's Digest*, 2016: https://www.rd.com/article/the-stranger-who-changed-my-life-my-enemy-my-friend/; Documentary, *Second Meeting*, op. cit.; 'Foes Now Friends': https://www.bbc.co.uk/news/

world-europe-20209770

2 Tony Gaunt, Martin-Baker Aircraft Company.

3 This is a very basic explanation of a very complex system which varies depending on the type of aircraft and role, and described by Tony Gaunt at Martin-Baker.

4 Neil Wilkinson's account based on JN interview; *Inside Out*, BBC One (Yorks and Lincs), 16 January 2012.

5 Edited translation of the longer letter.

6 https://brsaerospace.com/

7 The post-accident reflections of Ashley Stevenson and Kate Saunders are based on JN interviews.

8 Mike and Glennon Schaffner's account based on JN interviews, and *Inside Out* documentary, *Disappearance of RAF pilot*, BBC One (Yorks and Lincs), 16 September 2002.

9 *Inside Out*, ibid.

10 Royal Air Force Aircraft Accident Report Lightning F6 XS894. Some quotes edited.

11 The full letter has been edited for brevity and clarity.

12 Tom Wolfe, *The Right Stuff* (Vintage, 2005), p. 33 – quote edited.

BIBLIOGRAPHY

Blackman, Tony, *Victor Boys: True Stories from Forty Memorable Years of the Last V Bomber* (Grub Street, 2009)

Blackman, Tony, *Vulcan Test Pilot: My Experiences in the Cockpit of a Cold War Icon* (Grub Street, 2019)

Blackman, Tony and Wright, Anthony, *Valiant Boys: True Stories from the Operators of the UK's First Four-Jet Bomber* (Grub Street, 2014)

Bowyer, Chaz, *Wellington at War* (Ian Allan Ltd, 1982)

Burden, Rodney, *Falklands: The Air War* (Arms & Armour, 1986)

Caygill, Peter, *Lightning Eject: The Dubious Safety Record of Britain's Only Supersonic Fighter* (Pen & Sword Aviation, 2012)

Coville, Chris, *Fighter Pilot: From Cold War Jets to Spitfires – The Extraordinary Memoirs of a Battle of Britain Memorial Flight Pilot* (Air World, 2021)

Dingemans, Peter, *My Incredible Journey from Cadet to Command* (Brewin Books, 2016)

Gledhill, David, *The Phantom in Focus* (Fonthill, 2012)

Gruters, Guy D., *Locked Up with God* (Guy Dennis Gruters, 2012)

Gunby, David, *First Out in Earnest: The Remarkable Life of Jo Lancaster DFC* (Fighting High Ltd, 2016)

Hall, Ian, *Tornado Boys* (Grub Street, 2016)

Hay, Doddy, *The Man in the Hot Seat* (Collins, 1969)

Jewell, John, *Engineering for Life: The Story of Martin-Baker* (Martin-Baker Ltd, 1979)

Jolly, Rick, *Doctor for Friend and Foe* (Osprey, 2019)

King, Col and Guthrie, Ron, *Escape from North Korea: The Ron Guthrie Story* (Meteor, 2002)

McConnell, Malcolm, *Into the Mouth of the Cat: The Story of Lance Sijan, Hero of Vietnam* (W.W. Norton & Company, 1985)

Middleton, Donald, *Tests of Character: Epic Flights by Legendary Test Pilots* (Airlife Publishing, 1995)

Napier, Michael, *Blue Diamonds: The Exploits of 14 Squadron RAF 1945–2015* (Pen & Sword Aviation, 2015)

Napier, Michael, *Tornado GR1: An Operational History* (Pen & Sword Aviation, 2017)

Nichol, John, *Tornado: In the Eye of the Storm* (Simon & Schuster, 2021)

Philpott, Bryan, *Eject! Eject!* (Motorbooks International, 1989)

Pike, Richard, *The Lightning Boys* (Grub Street, 2011)

Price, Alfred, *Avro Vulcan: Workshop Manual* (Haynes, 2010)

Rivas, Santiago, *Skyhawks Over the South Atlantic: Argentine Skyhawks in the Malvinas/Falklands War* (Helion & Co, 2019)

Seidov, Igor, *Red Devils Over the Yalu: A Chronicle of Soviet Aerial Operations in the Korean War 1950–53* (Helion & Co, 2013)

Sharman, Sarah, *Sir James Martin: The Authorised Biography of the Martin-Baker Ejection Seat Pioneer* (Patrick Stephens, 1996)

Smallwood, William, *Warthog: Flying the A-10 in the Gulf War* (Brassey's, 1993)

Tuttle, Jim, *Eject!: The Complete History of US Aircraft Escape Systems* (Motorbooks International, 2002)

Walpole, Nigel, *Seek and Strike: RAF Brüggen in War and Peace* (Air World, 2020)

White, Alan, *Lightning Up: The Career of Air Vice-Marshal Alan White* (Pen & Sword Aviation, 2009)

White, Rowland, *Harrier 809: Britain's Legendary Jump Jet and the Untold Story of the Falklands War* (Bantam Books, 2020)

Yeager, Chuck and Janos, Leo, *Yeager: An Autobiography* (Bantam Books, 1985)

PICTURE CREDITS

INTEGRATED

Illustrations © Agnes Graves
p. 386 map © Bill Donohoe

PLATE SECTIONS

1. Low-level ejection by SSGT Bennie J. Davis III; both Bernard Lynch photos © Martin-Baker
2. AW52 and Lancaster: Jo Lancaster; Lancaster with the author: John Nichol
3. Both Hay photos: Hay family archive; Martin and Baker © Martin-Baker
4. Author with Mk1: John Nichol; Lightning ejection © Jim Mead; author with James and John Martin: Tony Gaunt
5. All images: Schaffner family archive
6. Super Sabre © APFootage/Alamy Stock Photo; Sijan at USAF and Gruters in cockpit: Guy Gruters
7. All images: Guy Gruters
8. Guthrie in cockpit: Ron Guthrie; test ejection with sled © Martin-Baker
9. Wilkinson: Neil Wilkinson; *Coventry* © Royal Navy Official Photographer, from Fleet Photographic Unit, HMS *Excellent* collection; Lucero and Velasco: Oscar Arredondo via Santiago Rivas

10. Castro Fox: Santiago Rivas; Wilkinson at crash site and Wilkinson with Velasco: Nicola Rees

11. Griggs's Jaguar © Ian Black; Griggs with ejection handle: Steve Griggs; Phantom © Ian Black

12. Saunders with Harrier and in hospital bed: Kate Morris; Stevenson and Saunders reunited: Ash Stevenson

13. Both images: Ian McLean

14. Tornado and Jaguar © Ian Black; author with Johnston: Neil Johnston

15. Both images: Tom Moloney

16. Tornado test flight © BAE Systems; author and Peters: John Nichol

17. Jones rescue: unknown; Jones and Slade: Larry Slade; Mk10 seat © Martin-Baker

18. Top: W. Gilson; bottom: Associated Press/Alamy Stock Photo

19. Top and middle: Associated Press/Alamy Stock Photo; bottom: Kurt's Kustom Photography

20. Richardson and Costello at crash site: Mark Richardson; Weaver with daughter: Ian Weaver; Tornado F3s © Ian Black

21. Both images: Ian Weaver

22. Zelko with flag: Dale Zelko

23. Civilians on wing © AFP Contributor/Contributor; Zelko with Dani © Andrej Isakovic/AFP via Getty Images

24. Test ejection © Martin-Baker

INDEX

Page references in *italics* indicate images.

*Read on for an extract from
John Nichol's upcoming book*

THE UNKNOWN WARRIOR

A Personal Journey of Discovery
and Remembrance

PROLOGUE

I am standing outside Westminster Abbey with a group of my
fellow ex-prisoners of war from the 1991 Gulf War. We are suited
and booted, medals glinting in the July sunshine. The flags flutter
in the breeze behind us, across Parliament Square. I trace the stiff
outline of the invitation in my jacket pocket with my fingertips as
we wait to be ushered inside. It feels like something of a miracle
to be here. Everyone who has fought in a foreign war has one
dream above all others: coming home. And now here we all are –
senior officers, junior airmen and airwomen, test pilots, chaplains,
princes, dukes, Second World War veterans in wheelchairs,
widows with determined smiles, and humdrum ex-Tornado navigators like me. All lining up in pressed shirts and polished shoes
for a service to commemorate 100 years of the Royal Air Force.

It's also supposed to be a celebration. But everyone attending will
have had comrades who lost their lives in war or in training, including some whose bodies have never been found. There is something
about these great ceremonial occasions, with their silences, their
symbolism and their ancient language as familiar as a pre-flight
checklist, that can make the grandest public event feel deeply personal, too. Sometimes uncomfortably so, if you are catapulted into
the past.

Edward I climbed these three low steps to the Great West Door for his coronation in 1274. So has almost every monarch since. Many were later carried up feet-first for their funerals.

Time after time, century after century, sombre crowds have gathered here, each man and woman connecting privately with a public sense of loss. In 1603, six knights carried the body of Queen Elizabeth I past this very spot, straining beneath the weight of a lead casket topped with her robed effigy. 'Westminster was surcharged with multitudes of all sorts of people in their streets, houses, windows, leads and gutters, that came out to see the obsequy,' wrote the chronicler, John Stow, 'and when they beheld her statue lying upon the coffin, there was such a general sighing, groaning and weeping as the like hath not been seen or known in the memory of man.'[1] There is nothing starker than a coffin – a body in a box – carried with great solemnity, to take the doubt out of death. And nothing more evocative than the field of tiny wooden crosses, each one bearing a poppy, that blooms here every Remembrance Day.

We shuffle out of the sunlight and into the pale stone magnificence within. Somewhere in the distance is the tomb where they laid Elizabeth I to rest. Elsewhere lie Edward the Confessor, Sir Isaac Newton, Chaucer, Dickens and Stephen Hawking. Since 1066, it has been the location of the coronations of forty English and British monarchs, a burial site for many of them, and the venue of at least sixteen royal weddings since 1100.

Sir Winston Churchill was laid out beneath an etched grey slab at our feet. My eye is drawn to the shiny rectangle of black marble beyond him, guarded by a neat rampart of poppies, which marks the grave of the 'Unknown Warrior'. As we hover in the area chatting, a more knowledgeable friend tells me that it is the only memorial stone in the abbey which no one ever walks on. And that when the Duke of York (later King George VI) married Elizabeth Bowes-Lyon there in 1923, she placed her bouquet on it in memory of her brother Fergus, killed eight years earlier during the Battle of Loos, who at that time had no known grave.[2] The Queen Mother's spontaneous act of remembrance was echoed by her daughter after her wedding to the Duke of Edinburgh in 1947. Princess Anne followed suit in 1973, as did Sophie, Countess of Wessex, in 1999. Kate, Duchess of Cambridge, continued the tradition in 2011, as have many royals since.

I am embarrassed to admit that, despite attending countless services in Westminster Abbey, I have never properly examined the grave before. So today, surrounded by so many Known Warriors, I feel the need to stop and read the inscription. But it's too late; the ushers are tapping their watches. *Please move inside, ladies and gentlemen; the royal family will be here soon.*

Taking my seat, I glance across the aisle at the uniformed figures lined up opposite us; celebrated military figures, the weight of their service measured out in medals and gold braid. So many medals. So much service. I wonder if I deserve to be here, a fossil dug up from the sands of a foreign conflict fought and forgotten long ago. Some of the attendees are too young to remember that first Gulf War in 1991 when I was shot down and captured, let alone the Falklands in 1982.

In the distance to our left, the minor royals are already beginning to file into the abbey, sidestepping the Unknown Warrior almost without a second glance, before turning to make dutiful small talk with the beaming clerics and officers lined up to greet them. Across the aisle, I wave at a familiar Bomber Command veteran in a wheelchair. With his hair as white as parachute silk and long row of medals on his chest, he is an old friend with whom I have shared many cups of tea – and glasses of red wine – over the years as we discussed the Allied bombing campaign in the Second World War. Today, he has asked me to help him leave the abbey at the end of the service when we make our way to Horse Guards for the more convivial rituals of the day. No doubt more tea and red wine will be involved.

Westminster Abbey must have known many sombre silences in its thousand-year history. But today, thankfully, the Great West Door has been flung open, allowing the dazzling sunlight and the excited chatter of the crowds to venture in on the breeze.

The BBC is broadcasting live coverage of the whole day, featuring archive footage of past conflicts, as well as interviews with veterans old and young, widows and serving warriors too. As more members of the royal family arrive, ladies in pastel colours bob and curtsey. Princes hand their peaked caps to equerries. Politicians attempt to look as if they do this every day, though the darting eyes of some give them away. Prince William and the Duchess of Cambridge arrive. More smiles; more small talk. The Prince of Wales walks in and steps smartly around the grave of the Unknown Warrior, while Camilla clasps her hands, looking fiercely cheerful beside him.

Now a fanfare of burnished trumpets cuts through the thick air. We all rise and turn our heads, staring – while pretending not to stare – towards the Great West Door. Led by the Dean of Westminster Abbey, the Queen – looking neat and spry in her suit of blue-and-turquoise silk – walks in without the aid of a stick. Out of the corner of my eye, I watch her careful footwork on the abbey's polished stones as she steers a path around the poppies surrounding the black slab which blocks her processional route.

Once again I find myself strangely drawn by this mysterious grave, somewhat ashamed to realise how little I know of its origins.

* * *

Later, when the sounds of the organ voluntary are tumbling out from the decorated pipes above our heads, I collect my wheelchair-bound veteran and push him slowly down the aisle, chatting about the service, the royals, and the next part of the day. We are both keen to move on to the red wine aspect of the celebrations.

As we reach the grave of the Unknown Warrior he raises his hand. 'Let me just take a moment, John,' he says, quietly.

Many of his Second World War comrades are still missing, decades after the end of the war. Of the 125,000 Bomber Command aircrew who took to the skies, 55,573 were killed. Around a meagre 50:50 chance of survival. Countless numbers were never seen or heard of again. 'As long as you need,' I reply. I have lost quite a few friends myself, but for me, it is the haunting notes of the Last Post that normally bring sad memories to the fore.

While the rest of the congregation sallies out into the sunlight, and my white-haired companion bows his head in silence, I'm now able to closely examine the gilded inscription carved into the jet-black slab of Belgian marble:

BENEATH THIS STONE RESTS THE BODY

OF A BRITISH WARRIOR

UNKNOWN BY NAME OR RANK

BROUGHT FROM FRANCE TO LIE AMONG

THE MOST ILLUSTRIOUS OF THE LAND

AND BURIED HERE ON ARMISTICE DAY

11 NOV: 1920, IN THE PRESENCE OF

HIS MAJESTY KING GEORGE V

HIS MINISTERS OF STATE

THE CHIEFS OF HIS FORCES

AND A VAST CONCOURSE OF THE NATION[3]

I have *read* it before, but today is the first time that the words truly resonate. I suppose I'd always just thought it was symbolic, like the Cenotaph. Embarrassingly, the penny only now drops. There really is the *body* of an unknown soldier under there. I feel I should have known this. How did this man get from the killing fields of the Western Front to central London? Did they bring him straight from a First World War battlefield? Which one? Or did they dig up a body after the Armistice? And what makes him a warrior, rather than a soldier?

'He could have been an airman or a sailor,' murmurs the mind reader in the wheelchair. 'Nobody knows who he was.'

As I think about this, he points out a ship's bell, embossed with 'HMS VERDUN', hanging silently on the pillar beside us. '*Verdun* was the destroyer that brought back his body in a special coffin from France,' he says, 'with an escort of six other ships, as if he were royalty.' I stare at the grave while my 94-year-old friend begins to tell me about the large crowds that filled London that day in 1920; about the special places reserved for the widows whose husbands and *all* of their sons had been killed in the war. About the honour guard of soldiers wearing the Victoria Cross, our country's highest award for courage in the face of the enemy, who lined the aisle. Again, why don't *I* know any of this story?

He gestures towards the translucent Union Jack which hangs high and limp and almost unnoticed above the throne of Edward the Confessor, in the alcove to the left of the Great West Door. 'The Padre's Flag,' he says, reverently. 'Still stained with the blood of the fallen, it was draped across the coffin . . .' His voice falters and stops. Is he thinking about *this* Unknown Warrior, or the hundreds he must have known personally during his own war years?

As I grapple with the enormity of what it represents, an idea is forming in my mind.

Sometimes the past can seem so remote; sometimes it seems like yesterday. Here, today, this humble Unknown Warrior is challenging me to connect the two; to join the dots between our lives today and all those innocent young men blown into fragments no more substantial than the poppies we pin to our coats each year.

So many names without bodies. So many bodies without names.

How did an anonymous corpse from the First World War come to be carried up the steps just in front of us, to be buried among the kings? Why did such vast crowds turn out to witness the passing of an unknown body in a box? And how does the Unknown Warrior still exert such a powerful hold upon us today?

The idea of placing myself in his boots is not a comfortable one. But as we walk slowly away from Westminster Abbey, amid the churning traffic and the tourists wandering through Parliament Square, I am determined to find out more and, if possible, retrace the life and journey of the Unknown Warrior.

The worldwide Covid pandemic halted my search for answers. As did writing a book about a more modern conflict involving my colleagues on the Tornado force during the 1991 Gulf War, and another about those whose lives have been saved by ejection seats. But I have had some time to research a little more about the First World War. I was staggered to discover the stark reality of the casualty figures: around 10 million military deaths, nearly 7 million civilian; 21 million military personnel wounded.[4] One of the heaviest losses of life on a *single* day, 1 July 1916, occurred during the Battle of the Somme, when (although figures vary) the British Army alone suffered around 57,000 casualties – dead, wounded and missing.[5]

These losses are on a scale almost too terrible to comprehend. But it was another fact which rocked me.

Commonwealth War Graves Commission files show that 526,816 British and Commonwealth soldiers of that 'Great' War have no known resting place. Of those, 338,955 have never been buried at all, while 187,861 do have graves but have not been identified.[6] What scale of devastation, what level of brutality, could result in more than *half a million* men having no known resting place? All those families unable to mourn properly, with no tangible location to focus their grief.

* * *

Following the death of the Queen in September 2022, I was honoured to be invited to pay a short tribute during the Royal British Legion's November Festival of Remembrance. In front of the new King, and a TV audience of millions, I had the privilege of talking about his late mother's dedication to a life of service for our nation.

Then, on the fortieth anniversary of the conflict, we went on to remember the sacrifice of those who lost their lives during the 1982 Falklands War. I watched as Mary Fowler, dressed in black and standing alongside an image of her father projected onto the floor, explained how, when she was only fourteen, he was killed aboard HMS *Coventry* during an Argentinian air attack. With thousands in the hall, and countless more watching around the world, she read aloud from the final letter she ever received from him, her voice cracking as she struggled to contain her emotions.

> *Dear Mary,*
>
> *By the time you get this letter, your birthday will have come and gone. I'm sorry I can't be with you, but I shall be thinking of you.*
>
> *I suppose your mum is a little upset so I would like you to look after her.*
>
> *I'm afraid I do not know when I shall be home again. But let's hope it's not too far away.*
>
> *Well love, that is about all for now, so till next time, God bless and take care of yourself.*
>
> *Love,*
>
> *Dad xxx*[7]

They were his last words home. I didn't bother trying to contain my own emotions; tears were streaming down my face.

Mary told me afterwards that her pain was made infinitely worse by the fact that when he and many of his friends went down with the ship, their bodies were never recovered. She had no grave to visit, no place to contemplate a life without him, nowhere to grieve. As a young serviceman, I had watched that war unfold on TV, seen ships sinking into the freezing waters. I had even been deployed on a Task Force vessel later that year. I now realised that though I knew about the death toll, I hadn't really thought about its wider resonances.

The similarities between Mary's experience of loss in 1982 and that of the relatives of the Great War's Unknown Warriors in 1918 thrust my earlier promise to investigate the story more sharply into focus.

Sometimes the past can seem so remote; sometimes it seems like yesterday . . .

ENDNOTES

Prologue

1 https://www.tudorsociety.com/28-april-1603-elizabeth-funeral/
2 In 2011 Captain Bowes-Lyon's grandson produced information that his grandfather had actually been buried in Quarry Cemetery, Vermelles. The evidence was accepted and a special memorial headstone to Captain Bowes-Lyon was erected, inscribed 'Buried near this spot' as there is no certainty about the precise location of his remains within the cemetery: https://www.westernfrontassociation.com/on-this-day/27-september-1915-captain-fergus-bowes-lyon-8th-black-watch
3 First part of the inscription.
4 https://www.census.gov/history/pdf/reperes112018.pdf
5 These figures vary depending on sources and can change as bodies are identified: https://www.britannica.com/event/World-War-I/Killed-wounded-and-missing
6 https://blog.forceswarrecords.com/the-unknown-warrior-gone-but-never-forgotten/
7 Michael Fowler letter to Mary Fowler, written 17 April 1982. Edited.

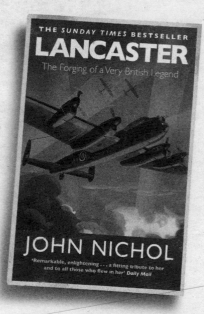

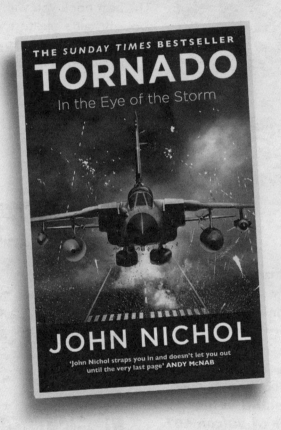